In the Mind's Eye

*Multidisciplinary Approaches to the
Evolution of Human Cognition*

**edited by
April Nowell**

INTERNATIONAL MONOGRAPHS
IN PREHISTORY

Archaeological Series 13

© 2001 by International Monographs in Prehistory
All rights reserved

Printed in the United States of America
All rights reserved

ISBN 1-879621-30-4 (Paperback)
ISBN 1-879621-31-2 (Hard Cover)

Library of Congress Cataloging-in-Publication Data

In the mind's eye : multidisciplinary approaches to the evolution of human cognition / edited by April Nowell

     p. cm. — (Archaeological series ; 13)
  Includes bibliographical references.
  ISBN 1-879621-31-2 (alk. paper). — ISBN 1-879621-30-4 (pbk. : alk. paper)
  1. Genetic psychology. 2. Cognition. I. Nowell, April, 1969- II. Archaeological series (Ann Arbor, Mich.) ; 13.
BF711 .I5 2001
155.7—dc21

                                                                                  2001026458

This book is printed on acid-free paper.

International Monographs in Prehistory
P.O. Box 1266
Ann Arbor, Michigan   48106-1266
U.S.A.

Table of Contents

List of Contributors iv
Forward
    Philip Tobias vii
Acknowledgments xiii

Introduction
    April Nowell 1

Part I. Archaeology and Cognitive Science
    1. The Role of Archaeology in Cognitive Science
        Thomas Wynn 9
    2. The Re-Emergence of Cognitive Archaeology
        April Nowell 20

Part II: On the Ground: Interpreting Material and Non-Material Artifacts
    3. Memories out of Mind: The archaeology of the oldest memory systems
        Francesco d'Errico 33
    4. A Pragmatic View of the Emergence of Paleolithic Symbol Using
        Martin Byers 50
    5. Nonmaterial Artifacts: A Distributed Approach to Mind
        Shirley Strum and Deborah Forster 63

Part III: Paleoneurology
    6. Archaeological Implications of Paleoneurology
        Harry J. Jerison 83
    7. Intellectual Surplusage: The Role of Bipedalism
        Sean C. Hogan and Gordon G. Gallup, Jr. 97
    8. Before or After the Split? Hominoid Brain Structures and the Evolution of the Human Mind
        Katerina Semendeferi 107

Part IV: Information Processing in Human Evolution
    9. Multilevel Information Processing, Archaeology and Evolution
        Philip Chase 121
    10. Behavioral Response to Variable Pleistocene Landscapes
        Richard Potts 137
    11. The Fossil Evidence for the Evolution of Human Intelligence in Pleistocene *Homo*
        Anne Weaver, Trenton W. Holliday, Christopher B. Ruff and Erik Trinkaus 154

Part V: A Final Word: The Origins of Language
    12. On the Neural Bases of Spoken Language
        Philip Lieberman 172
    13. Discovering the Symbolic Potential of Communicative Signs—The Origins of Speaking a Language.
        William Noble and Iain Davidson 187

List of Contributors

A. Martin Byers
Department of Anthropology
McGill University
Room 717, Stephen Leacock Building
Montreal, Quebec
Canada H3A 2T7
ambyers@sprint.ca

Philip Chase
University of Pennsylvania Museum of Archaeology and Anthropology
University of Pennsylvania
33rd and Spruce Sts.
Philadelphia, PA 19104-6324
pchase@mail.sas.upenn.edu

Iain Davidson
Department of Archaeology and Palaeoanthropology
University of New England
Armidale, New South Wales
Iain.Davidson@une.edu.au

Francesco d'Errico
Chargé de recherche au CNRS
UMR 5808 du CNRS,
Institut de Préhistoire et de Geologie du Quaternaire
Av. des Facultés,
33405 Talence, France
f.derrico@iquat.u-bordeaux.fr

Deborah Forster
Department of Cognitive Science
University of California, San Diego
Mailcode 0515
9500 Gilman Drive
La Jolla, CA 92093-0515
forster@cogsci.ucsd.edu

Gordon G. Gallup, Jr.
Department of Psychology
Social Science Building 112
State University of New York at Albany
Albany, New York 12222
Gallup@csc.albany.edu

Sean C. Hogan
Department of Psychology
Social Science Building 112
State University of New York at Albany
Albany, New York 12222
hogan1858@hotmail.com

Trenton W. Holliday
Trenton W. Holliday
Dept. of Anthropology
Tulane University
New Orleans, LA 70118 USA
e-mail: thollid@mailhost.tcs.tulane.edu

Harry Jerison
Department of Psychiatry and Biobehavioral Sciences
UCLA Medical School
Los Angeles, CA 90024
hjerison@ucla.edu

Philip Lieberman
Cognitive and Linguistic Sciences
Box 1978
Brown University
Providence, Rhode Island 02912
philip_lieberman@brown.edu

William Noble
Department of Psychology
University of New England
Armidale, New South Wales
wnoble@metz.une.edu.au

April Nowell
Department of Anthropology
University of Victoria
P.O Box 3050, MS 7046 Victoria, BC
Canada V8W 3P5
Victoria, BC, Canada
anowell@uvic.ca

Richard Potts
Department of Anthropology
National Museum of Natural History/National Musuem of Man
Smithsonian Institution
Washington, DC 20560
MNHAN064@SIVM.SI.edu

Christopher B. Ruff
Department of Cell Biology and Anatomy
John Hopkins University School of Medicine
725 North Wolfe Street
Baltimore, MD 21205
cbruff@jhmi.edu

Katerina Semendeferi
Department of Anthropology
University of California, San Diego
9500 Gilman Drive
La Jolla CA 92093-0532
ksemende@weber.ucsd.edu

Shirley C. Strum
Department of Anthropology
University of California, San Diego
9500 Gilman Drive,
La Jolla, CA 92093-0532
USA
1-619-534-4145 (anth dept)
sstrum@ucsd.edu

Phillip Tobias
Department of Anatomy and Human Biology
Palaeoanthropology Research Unit
University of Witwatersrand
South Africa
055SHAD@chiron.wits.ac.za

Erik Trinkaus
Department of Anthropology
Campus Box 1114
Washington University
St. Louis, MO 63610
trinkaus@artsci.wustl.edu

Anne H. Weaver
Department of Anthropology
University of New Mexico
Albuquerque, NM 87131
oneweaver@aol.com

Thomas Wynn
University of Colorado at Colorado Springs
Colorado Springs, Colorado 80933-7150
tgwynn@uccs.edu

Forward:
The Archaeology of Intelligence
Phillip V. Tobias O.M.S.G., F.R.S.

To relate the findings and interpretations of archaeology to the evolution of the mind, as this work does, is to make a forthright statement—that the techniques and symbols explicit and implicit in the archaeological record are directly related to the cognitive abilities, mental competences and intelligences of evolving human beings. However, the book does not stop there. By including chapters on human biology, primates and behaviour, it offers a further positive affirmation: it is not only from the archaeological record that we may glean evidence on the evolution of hominin intelligence. The endocranial casts of ancient hominins, the analogy and in a hopeful mood the homology between the brains and behaviours of modern humans and of non-human primates, and those of our remote ancestors, all add grist to the mill of scholars probing the origins and stages of hominid and hominin intelligence. Bearing witness to these two declarations, Dr. April Nowell, the Editor, has brought together just such a diverse array of scholars and topics. She succeeds in focusing their respective insights, as under a fierce and multi-facetted spotlight, on the awakening of human intelligence.

I first had the pleasure of exchanging views with April Nowell on these matters in the Department of Anthropology at the University of Pennsylvania, where in the early 1990's she was a doctoral student and I a visiting professor. Already her interest in different lines of evidence on the emergence and evolution of the human intellect was apparent to me. It is especially pleasing and worthy of congratulation that some of the roots and shoots which she tended so lovingly then have borne fruit in this book. Like hers, my own interest in the mind of early humans was sparked by archaeological reflections rather than by the endocasts with which I later became involved after the first cranium of *Homo habilis* was discovered in 1960. Perhaps I may be forgiven a brief retrospective glance at my personal odyssey in quest of the roots of human intelligence.

Like Bees Foraging in the Garden of Memory

As a schoolboy in Durban, Natal, during the Second World War, I spent many hours studying a superb exhibit of stone tools arranged by E. C. Chubb, the Curator of the Durban Museum. They had been excavated from a cave near the mouth of the Mgazana River on the south coast of Kwazulu-Natal. The tools exhibited in a fine self-teach exhibition were of the Coastal Smithfield Industry and of the Middle Stone Age (on the classification of Sub-Saharan industries devised by John Goodwin of the University of Cape Town and Clarene van Riet Lowe, one of my mentors at the University of the Witwatersrand). The youthful wonderment excited by these superbly crafted, symmetrical arrowheads and spearpoints set me pondering on the mentality of the prehistoric people who had made and used them. Of course, those on display were the pick of the bunch and I had no idea how many thousands of indifferent specimens had been left out of the public exhibit. In exploring the boundaries of mentality, I suppose that one should judge the achievements of an aeon by the best of which its practitioners were capable, as Clive Bell seemed to suggest in his work on *Civilization*.

What was running in the mind of a fifteen year old in a South Africa which was racially divided and firmly based on segregation and discrimination? It was the revelation that these works of art in stone were made by prehistoric people long before there came to southern Africa European colonists with their technology, artistry and craftsmanship, and an overweening belief in their own superiority. So that archaeological exhibit in the Durban Museum imbued my thoughts with two things: the length of time for which the human intellect had been capable of wonderful accomplishments—and the hubris and prejudice of which the same human mind was capable.

The state of wonder was enhanced in 1946 when I carried out the first excavation of an archaeological deposit in the Transvaal Province. Mwulu's Cave is not far from the Makapansgat Limeworks which eighteen months later, were to yield fossils of *Australopithecus africanus*. The Mwulu's Cave deposit gave us a rich haul of artefacts from three phrases of the Pietersburg Industry.

Among them we found the earliest signs of artistic activity in the South African Middle Stone Age. These were fragments of red and yellow ochre, a small stone pestle which we assumed had been used to crush and grind the ochre, and a carefully fashioned specularite pencil. Later dating showed that such industries extended back in time earlier than 100,000 years before present! It seems that the artistic tradition had been alive in South Africa long before the making of those more recent, exquisite rock paintings and engravings which are preserved in more than 20,000 South African prehistoric sites. So it dawned on me when I was twenty, that advanced techniques and artistic creations were part of the human capacity in Africa—once haughtily spoken of as the Dark Continent—at least as far back as the later days of the Middle Pleistocene.

During these early adventures, I was a medical student at the University of the Witswatersrand, Johannesburg, where I was to spend over half a century of my adult life. My anatomy studies took place in the department presided over by Raymond A. Dart, whose first and abiding love had been the brain, and certainly not fossils! One of my earliest undergraduate projects was a study of the brains of reptiles and mammals, which I carried out under the supervision of Michael K. Wright. Always lurking in my thoughts was the question of the size and shape of the brain and their possible relevance to human achievements. In the 1950's, as a teacher of anatomy under Dart, my medical science and honours students often discussed these questions with me. I read avidly Dart's prescient paper, "The relationships of brain size and brain pattern to human status" (1956).

Half a Litre of Brains

The problem became a real challenge to me when I was invited by the Leakeys, in 1959, to make the definitive study of the astonishing cranium that Mary Leakey discovered in the Olduvai Gorge, Tanzania. This was the type specimen of what Louis at first called *Zinjanthropus boisei* and which I later named *Australopithecus (Zinjanthropus) boisei*. It was not difficult to make an endocranial cast of this specimen and then, by water displacement, to determine its volume. It turned out to have an endocranial capacity of some 530 c.c., a little larger than the capacities of the South African specimens of *Australopithecus africanus*, and only slightly more than one third of the average capacities of modern human populations.

It was already clear that brain size was not everything: nevertheless, it did pose some tricky questions. From the studies of Ralph Holloway and of myself, it emerged that the australopithecines had brain sizes in the 400s and low 500s of cubic centimeters, whilst the Asian fossils of *Homo erectus* from Java and China had capacity values which ranged from 750 c.c. to 1225 c.c. There was a yawning chasm between these two groups—a hiatus from 530 c.c. to 750 c.c. Was this strange distribution the freakish result of sampling, or did something else, an as yet undescribed species, have capacity values which would fill this gap?

We did not have long to wait for the beginnings of an answer. In November 1960 Jonathon Leakey (the eldest son of Louis and Mary) discovered the parietal bones, mandible and other remains of Olduvai hominid 7—which was to become the type specimen of *Homo habilis* (Leakey, Tobias and Napier 1964). I devised a method for estimating the endocranial capacity when one had only part of a calvaria and, thus, only a partial endocranial cast. The capacity of this and subsequently discovered crania of the same species turned out to lie in the no-man's-land between the upper limit of the sample range for *Australopithecus africanus* and the lower limit of the range for *Homo erectus*. It took nearly as long to convince the world of the validity of *H. habilis* as a new species within the genus *Homo*, as it had taken Raymond Dart and Robert Broom to convert their fellow-scientists to an acceptance of *A. africanus* as a species which was trembling on the brink of humanity. Elsewhere I have explored such instances of premature discoveries and their significance in the history and philosophy of science. It is not necessary to develop the theme here.

For decades it was fashionable to decry studies on brain size. What was really important, it was averred, was the quest for signs of morphological reorganization of the brain during phylogeny, the expansion of some regions of the brain but not of others. Certainly this approach is highly important in any study of the evolving functioning brain, including the archaeology of mentation and intelligence. Yet, crude as mere size may seem to be, there are some valuable and interesting insights sill to be wrested from the analysis of brain sizes.

Firstly, it is often assumed that a population with larger adult brain sizes would have a larger mean brain size at birth. This is not necessarily so. It is questionable whether the capacity of the pelvis or birth canal of modern human females is three times as great as that of gorillas, chimpanzees or *A. africanus*, although the respective adult brain-sizes are on average about 3 to 1. It is instructive to compare the brain size at birth with that in adulthood. In living humans, the brain size at birth is approximately 25% of that in the adult; in gorillas and chimpanzees, the corresponding figure is of the order of 60%. If the small brained, ancestral australopithecines resembled the living great apes in this respect, it would follow that during the evolution of the genus *Homo* from *Australopithecus* and during the further evolution of some of the species within the genus *Homo*, there must have occurred a series of diminutions in the percentage of the adult brain size which was already present at birth. Thus, it is reasonable to surmise that the percentage dropped from 60% by steps and stages until the modern human value of about 25% was attained. It would follow that, during encephalization, most of the hominins' increase in brain size must have been largely post-natal. If this reasoning is true, most of the trebling in brain size that has occurred in hominins during the last two million years has taken place when the developing foetus was no longer ensconced within the womb and no longer dependent on maternal nutrition.

Secondly, an important implication of this ontogenetic timing of brain enlargement is that the encephalizing processes—or at least a most significant proportion of them—took place post-natally in a creature which was subject to the vicissitudes of the environment and fluctuating food-resources. In other words, encephalization in the hominins (although not necessarily in monkeys and baboons, nor in the cetaceans) was a highly vulnerable process. Since Loren Eiseley touched on this subject in *The Immense Journey* (1957), and I (1971) and H. T. Epstein developed it, a number of more recent studies, such as those of Robert Martin, have explored the possible metabolic constraints on human brain weight at birth and post-natal growth.

Thirdly, another aspect of the problem of the vulnerability of the brain is that this factor might be expected to pose a more serious challenge to development and survival at times when hominins were living in more straitened environmental circumstances. This has suggested to Sergio Dani and myself a possible explanation of the extraordinarily high juvenile mortality which has been found in certain of the early hominin populations. When the climate of Africa underwent sharp deterioration from 2.6-2.5 million years onwards, we find much higher proportions of immature individuals among the fossilized hominin populations than in the more salubrious or *mesic* conditions before than period. Thus, *Australopithecus africanus* populations from 2.8-2.6 myr showed a "lower risk" demographic pattern than the more recent taxa, *A. (Paranthropus) robustus* and *Homo Habilis* and *H. erectus*. I have recently postulated that changes in the demographic pattern depended upon changes in the prevailing environment—and this hypothesis is supported by the palaeo-ecological data of the fossil record.

Fourthly, some unanswered questions remained from this analysis. Why was it that of the three taxa (*A. robustus*, *H. habilis* and *H. erectus*) that lived in more exacting conditions, after the 0.6-2.5 myr deterioration, the two *Homo* species showed even higher risk demographic patterns (with no fewer than 73% of juveniles among the available samples) than did the robust australopithecines? The answer appears to lie in the degree of encephalization, as S. Dani and I have recently been demonstrating. The reasoning goes like this:

Increasing brain size or encephalization might be expected to have made extra demands upon the body's nutritional reserves and the population's resource base. This is because, of all the tissues in the body, neural tissue has the greatest nutritional requirements and constitutes a major metabolic burden to the ontogenetically developing organism. It might therefore be expected that, at a time of minimal encephalization, a population's demands on the resource base would be minimal; during a phase of greater encephalization, these demands would be enhanced; whilst a population undergoing strong encephalization e.g. the earlier species of *Homo*, would be subject to an appreciable increase in its nutritional demands. On testing this hypothesis by reference to four fossil hominin species, our evidence appears to show that different demographic patterns among early hominins had two principal determinants. One was the circumambiant climate which could explain part of the differences, namely those between *A. africanus* on the one hand, and *A. robustus* and *H. habilis* on the other. The second factor was the intensity of encephalizing pressures to which the taxa were exposed and this appears to account for the differences between the demographic patterns of the synchronic taxa, *A. robustus* and *H. habilis*.

It seemed reasonable to conclude that, while the increased metabolic burden of a larger brain would be a handicap under any circumstances, its presence in the face of an adverse environment would make survival even more difficult and might be expected to lower the average expectation of life. On the other hand, this metabolic and demographic disadvantage must have been more than outweighed by the selective advantages that greater encephalization brought in its wake. The sharp change to a high risk demographic pattern went hand in hand with an enhancement of selection pressures favouring encephalization.

Crude as brain size may once have seemed to some investigators, the analysis of such data in the light of life-cycles, demographic patterns, metabolic constraints—and many other factors, perhaps even problems of brain cooling, such as Dean Falk has been contemplating, has provided many more subtleties in our toying with the dramatic changes in brain size that have occurred in human lineage.

A Certain Quality in Human Lives

If this is so for quantitative encephalization, it is true also for the dissection of the quantitative changes in the hominin brain. These have been studied from more critical analyses of the surface features of the endocranial casts of ancient and modern hominins, and of the brains themselves of modern apes and humans. Exciting developments have been the demonstration of prominences on the endocasts of *H. habilis* and of later hominins in the regions occupied in modern humans by the areas of Broca and Wernicke: which nearly twenty years ago, first led me to claim that the most salient areas (though not the only ones) subserving spoken language were present already in ancestral *Homo* populations as long ago as 2.0 my! When supplemented by other evidence, this seminal observation led me to claim that *H. habilis* was the world's first speaking Primate: it was just one hundred years after the death of Broca.

It has long been held that asymmetry of the *planum temporale*, a small part of the upper surface of the temporal lobe of the cerebral cortex, which is regarded as part of Wernicke's Area, was associated with spoken language. Yet Patrick Gannon and his associates have just shown that in 17 out of 18 chimpanzee brains the *planum temporale* was larger on the left than on the right, as in most modern human brains. Does this tell us something about "chimp-speak", or does it tell us that the view that asymmetry of the *planum temporale* was associated with spoken language was simply not correct?

When Dart claimed seventy years ago that he could recognize the lunate sulcus on the endocast of the Taung child (type specimen of *A. africanus*), he could not have foreseen the lengthy controversies that his claim would generate, almost to the end of the century! It was not a simplistic question of whether he was or was not making a correct anatomical observation. The importance of his claim lay in the understanding that this sulcus marked the posterior boundary of one of the most advanced parts of the human brain. I refer to the parietal lobe of the cerebrum, sometimes spoken of as "the association of association cortices"! If that junction lay far to the front, the possessor would have had a relatively small parietal lobe, as in modern ape brains. On the other hand, if that sulcus lay far back on the surface of the cerebrum and if it truly (functionally) marked the posterior termination of the parietal lobe, then it would have signified a human-like brain, for all is small size. So much hinged on the correctness of the identification of the lunate sulcus and its position. Many of us have long since given up the attempt to identify the sulcus and have concluded that it is simply not possible to determine the position of the lunate sulcus in the australopithecine endocast with precision. This is an unsatisfying "conclusion", but it serves as a healthy corrective to the view of some cynics that no detail can be read on the surface of endocasts and that those who aspire to do so are latter-day soul mates of the phrenologist Franz Gall. On the contrary, a great deal of the gyral and sulcal patterns can be read from endocasts—as is confirmed by comparisons of modern primate brains with the artificial endocasts made in the corresponding brain-cases—but the lunate sulcus is not among them.

Several decades ago both Tadeusz Bielicki and I claimed that a self-reinforcing, positive feedback relationship existed between hominin brain evaluation and culture. Hands and eyes were the intermediaries and, I proposed in the 1970's, spoken language was a critical component of the feedback system that ensured its trans-generational transmission. The entire loop, it seemed, was subject to natural selective pressures. Although, at one time, this feedback concept appeared not to have had any substantial impact on studies in brain evolution, it is interesting to note that the idea has recently been reviewed by D. P. Agrawal and Sheela Kusumgar at Ahmadabad, India (1997): "humans stand apart from other animals by virtue of the complex material world that they have created. So the material world, tools and other artifacts...were a powerful stimulus, also a feedback, to the evolution of the human brain, mainly to its language and co-ordination areas (op. cit., p. 62)."

Many of these musings are featured in this book; some are not. Humans cannot and must not pretend to learn about themselves from contemplating their encephalic navel. Let us not forget—and April Nowell's book does not do so—what Earnest Hooton called *Man's Poor Relations*.

Why Men Behave like Apes and Vice Versa (Hooton)

The inclusion in this work of research in primatology and psychology testifies to a growing interest in the capabilities of chimpanzees and other apes. When Dart launched the Taung child upon the world, little was known of the behaviour of apes in the wild. Such studies as those of Wolfgang Kohler in the Canary Isles, N. Kohts in Moscow, and Robert Yerkes in Florida, had suffered from the shackles of ape life constrained in captivity. It took studies of primates in the wild by Adriaan Kortlandt, Jane Goodall, Diane Fossey, Birute Galdikas—to make critical advances possible in our understanding of how apes behave. These studies followed Eugene Marais' and Raymond Dart's own display of the carnivorous propensities of the chacma baboons in the wilds of southern Africa.

In the last 30 or 40 years, there has been a veritable explosion of knowledge on the behavioural traits of chimpanzees in the wild. We now know a great deal about inter-individual, intrafamily and intergroup relations, about child-rearing and sex habits, communication and problem-solving among the living apes. How human they have come to be seen—or should that be, how apish human ancestors—which we used to call hominids—must have been. Thomas Henry Huxley was faced with the urging of some of his contemporaries that humans should be placed in an order of mammals distinct from all other orders. Huxley had to show—and by morphological comparisons he was able to show—feature by feature, that "the structural differences which separate Man from the Gorilla and the Chimpanzee are not so great as those which separate the Gorilla from the lower apes". Therefore, he concluded, Man's place in Nature was not in a distinct order of mammals, but in the same order as the other Primates. Huxley was not to know that the closeness of humans to the anthropoid apes would be substantiated more than a century later when cytogenetics and molecular genetics were added to the armamentarium of the evolutionist.

So biological evidence has progressively closed the gap between humans and apes—that very gap that was once believed to be filled by the now effete "missing link"—until for several decades past far-sighted biologists and systematists have claimed that it is no longer justified to keep them apart in two different families. The old apartheid which had conceitedly dragooned us—who are higher primates—into Hominidae and Pongidae has been swept away. An integrated society has emerged in which, without churlishness, apes and humans are viewed as members of the same family!

The separation between them is relegated to lowlier levels of the taxonomic hierarchy. That is why, in this introductory essay, I have used the term *hominins* to refer to the erstwhile hominids; for hominins means members of a *tribe* (in the taxonomic sense), that is a group between the subfamily and the genus in the ladder of systematic categories.

Such are the lessons we have learned from the morphological and the biochemical appraisal of apes and humans. What have the behavioural studies taught us?

As with the morphology, so has the ethology shown us how very close are the resemblances between apes and humans, especially chimpanzee and bonobo. This is not the place to review all of the burgeoning evidence which has made this conclusion irresistible. However, in the context of the archaeology of mind, even the apparently sacred human preserve of tool-making and tool-using, which had led Kenneth Oakley (1949) to speak of *Man the Tool-Maker*, has been toppled from its pedestal. Although it is only just over thirty years since Oakley could write, "... it is evident that man may be distinguished as the tool-making primate...", barely five years ago Ian tattersall, in his customarily pungent manner, coould scold palaeo-anthropologists for allowing themselves to be mesmerized into classifying the makers of any stone tools, however crude, in the genus *Homo*. Yet, in the 1980s and the 1990s, such investigators as C. and H. Boesch, William McGrew and Frédéric Joulian have thrown a flood of new light on the implemental activities of wild chimpanzees. It is not simply a question of tool-usage but of tool-making that has shown how astonishingly human-like these apes of the genus *Pan* are proving to be. Joulian has even compared Oldowan stone tools made, most people believe, by *Homo habilis* in the dying days of the Pliocene and at the dawn of the Early Pleistocene—with what he dubs analogous tools made by chimpanzees of today. These observations have led to him to challenge such statements as those of N. Pioget and J. Pelegrin, to the effect that the mastering of a platform and of the conchoidal fracture "are highly significant characteristics which prove early hominid abilities to be superior to those of chimpanzees".

Joulian has even gone so far as to contest the long cherished paradigm that "culture" is an exclusively human realm: for the pursuit by various groups of West African chimpanzees of nut-cracking with some populations, but not within other interspersed groups of the same species, has strongly suggested the transmission of such behavioural traits by epigenetic means, in a word by learned behaviour of a kind which has conventionally been assigned to human cultural behaviour. These are early days for so dramatic a change of paradigm: yet, however old we may be in mind or body, however deeply mired in old and traditional ides, we must forever remain open to the likely impact of new evidence and new ideas.

In a sense this book is dedicated to a review of the archaic and an exploration of the novel. It strikes fearlessly at the worn out, the cliched and the antediluvian: it sets forth boldly to espouse the novel, the root and branch surprise—in a field of study where received wisdom is still fresh and often ill-based.

References Cited

Dart, Raymond
    1956   The relationship of brainsize and brain pattern to human status. *South African Journal of Medical Science* 21:23.

Eiseley, Loren
    1957   *The Immense Journey*. Random House, New York.

Leakey, L. S. B., P. V. Tobias and J. R. Napier
    1964   A new species of the genus *Homo* from Olduvai Gorge. *Nature* 202:7-9.

Oakley, Kenneth
    1949   *Man the Tool-Maker*. British Museum, London.

Tobias, P. V.
    1971   *The Brain in Hominid Evolution*. Columbia University Press, New York.

Acknowledgements

In the spring of 1996 I was fortunate to organize a symposium, titled *The Archaeology of Intelligence*, which formed part of the Society for American Archaeology meetings held in New Orleans. The symposium invited speakers from archaeology, physical anthropology, primatology, psychology and biology to share their research and ideas on the evolution of the human mind. The quality of the papers presented and the engaging nature of the discussions which followed the symposium prompted the formulation of this book. As the subject of the human mind becomes of greater interest to archaeologist it is crucial to clearly outline the state of our knowledge on this topic and to point out the direction future research should be taking. It is equally important to clearly define the contribution that archaeologists can make to the larger study of human intelligence. Because this book presents both review articles and new work it is hoped that researchers will turn to this volume as both a means to familiarize themselves with current lines of investigation as well as a general resource for information on this topic.

The success of this volume and the symposium upon which it is based is due to the efforts of a number of people. In particular, I would like to thank Harold Dibble for the initial suggestion to organize a conference on the evolution of intelligence and for his continued advice and enthusiasm for the project. I would also like to thank Philip Chase, Leslie Aiello, Merlin Donald, Stuart Lipkin, Paul Mellars and Anthony Miller for helpful commentary and Steve Miller for editorial suggestions. I wish also to acknowledge all of the work put in by the authors in this volume. I truly appreciate their diligence and patience. I am also grateful to Robert Whallon for his tremendous effort in making this a successful project. Finally, I would like to thank my husband, Jon, for his advice, support and good humor throughout the seemingly never-ending process from idea to fruition.

Introduction

April Nowell

Part I: Archaeology and Cognitive Science

The last decade has witnessed a sophistication and proliferation in the number of studies focused on the evolution of human cognition both within and outside anthropology. This movement reflects a renewed interest in the evolution of the human mind that is taking place in many disciplines such as cognitive ethology and evolutionary psychology (Mithen 1995:306). Cognitive archaeology, particularly as it is applied to the Paleolithic period (see Wynn, this volume; Nowell, this volume), focuses on the evolution of the development of human cognition, language and symboling. Archaeologists within this area of study draw upon biology, primatology and psychology among other disciplines. They integrate these datasets and perspectives with archaeological data in order to coherently study and explain the evolution of the human mind. Cognitive archaeology represents an exciting and, in some ways, a new direction for archaeology as a whole by extending the boundaries of traditional archaeological inquiry.

Opening the volume is the work of Thomas Wynn, arguably one of the first researchers to systematically study how the archaeological record could be used to explore questions of cognition on an evolutionary scale (see Nowell, this volume). Drawing on data from psychology and cognitive science, Wynn looks to the archaeological record to study the evolution of sex differences in spatial cognition. Through a study of projectile technology and subsistence practices he concludes that these differences evolved very late in hominid evolution and were merely the by-product of natural selection acting on prenatal androgens.

Nowell (this volume) looks at the role of archaeologists within the cognitive sciences and, in particular, at the rise of cognitive archaeology. Drawing on some of the themes introduced by Wynn, she explores the possibilities and limitations involved in studying the mind through the archaeological record.

Part II: On the Ground: Interpreting Material and Non-Material Artifacts

One important goal of Paleolithic archaeology is to explore the nature of the relationship between material culture and hominid cognition. In their study of this relationship many of the authors in this volume base their work or test their hypotheses to some degree on the lithic evidence (e.g. Byers, Chase, d'Errico, Strum and Forster, Potts, Wynn etc.) and there are good reasons for doing so. First, lithics represent the most abundant data source of this period. Because of their durability, lithic artifacts have at least a two and a half million-year record and an immense geographic distribution (see, for example, Barton 1990). This situation permits regional and temporal comparisons. In addition, their abundance makes them amenable to statistical testing. Arguably, the archaeological record is far more complete than the human paleontological record (Wynn n.d.:4).

Most importantly, lithics are an excellent subject of study because they are the direct product of hominid behavior. As Wynn (1996:263) notes, "tools typically carry information about the user beyond the mere mechanical task performed." While it is still a considerable source of debate whether or not individual flake removals were intentional in the sense of being preconceived (i.e. forming part of a *chaîne opératoire*, see below), there is an enormous amount of information to be gleaned from lithics about the choices made by hominid knappers—platforms have been prepared, flakes have been removed, some flakes have been modified and put to use while others have been put to use without modification and still others have not been used at all.

What this means is that archaeologists can isolate knapping choices. The next step is to investigate whether or not these choices can provide archaeologists and other researchers with information about technological strategies, cultural differences and, possibly, even cognitive differences among our hominid ancestors (see, for example, Robson Brown 1993:233). As Barton (1990) writes, while lithics "probably represent accumulations of artifacts discarded over unknown time spans rather than residues of single occupations by coherent social groups, they still provide information about variation in artifacts and associated human behavior within the context of space, time, and environment (see also Barton 1988)."

It is possible to argue that since non-human primates use and make tools that the value of these behaviors for the study of human evolution is lessened but as Gowlett (1984:175) writes, "tool modification by other primates does not erode the significance of human tool-making; it serves, if anything, to highlight how much further we have gone in that direction than any other species," and "it is increasingly plain that human beings of the present day are distinguished above all by the *complexity* of their tool making (emphasis in the original)" (but see discussion in Wynn and McGrew 1989; Joulian 1996).

There are, however, many factors that complicate attempts to study lithic artifacts. First, it has been observed that lithics may not necessarily be the most appropriate way of judging hominid cognitive abilities (Toth 1985b, see also Toth and Schick 1986). In fact, it has been suggested that they may only indicate the *minimum* capabilities of early hominids that were necessary to use and manufacture stone tools (Wynn 1981, 1998; Wynn and McGrew 1989). They may have employed a more complex behavioral repertoire in other aspects of their lives such as subsistence strategies, patterns of land use and so on.[1] For this reason the lithic data must always be studied in conjunction with data from other parts of the archaeological record and from human paleontology and the cognitive sciences. It is important to note, however, that the correlate of this argument is that if it can be demonstrated that early hominid stone tools are the product of complex cognitive behavior then a case might be made for extending this complexity to other behavioral realms (but see Mithen 1996).

A final point to consider is that in the Paleolithic it is far from clear exactly what a "tool type" means. This is an issue of fundamental importance because many of the studies that focus on lithics are based, at least in part, on assumptions concerning artifact morphology (e.g. Holloway 1969, Calvin 1993). The traditional approach has been to consider these types as real entities in the sense of Spaulding (1953) and as the product of mental templates.[2] This assumption is what Davidson and Noble (1993a) have referred to as the "*finished artifact fallacy*" as it is becoming increasingly clear from a number of studies that many factors that have little or nothing to do with cognition influence tool morphology (see also Barton 1990:69). Tools uncovered in the archeological record have a longer and more complicated history than was once thought and the final form of the artifact often has little to do with the specific mental template of a desired end product. As Barton (1990:70) notes, "virtually all lithics found at sites entered the archaeological context because they were no longer of value to the makers and users." The findings of these studies have important ramifications for how archaeologists and others using archaeological data need to approach the study of the hominid mind.

While it is clear that despite these issues lithics are an indispensable source of data for researchers there is much more to the archaeological record, even of the Paleolithic, than accumulations of stones. Accordingly, each of the authors in this section approaches the archaeological record from a different perspective. D'Errico (this volume) is interested in the ways in which information was stored in the Paleolithic. Observing that only humans create devices to store thoughts and memories, he argues that these devices permit the later retrieval of information by other members of a group. He refers to these devices as Artificial Memory Systems or (AMS). In the Upper Paleolithic there are numerous sequentially marked and notched bones that have been variously described as calendars or "*marques de chasses*." D'Errico suggests that these bones may be better understood as AMS. He turns to the ethnographic record to discern the general principles by which information is stored other than through writing. He then offers detailed criteria for recognizing AMS in the archaeological record.

Byers takes a more philosophical approach to discerning symboling behavior in the archaeological record. He focuses on the question of whether the use of symbols developed gradually or suddenly. Byers arguing for a sudden appearance of symboling behavior in the early Upper Paleolithic, suggests that for a symbol using population all of its material culture is symbolic including what he terms "utilitarian" artifacts such as stone tools. For this reason, the archaeological assemblages of non-symboling hominids will look different from those of symbol-using hominids. Byers refers to this as Style 1 and Style 2 respectively. Byers argues that when symboling becomes the dominant means of processing information there will be a rupture in the archaeological record—an abrupt change from Style 1 to Style 2. He concludes that such a rupture characterizes the Middle to Upper Paleolithic transition.

Finally, Strum and Forster (this volume) provide the primate perspective on the evolution of the human mind and the nature of non-material artifacts. As they note, "to define what is human, we need to know what is not" (see also Semendeferi, this volume). In some ways similar to Weaver et al. (this volume, see below), they focus on the relationship between the social, cognitive and technical/material-culture realms of life. Drawing on their experience and knowledge of baboon behavior, these researchers develop a number of implications for archaeologists. They argue that future models of the evolution of the human mind have to incorporate significant shifts in the meaning and focus of these realms in order to grasp in a more complex way the limits and possibilities of both human and non-human primate material culture, sociality and cognition and their interrelationships.

Part III: Paleoneurology

Without a doubt one of the most dramatic trends in human evolution is a marked increase in cranial capacity. In fact, modern humans have trebled their brain size relative to their earliest ancestors. This increase in size is often related to an increase in hominid intelligence. With approximately 100 billion neurons, the human brain is the largest of all of the primates and certainly one of the largest in the animal kingdom (Wolpoff 1999). Furthermore, while representing only 2% of a human's body weight, the brain accounts for 20% of the body's metabolic requirements (see Weaver et al., this volume). While it has not been demonstrated that intra-species differences in cranial capacity (beyond a certain minimum threshold) are related to differences in intelligence (Jerison 1973), within the catarrhini, brain size and intelligence differences *between* species do seem to be correlated (e.g. baboons, chimpanzees and humans) (Wolpoff 1999; see also Jerison 1973).

For these reasons many researchers are interested in exploring possible relationships between brain size and intelligence and there are a number of ways of doing so. First, one can look at *absolute* brain size as an indicator of intelligence but it seems that this measure alone cannot account for human intelligence since less intelligent animals such as the elephant and the sperm whale have brains that are at least as large. Although there is a relationship between brain size and body size (larger individuals on average tend to have larger brains) it seems that allometry is not a sufficient explanation of cranial capacity. According to Wolpoff (1999:143), a classic example is that of the 5000-pound stegosaur with a brain no larger than a "grapefruit." He notes that "absolute brain size alone may be an indicator of intelligence but it is an ambiguous one."

For this reason, many researchers consider relative brain size to be a more reliable measure of intelligence and prefer this approach. To continue with the above example, in humans, the brain accounts for 2% of body weight while in sperm whales this figure drops to 0.02%. Interestingly, some prosimians have even higher brain to body weight ratios than humans which suggests that relative brain size may become important only after a minimum threshold of absolute brain size is exceeded (Wolpoff 1999). But even this may not be enough as other species, such as the porpoise, have ratios that are equal to humans but do not exhibit the same degree of behavioral complexity.

A third way to approach brain size is to look at what has been termed "excess neurons" based on Jerison's (1973) encephalization quotient or EQ. Jerison's EQ is the ratio of actual brain size to expected brain size with expected brain size defined by the average brain size for living mammals taking body size into account. By using the EQ, any living or fossil mammal may be compared to any other mammal with respect to brain size. But a comparison of the EQ's of members of the Anthropoidea shows that chimpanzees place way above the mammalian average while gorillas fall far below it suggesting to Wolpoff "behavioral variation far beyond that observed (1999:144)" between the two species.

It is for these and other reasons that many prefer to study cranial reorganization in addition to, or instead of, absolute and/or relative brain size (see Semedeferi, this volume). There are two difficulties associated with the kind of study. One is that researchers only understand the neurology of different animal species in the most general way. According to Wolpoff (1999:144) the likelihood of conducting "the kind of experiments needed to trace human neural circuitry or complex human behaviors in the future" is remote. Most studies of the brain remain "of the 'black box' type of how brains *should* function."

The second problem is that in fossil hominids, both cranial capacity and cranial reorganization are estimated from endocranial casts (Aiello and Dean 1990). Using these casts researchers can identify sulci, gyri, sutures and pathologies. While endocasts represent a significant data source, there are many difficulties associated with studying them (see Aiello and Dean 1990:176-195). First, non-human primates are often employed as comparative models but human brains have developed more secondary sulci than apes or monkeys and these secondary sulci often obscure primary ones. In addition, the shapes of the primary sulci are quite variable over the cortex rendering identification difficult (ibid.). Other problems include the fact that hypothesized functions of different regions of the cortex remain conjectural because they cannot be tested with architectonic maps or electrode stimulation and that the assumptions of homologous features between modern humans and fossil hominids remain, of necessity, speculative as well (ibid.). It should also be noted that the number of endocasts is rather small. Thus, paleoanthropologists have to make inferences about a species from a limited sample. Factors such as age, sex, height, nutrition, disease and degree of intellectual stimulation in childhood will introduce variation in living humans and it may be that some of these variables affected early hominids as well (Tobias, pers. com.).

Not withstanding these difficulties, at a minimum it can be said that there is a relationship between brain size and intelligence but the nature and details of this relationship remain unclear (Wolpoff 1999). It is within this context that the authors of this section grapple with understanding hominid brain evolution. Jerison begins his paper with a detailed summary of the evolution of the mammalian brain in terms of its organization, function and encephalization. Like Chase (this volume; see below), Jerison believes the brain's capacity to process information is of central importance in understanding human evolution. Essentially, Jerison argues that the evolution of the human mind is the story of the evolution of the "brain's capacity to perceive and know a real world." Following from this, he believes that language evolved first as a means of processing information about the environment and not as a means of communication. He explains the expansion of the hominid brain as a result of the need for additional information-processing functions. According to Jerison, human language operates by sharing knowledge through the creation of images in the minds of others. For Jerison, then, language is fundamentally a cognitive system. Similar to Chase, he further suggests that this type of system can only in part be genetically determined. Like Potts (this volume), Jerison believes that behavioral flexibility is crucial to a successful adaptation in changing environments.

Hogan and Gallup, much like Jerison, Weaver and others in this volume, are interested in explaining the increase in cranial capacity that characterizes Pleistocene hominid evolution. Pointing out that modern humans use their brains for all sorts of tasks that could not have been anticipated millions of years ago, they ask a provocative question—"What mechanism could favor the evolution of abilities not directly necessary for our day to day adaptation and survival?" For these researchers, the answer lies in the relationship between the anatomical restructuring of the primate form to accommodate bipedalism resulting in a narrowing of the pelvic inlet and the potential for infant head trauma during birth. According to Hogan and Gallup, hominids developed surplus brain tissue during the birth process. They contend that the side effect of this surplusage is increased intellectual activity.

Finally, Semedeferi, focuses on the extent and timing of human cranial reorganization. By taking a comparative approach, she attempts to identify uniquely human neural substrates of behavior. Semedeferi's analysis of large sectors of the hominoid brain points to differential sizes for some of the sectors of the human brain when compared to the rest of the hominoids. She argues that some of the neural circuits involved in cognitive and emotional processes were present in the African hominoid stock prior to the split of the hominoid line, while others constitute real novelties. The implication of this work for archaeologists is that it should be possible for them to search for the expression of these novel neural circuits in the archaeological record.

Part IV: Information Processing in Human Evolution

For a growing number of researchers, such as Jerison (this volume), information processing is key to understanding human evolution. Information processing refers to how a species perceives, interprets and analyzes its environment and produces a corresponding behavior (see Chase, this volume). One researcher that has explored this topic in detail is Kathleen Gibson. For Gibson (1993), there are several differences between modern human and ape information processing capacities. If it is valid to think of humans and apes as lying on opposite ends of a continuum then it is possible to use these differences in processing capacities as a means of understanding the evolution of these abilities in the hominid line. Specifically, she argues (1993:251) that humans possess a greater degree of "brain-size mediated information processing capacity" that they are able to apply to tool-use, language and social behavior. She further argues that in human beings, these behaviors are mutually interdependent, that at least tool-use and language abilities are strongly canalized and that the development of all three of these behaviors necessitated quantitative increases in brain size with a concomitant increase in information processing capacities.

What all of this means for Gibson (1993) is that modern information processing capabilities along with tool use, language and social behavior must have evolved gradually over time as a function of the relatively gradual expansion of the hominid brain. She also suggests that as more information could be processed and held in the mind simultaneously humans were capable of creating increasingly complex mental constructs. In Gibson's view then it is this complexity, for instance, that is responsible for the differences between human language and ape symbolic communication. Humans are able to break down phenomena into components and recombine them in more complex ways than the apes. For Gibson (1993:252), in terms of syntax and content, ape communication "displays information processing capacities and mental constructional skills similar to those displayed by very small children."

Interestingly, Chase (this volume) argues that while there remains a genetic component to human behavior, hominid evolution is characterized by a *lessening* of the degree to which this behavior is strictly determined by genetics. He uses the archaeological record to trace this trend. Over time, he argues, learning becomes a more important factor in information processing leading to greater behavioral flexibility. According to Chase, the appearance of stone tools in the archaeological record at 2.5 mya represents modifications at the genetic level resulting in the development of new skills not shared by other primates (contra Wynn and McGrew 1989 and Joulian 1996). By contrast, the great distance that non-utilitarian items were transported in the Upper Paleolithic in comparison with the Lower and Middle Paleolithic is a behavior that is the result of information processing at the highest level and is not genetic in nature.

Potts (this volume) is also interested in behavioral flexibility but focuses on the role of the environment in the evolution of the human brain. According to Potts researchers usually explain the increase in brain size and complexity that characterize hominid evolution by suggesting that a limiting factor such as available metabolic energy or heat stress was overcome or that for adaptive reasons, a larger, more complex brain was selected for through natural selection (see discussion in Weaver et al., this volume). For Potts, the weakness of these approaches is their lack of context. Drawing on evolutionary biology, paleoecology, geology, primatology and evolutionary psychology he argues that hominid land use is a more appropriate way of assessing behavioral responsiveness and can be used to supplement other approaches. He then compares the environments of five Pleistocene sites including the distribution of stone tools across the landscape and concludes that there is evidence to support the notion of greater behavioral flexibility in response to increasingly variable habitats over time.

Concluding this section is the work of Weaver, Holliday, Ruff and Trinkaus. They focus on hominid evolution in the Pleistocene because it is during this period that there is a tremendous increase in cranial capacity, and a concomitant increase in the complexity of the archaeological record. Following a modular view of hominid intelligence they identify four types of intelligence that may have developed during the Pleistocene and which may explain the expansion of the hominid brain. These are object-oriented, conceptual, social and linguist intelligences. They argue that it is only through the combined effort of paleoneurology, neuroanatomy, biology and archaeology that these types of intelligences will be identified in the fossil record and their behavioral implications understood.

Part V: A Final Word: The Origins of Language

While language is of considerable antiquity, writing—the tangible expression of language—is a much more recent phenomenon and for this reason the origin and evolution of linguistic abilities remains controversial. While it may not be possible to "infer linguistic ability directly, it may at least be possible to detect correlated processes

Introduction

(Graves 1994:158)." This situation has prompted some researchers to turn to the archaeological record to look for linguistic clues. As Wynn (1991:191) notes, "because archaeological evidence is behavioral in nature, as opposed to the anatomical evidence of fossils, some consider it a more likely informant about language." While a discussion of this topic is beyond the scope of this introduction, many have debated the degree to which stone tools may be able to inform archaeologists about the origins of language (see, for example, Holloway 1969; Hewes 1973; Foster 1975; Kitahara-Frisch 1978, 1980; Dibble 1989; Chase 1991; Noble 1991; Calvin 1993; Noble and Davidson 1996 and several papers in Gibson and Ingold 1993).

Others have chosen to study endocranial casts to explore possible relationships between any apparent cranial reorganization and language (e.g. Holloway 1981a, 1981b, 1983a, b). Most often these kinds of studies have focused on the expansion of the parietal lobes, the development of cerebral asymmetries (or, specifically, right-fronto petalias, a condition seen in most modern humans), identifiable Broca's areas (but see Lieberman, this volume) and the position of the lunate sulcus among other possible evidence of reorganization. Within this context several hypotheses have been generated such as Wilkens and Wakefield's (1995) Reappropriation Hypothesis which holds that the linguistic areas of the brain were initially responsible for *manipulative activities*. Selection pressures acting on the hand for a precision grip were simultaneously acting on the motor cortex. Thus, it is argued that the association cortex is a result of the selection pressures placed on the neural control of the evolving hand (see also Frost 1980).

A tack taken by other researchers has been to reconstruct the "anatomy of speech." There are several difficulties with this approach. First, the ability to learn language and to communicate using sign language or speech cannot be inferred simply from reconstructing hominid vocal tracts (Aiello and Dean 1990:232). Parrots have the ability to mimic human vocalizations but it seems that they do not have the capacity for human language. Second, early hominids may have used sign language or some other non-verbal means of communication and this cannot be known from looking at vocal tracts. Third, much of the anatomy of the vocal tract is composed of soft tissue with does not survive in the archaeological record. Nonetheless, it is still possible to investigate to some degree whether or not hominids had the mechanisms necessary for language. These sorts of studies center on the hyoid bone from Kebara (Falk 1975; Arensburg 1989; Duchin 1990; Lieberman 1993) and basicranial flection.

A related topic is research that has been conducted on the mostly complete *Homo erectus* skeleton from Nariokotomi. The vertebral canal of the thoracic vertebrae in this specimen (WT-15000) are argued to be more ape-like in size (see Weaver et al. this volume) suggesting that this species did not have ability to control breathing in the manner necessary for speech (Boyd and Silk 1997). Specifically, the enlarged canal of modern humans accommodates nerves that enervate the muscles of the rib cage and the diaphragm which are necessary for control of the complex breathing used in speech.

It is within this context that the authors of this section approach the origins of language. While Lieberman (this volume) agrees with Jerison that there is a strong relationship between the anatomical structures involved in language and cognition, he argues that language evolved primarily as a means of communication. Through a careful presentation of the neurological literature, Lieberman suggests that a Functional Language System or FLS exists in the brain to regulate "the production and comprehension of spoken language" and it is the FLS that renders the human brain unique. According to Lieberman early anatomically modern humans possessed modern human speech anatomy. From this observation he infers the existence of an FLS and probably fully modern cognitive abilities as well. He contrasts Neanderthals to early anatomically modern humans. Lieberman argues that while capable of vocalizing, the speech anatomy of Neanderthals suggests that they could not make use of the full range of sounds typical of modern human languages. He further notes that genetic studies indicate that differences in dialect can act as an isolating mechanism. Lieberman argues that the implication of his study in conjunction with reconstructions of Neanderthal speech anatomy is that their differences in language ability may have led to their extinction.

Concluding the volume is the work of Noble and Davidson who return to the importance of information processing in human evolution but specifically in the context of language abilities. Like so many authors in this volume they argue that language evolved as part of a bio-behavioral package that "conferred advantages in terms of information control." They support their positions through a series of experiments with both human and non-human primate subjects.

Notes

[1] It is possible to imagine making a similar argument for non-human primates. The material culture of Bonobo chimpanzees does not necessarily reflect, for example, the complexity of their social organization.

[2] A brief note about the usage of the term "mental template" is warranted. The term mental template is used to suggest a symbolic ideal or type. Drawing on Deetz (1967), Gowlett (1984:177) defines a mental template as a set "of linked abstractions, held in the mind, which serve as a pattern for activities in the outside world." This set of abstractions is rigid enough so that the mind can retain it but "flexible enough to adjust to the manufacture of the individual tool (ibid.)."

References Cited

Aiello, L. and C. Dean
    1990 *An Introduction to Human Evolutionary Anatomy*. Academic Press, London.

Arensburg, B.
    New Skeletal Evidence Concerning the Anatomy of Middle Paleolithic Populations in the Middle East: the Kebara Skeleton, in *The Human Revolution: Behavioural and Biological Perspectives on the Origins of Modern Humans*, edited by P. Mellars and C. Stringer, pp. 165-71. Edinburgh University Press, Edinburgh.

Barton, M.
    1988 *Lithic Variability and Middle Paleolithic Behavior: New Evidence from the Iberian Peninsula*. BAR, Oxford
    1990 Beyond style and function: A view from the Middle Paleolithic. *American Anthropologist* 92:57-73.

Boyd, R. and J. Silk
    1997 *How Humans Evolved*. W. W. Norton and Company, New York.

Calvin, H.
    1993 The unitary hypothesis: A common neural circuitry for novel manipulations, language, plan-ahead and throwing? In *Tools, Language and Cognition*, edited by K. Gibson and T. Ingold, pp. 230-250. Cambridge University Press, Cambridge.

Chase, P.
    1991 Symbols and Paleolithic artifacts: style, standardization, and the imposition of arbitrary form. *Journal of Anthropological Archaeology* 10:193-214.

Davidson, I., and W. Noble
    1993a On the evolution of language. *Current Anthropology* 34:165-166.

Dibble, H. L.
    1989 The implications of stone tool types for the presence of language during the Lower and Middle Paleolithic. In *The Human Revolution: Behavioral and Biological Perspectives on the Origins of Modern Humans*, edited by P. Mellars and C. Stringer, pp. 415-432. Edinburgh University Press, Edinburgh.

Duchin, L. E.
    1990 The evolution of articulate speech; comparative anatomy of the oral cavity in *Pan* and *Homo*. *Journal of Human Evolution* 19:684-695.

Falk, D.
    1975 Comparative anatomy of the larynx in man and the chimpanzee: implications for language in Neanderthals. *American Journal of Physical Anthropology* 43:123-132.

Foster, M.
    1975 Symbolic sets. Paper presented in the symposium Toward an Ideational Dimension in Archaeology. Meetings of the Society for American Archaeology, Dallas.

Frost, M.
    1980 Tool behavior and the origins of laterality. *Journal of Human Evolution* 9:447-459.

Gibson, K.
    1993 Tool use, language and social behavior in relationship to information processing capacities. In *Tools, Language and Cognition in Human Evolution*, edited by K. Gibson and T. Ingold, pp. 86-108. Cambridge University Press, Cambridge.

Gibson, K. and T. Ingold (eds.)
    1993 *Tools, Language and Cognition in Human Evolution*. Cambridge University Press, Cambridge.

Gowlett, J.
    1984 Mental Abilities of Early Man: a look at some hard evidence. In *Hominid Evolution and Community Ecology*, edited by R. Foley, pp. 167-92. Academic Press, London.

Graves, P.
    1994 Flakes and ladders: what the archaeological record cannot tell us about the

origins of language. *World Archaeology* 26:158-171.

Joulian, F.
- 1996 Comparing chimpanzee and early hominid techniques: some contributions to cultural and cognitive questions. In *Modeling the Early Human Mind*, edited by P. Mellars and K. Gibson, pp. 173-190. McDonald Institute Monographs, Cambridge.

Kitahara-Frisch, J.
- 1978 Stone tools as indicators of linguistic ability in early man. *Kagaku Kisoron Gakkai Annals* 5:101-109.
- 1980 Symbolizing technology as key to human evolution. In *Symbol as Sense: New Approaches to the Analysis of Meaning*, edited by M. Foster and S. Brandes, pp. 211-224. New York: Academic.

Hewes, G.
- 1973 An explicit formulation of the relation between tool-using and early human language emergence. *Visible Language* 7:102-27.

Holloway, R.
- 1969 Culture: a human domain. *Current Anthropology* 20:394-412.
- 1981a Revisiting the South African *Australopithecus* endocasts: results of stereoplotting the lunate sulcus *American Journal of Physical Anthropology* 56:43-58.
- 1981b Culture, symbols and human brain evolution. *Dialectical Anthropology* 5:287-303.
- 1983a Cerebral endocast pattern of *Australopithecus afarensis* hominid. *Nature* 303:420-422.
- 1983b Human brain evolution: a search for units, models and synthesis. *Canadian Journal of Anthropology/Revue Canadienne d'Anthropologie* 3:215-230.

Jerison, H.
- 1973 *Evolution of the Brain and Intelligence*. Academic Press, London.

Lieberman, P.
- 1993 On the Kebara KMH 2 hyoid and Neanderthal speech. *Current Anthropology* 34:172-175.

Mithen, S.
- 1995 Paleolithic archaeology and the evolution of mind. *Journal of Archaeological Research* 3:305-332.
- 1996 *The Prehistory of the Mind*. Thames and Hudson Ltd., London.

Noble, W.
- 1991 The evolutionary emergence of modern human behavior: Language and its archaeology. *Man* 26:223-253.

Noble, W. and I. Davidson
- 1996 *Human Evolution, Language and Mind*. Cambridge University Press, Cambridge.

Robson Brown, K.
- 1993 An alternative approach to cognition in the Lower Paleolithic: the modular view. *Cambridge Archaeological Journal* 3:231-245.

Spaulding, A.
- 1953 Statistical techniques for the discovery of artifact types. *American Antiquity* 18:305-313.

Toth, N.
- 1985 The Oldowan reassessed: a close look at early stone artifacts. *Journal of Archaeological Science* 12:101-120.

Toth, N. and Schick, K.
- 1986 The first million years: the archaeology of protohuman culture. *Advances in Archaeological Method and Theory* 9:1-96.

Wilkens, W. and J. Wakefield
- 1995 Brain evolution and neurolinguistic preconditions. *Behavioral and Brain Sciences* 18:161-226.

Wolpoff, M
- 1999 *Paleoanthropology*. McGraw-Hill, Boston.

Wynn, T.
- 1981 The intelligence of Oldowan hominids. *Journal of Human Evolution* 10:529-41.
- 1991 Tools, grammar and the archaeology of cognition. *Cambridge Archaeological Journal* 1:191-206.
- 1996 The evolution of tools and symbolic behavior. In *Handbook of Human Symbolic Evolution*, edited by A. Lock and C. Peters, pp. 263-287. Clarendon, Oxford.
- 1998 Symmetry and the Evolution of Mind. Paper presented at The Hang Seng Center for Cognitive Science Workshop on Evolution of Mind

Wynn, T., and W. C. McGrew
- 1989 An ape's view of the Oldowan. *Man* 24:383-98.

1. The Role of Archaeology in Cognitive Science

Thomas Wynn

Abstract

Over the last twenty years students of human evolution, archaeologists included, have begun seriously to consider the evolution of human cognition. While archaeologists have employed a variety of approaches to this question, the most successful have employed concepts borrowed directly from cognitive science. Typically, the result is an enrichment of our understanding of traditional Paleoanthropological problems, the evolution of anatomically modern humans being an excellent example. Archaeology also has the potential to contribute to some of the research questions of cognitive science. Recently, for example, evolutionary psychologists have proposed a number of hypotheses to explain the sex differences in spatial cognition, the most popular of which focus on male hunting or female foraging. However, these proposed hypotheses fit neither the archaeological record for the evolution of spatial cognition nor the record for the evolution of hominid foraging. The archaeological record, therefore, necessitates a rejection or modification of hypotheses generated in cognitive science.

Cognitive Archaeology and Paleoanthropology

Cognitive archaeology today (1998) is a melange of very different approaches, theories, and interests that share only a conviction that the products of human action can inform us about the minds of the actors. It is united by a methodological reliance on material culture and a very general concern with minds as active shapers of that material culture, but beyond this there are fundamental differences. Many, perhaps most scholars who label their work as cognitive archaeology are interested in the content of past minds, particularly the symbolic meaning of objects or patterns, or the existence of meaning (see Renfrew and Zubrow [1994], for several examples). This approach is based largely on theories of human meaning derived from semiotics, symbolic anthropology, or social theory. It is not the subject of this essay. A second group of cognitive archaeologists is interested in the evolution of the structure of the human mind. The concern is not to understand individual actors (though individual actions constitute the data base) but to reconstruct patterns and trends in human evolution. This approach is allied with the general program of paleoanthropology and, in some cases at least, is based in theories of cognitive science. But within this general approach there are also differences in interest and method. It is convenient to separate these into three groups: an "implicitly" cognitive approach, the "*chaîne opératoire*", and cognitive science.

In implicit approaches "cognition" is more or less interchangeable with "intelligence." It is treated as a unitary phenomenon that evolved, by means of natural selection, as a component of the hominid adaptive niche. Certain cognitive abilities favored some hominids over others, leading to greater reproductive success, and overall evolutionary preeminence. Methodologically, this approach employs the same analytical units as normal paleoanthropology: Middle Paleolithic compared to Upper Paleolithic, or *Homo erectus* compared to Neanderthal. The approach rarely specifies cognitive variables or identifies how they would be recognizable in the archaeological record. Indeed, archaeologists often take their cue from fossils, especially brain size. Occasionally archaeologists have made arguments about cognition indirectly, as in Binford's argument for limited Neanderthal intelligence based on archaeological reconstructions of Neanderthal foraging (Binford 1982). When explicit about archaeological correlates with intelligence, archaeologists generally cite artifact complexity. Dennell (1997) provides a good recent example in a discussion of the Schoeningen spears:

> These represent considerable investment of time and skill — in selecting an appropriate tree, in roughing out the design and in the final

stages of shaping. In other words, these hominids were not living within a spontaneous 'five-minute' culture, acting opportunistically in response to immediate situations. Rather, we see considerable depth of planning, sophistication of design, and patience in carving the wood, all of which have been attributed only to modern humans. (Dennell 1997:768)

The problem with such implicitly cognitive approaches is their methodological naivete. "Planning", "sophistication of design", and "patience" are common-sense cognitive categories, but they are not rigorously defined cognitive abilities. Arguments based on such categories are reminiscent of arguments for the evolution of hunting, made thirty years ago, that were based on the simple association of tools and faunal remains. Since that time archaeology has developed a sophisticated set of concepts and techniques in faunal analysis that yields a much more subtle, and we hope, reliable picture of the evolution of hominid foraging. Archaeologists had to develop these concepts and techniques themselves (Binford 1981); they did not exist elsewhere. A similar theoretical vacuum is *not* true of human cognition. Perhaps the most surprising thing about much of cognitive archaeology is that it ignores an immense extant literature on human cognition. Psychologists and cognitive scientists abandoned or refined the "unitary" concept of intelligence decades ago, and yet many archaeologists who mention cognition still employ a vague, ill defined notion of intelligence and cognition. On the positive side, these implicit approaches to the archaeology of cognition have introduced an important perspective to the study of human evolution. It is essential that paleoanthropologists understand that prehistoric people had minds and that these minds structured their behavior. The actions of early hominids were not simply the passive consequences of foraging or social systems. On the down side, however, the implicit approach provides little real insight into the nature of cognitive evolution. It does not use "real" cognitive abilities and is largely ignorant of the major theories of cognitive science. It simply lacks the methodological rigor and theoretical power necessary to make reliable interpretations of past cognitive abilities.

Over the last fifteen years two approaches to the archaeology of cognition have appeared that are more sophisticated than implicit approaches, and which show the potential to provide a more detailed assessment of prehistoric cognition.

The first is the *"chaîne opératoire"* methodology, initially developed by Leroi-Gourhan and refined by Jacques Pelegrin and others at the CNRS (Pelegrin 1990). A *"chaîne opératoire"* is a reconstruction of the chain of decisions made in a technological activity. It is a more or less detailed description of an action based in sequential cognition. At its best, as in Nathan Schlanger's (1996) reconstruction of Marjorie's core from Maastricht-Belvedere, it provides a powerful description from which conclusions about cognition can be made. Marjorie's core is a refitted Levallois core from which Schlanger was able to describe a long sequence of knapping, and infer the chain of decision making. The knapper removed a series of Levallois flakes from the core, and for each episode had to modify the way he prepared the core in order to achieve a similar result. What is impressive is the description of an actual sequential cognitive task performed in prehistory, one that required flexibility in technique. However, Schlanger's interest is narrowly focused on reconstructing the "technique" of Levallois. He argues that it is arbitrary and incorrect to separate "mind" from "action". But while he is interested in the nature of technological action, and especially the way in which a knapper's intentions are affected by the ongoing process, he does not attempt to present his conclusions in terms of explicit cognitive abilities. Indeed, when he addresses cognitive evolution he falls back on concepts that are no better defined than those found in less sophisticated implicit approaches.

> The knapping of Marjorie's core cannot be said to have proceeded in an adventitious and responsive manner. In terms of wider implications, it cannot serve to restrict capacities of planning and consciousness to 'modern' humans, or to argue that flint knappers of 250,000 years ago were somehow more constrained by raw material or technology than were, for example, Upper Paleolithic blade makers. (Schlanger 1996:248)

Schlanger's analysis provides a much better understanding of Levallois (which was his goal), but does not inform us about well defined cognitive abilities. This is not a criticism of Schlanger's analysis. He and other proponents of *chaînes opératoires* are interested in technological action (technique) but not in the kinds of cognitive categories that are the focus of most of cognitive science. Such a description could provide a rich base for cognitive interpretation, but this is not the interpretive direction favored by its advocates. The *chaîne opératoire* technique remains a very useful comparative device, much better than the vague "complexity" of the implicit approach, but it has not yet provided much insight into the evolution of cognition.

The second formal approach to the archaeology of cognition borrows directly from the psychological and cognitive science literature. There have been several recent examples (Wynn 1993; Byers 1994, n.d.; Mithen 1994a, 1994b, 1996; Robson Brown 1994; Sinclair 1995; Steele 1995). As yet it is difficult to characterize this very eclectic approach, as the participants do not share specific concepts or methods. However, a few examples should provide a basic idea of the approach.

James Steele (1995) has applied concepts developed in neuropsychology. PET scans and research on aphasics indicates that the prefrontal lobes of the brain are instrumental in imaging and controlling sequential activities. In addition, neuroscience has discovered that much action is organized "peripherally," and not with a single central neural processor (i.e., there is modularity on a neural level). By analyzing videotapes of stone knapping, Steele hopes to identify the nature of the peripheral organization in stone knapping and also identify the role of the prefrontal lobes (using EEG, perhaps). If successful, this approach should allow interpretations of the neural processes underlying the reconstructed sequences of prehistoric stone knappers. It is also possible that Steele will be able to identify the elusive cognitive connection between stone knapping's sequential action and linguistic sequential organization, and provide a means for investigating the evolution of language.

My work (e.g., Wynn 1989) and that of Robson Brown (1994) have applied some of the concepts of developmental cognitive psychology. These are not based on neural evidence but on the actions and verbal reports of subjects, both children and adults. My work has been influenced most by Piaget and Inhelder's classic study of the development of spatial thinking (Piaget and Inhelder 1967). The applicable concepts consist of features of intuitive geometry, such as proximity, boundedness, perspective, and symmetry. Robson Brown has also focused on spatial concepts, but has used analytical concepts developed more recently in cognitive psychology, and which have a more rigorous experimental foundation. These include such specific cognitive abilities as mental rotation and visual frame independence (more on this later). Both my work and Robson Brown's have documented changes in the repertoire of spatial concepts used by early hominins. Our work has a much narrower focus than studies of "intelligence," but the results are much more reliable.

The archaeologist who has drawn most broadly from the cognitive science literature is undoubtedly Steven Mithen (1994a, 1994b, 1996). In particular, he has focused on the question of modularity and the evolution of "accessibility." Modularity refers to the idea that intelligence consists of many largely separate intelligences, like technical intelligence, social intelligence, and spatial intelligence. Here Mithen relies heavily on the work of Fodor (1983) and Gardner (1983), both highly influential cognitive psychologists. "Accessibility" refers to the degree of linkage between these various cognitive modules. Mithen cites Rozin and Schull (1988), who have argued that a key feature of human minds, compared to other primates, is our ability to apply knowledge gained in one module to experience usually controlled by another. Mithen uses these concepts to explain the advent of greater cultural complexity in the Upper Paleolithic. For example, Mithen points to the fact that Lower and Middle Paleolithic technologies were relatively insensitive to environmental differences and suggests that this reflects the modularity of technical and natural history intelligences. The two cognitive modules were not richly linked in the minds of *Homo erectus* and Archaic *Homo sapiens,* were therefore unable to "access" the technical module when faced with new or changing environmental conditions, and could not respond with technically innovative solutions. But technical and natural history intelligences *were* linked in the minds of early anatomically modern humans. This evolution of greater accessibility — a more general intelligence — allowed the developmental of specialized, environmentally specific adaptations in the Upper Paleolithic. It is readily apparent, I think, how much better an argument this is than the implicit examples cited earlier. It specifies the cognitive abilities in question and identifies archaeological correlates. More importantly, Mithen has illuminated an aspect of human evolution that had previously been only vaguely recognized. These examples illustrate a range of cognitive science theories applied to archaeological evidence. Following a long standing practice in archaeology, they have used concepts developed in another science to make sense of archaeological data. The key step in cognitive archaeology, as in any such borrowing, is methodological. Variables measured in modern experimental situations must be translated into variables that can be identified in material culture. Alternatively, an archaeologist can generate observable variables directly from an understanding of the general theory. In either case it is essential that the archaeologist have a full understanding of the theory (as Mithen clearly does, for example), not just a passing acquaintance or, worse, knowledge of only a few experiments.

Despite their theoretical grounding in cognitive science, the problems addressed in these examples are the same problems addressed by all Paleolithic archaeologists. Why was the Middle Paleolithic different from the Upper Paleolithic? How did the spatial concepts of *Homo erectus* differ from those of earlier *Homo*, or *Homo sapiens*? In other words, the units of interpretation are not new. Yes, this approach to prehistoric cognition adds much needed rigor and detail to studies of human evolution, but it has given little back to cognitive science. In the long run, I believe, cognitive archaeology will flourish or fade based not only on its contribution to paleoanthropology but also on its contribution to cognitive science as a whole.

Cognitive Archaeology as Cognitive Science

What can archaeology provide that is relevant to the current problems treated by the science of the human mind? I believe the answer to this question is "much," but it requires that archaeologists be willing to step away from their parochial concerns and categories, and attack some of the contemporary issues of cognitive science. Let me provide an example — sex differences in spatial cognition (see Wynn et al. 1996 for a fuller discussion).

One of the most provocative conclusions in all of the cognitive science literature is that there is a sex difference in spatial cognition. Men and women perform differently on standard tests of spatial thinking. Many individual studies have demonstrated a statistically significant sex difference on many spatial tasks (Halpern 1992). More telling, recent meta-analyses of these individual experiments have shown that the sex difference in performance is consistent and predictable (Voyer et al. 1995). Indeed, one spatial task, mental rotation, provides the largest sex difference in all of the cognitive science literature. From the perspective of cognitive psychology this sex difference in performance is real. The cause of this difference is less clear. Boys and girls grow up in different cognitive milieus, and have different experiences. However, when environmental variables are controlled (e.g., math scores, IQ, attitudes toward math and science, and so on), the sex difference in spatial ability remains as pronounced as ever (Halpern 1992). Most cognitive psychologists now maintain that the sex difference is physiologically based, tied in particular to prenatal hormonal levels (Geschwind 1987; Halpern 1992; Geary 1996), or less likely, adolescent hormone levels.

The cognitive psychological literature has identified four specific skills for which there are reliable sex differences. Tests of *spatial perception* require the subject to identify (or draw) the horizontal or vertical within a distracting frame. The most famous example is Piaget's "water level task," in which subjects attempt to draw the water level in a tilted drinking glass (Fig. 1). Underlying success on this task is the ability to ignore the orientation of the local spatial frame and attend to a separate

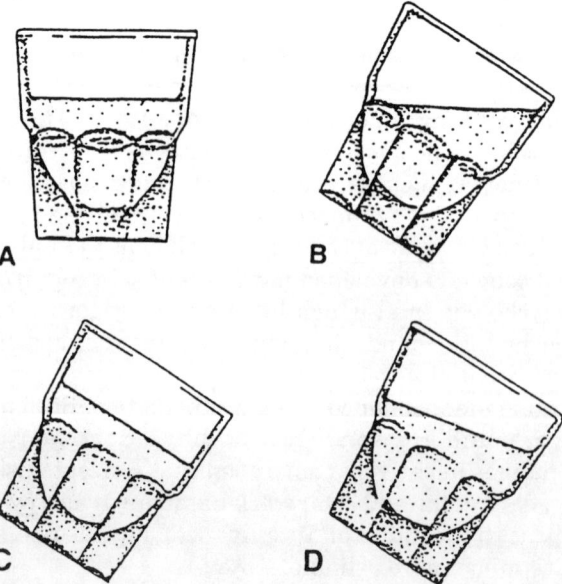

Fig. 1. The ability to correctly select the tilted glass with the appropriate water level (B) is a matter of *spatial perception*.

orientation. Even a significant percentage of college students fail this test of frame dependency (Thomas and Lohaus 1993), and there is a significant sex difference at all ages.

Mental rotation is the ability tapped in the block tests common on some intelligence tests. Typically, tests require the subject to view a complex figure and then pick the option that is a rotated version of the original (Fig. 2). Success on these tests hinges on the subject's ability to conceive of non-ego centered points of view and manipulate the image as it moves through them. This ability has consistently demonstrated the largest and most reliable sex difference in all of the cognitive psychological literature.

Tests of *spatial visualization* require subjects to pick simple patterns out of a complex background (often referred to as the embedded figures test—Fig. 3). Effect sizes for this test are the smallest of the four, and it is the only test in which women score higher than men (Silverman and Eals 1992).

Spatio-temporal ability is the ability to predict the time that a moving object will arrive at a designated point (Fig. 4). Tests commonly use a film or video of a moving object approaching a designated point (sometimes the viewer!). The video is blacked out, and the subject presses a button at the moment he or she thinks the object would have arrived. Although there is clearly a spatial component to this ability, there is also a time component, and the underlying basis of differential performance is harder to pinpoint. There is, however, a significant sex difference here as well.

The sex difference in spatial ability, and its neurological basis, are now widely enough accepted that evolutionary hypotheses have been proposed, and it is here that archaeology can make a direct contribution to an important issue in cognitive science. Several scholars have suggested that the sex difference in spatial cognition resulted from natural selection acting on a sexual division of labor. Jardine and Martin (1983), for example, suggested

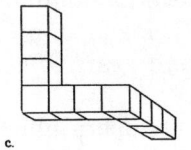 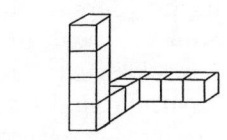

Fig. 2. The ability to judge (c) as equivalent to (a) is a matter of *mental rotation*.

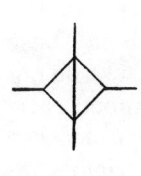 

Fig. 3. *Spatial visualization* is used to detect the figure on the left within the figure on the right.

Fig. 4. *Spatio-temporal* ability is used to predict the time of arrival of moving objects.

that throwing accuracy in hunting selected for male specific spatial abilities, and recently Geary (1996) has advanced a similar argument. Presumably, male hunters who could hurl projectiles more accurately would have greater reproductive success, and since females would not be under such selective pressure, over time selection would yield a consistent sex difference. As psychologists, Jardine and Martin and Geary relied on experimental evidence as the basis of their argument; their understanding of the anthropology of gender roles is clearly naive. Nevertheless, this hypothesis is a reasonable one and, more importantly, is amenable to testing through archaeological evidence. Archaeologists cannot see sex roles, especially in the Paleolithic, but we can see the evolution of spatial abilities and, less clearly, the evolution of hunting. If spatial cognition evolved as an aid to hunting, then there should be some clear correlations in the archaeological record. If spatial cognition in general did not evolve as an aid to hunting, then the sex differences in spatial cognition cannot have resulted from selection for hunting skill. As the evidence for the evolution of spatial thinking is the less well known, I will begin with it.[1]

The evidence of stone tools suggests that spatial cognition did not evolve as a package of linked abilities. Some of the abilities for which there are modern sex differences appeared early in the archaeological record, others very late. *Spatial perception,* the ability to see past local perceptual frames, appeared earliest. The rudimentary symmetry and regular diameters imposed on early bifaces and discoids (1-1.5 mya) indicate that the knapper could conceive a shape (two-dimensional) that was independent of the spatial frame of the blank (Fig. 5). This is analogous to the frame independence required in the water level task. Evidence for *mental rotation,* the ability to manipulate perspectives in the mind, appeared much later in the archaeological record. The three-dimensional, congruent symmetries imposed on late Acheulian bifaces required the ability to imagine invisible cross-sections and monitor many viewpoints simultaneously (Fig. 6; Wynn 1989; Wynn et al. 1996). This is directly analogous to the rotated polygons of mental rotation tests. Some late Acheulian knappers were apparently very good at mental rotation, as evidenced by the beautiful twisted profile handaxes present in many late Acheulian assemblages. The advent of the application of mental rotation to stone tools is hard to pinpoint. It was certainly present 300,000 years ago, but may well have appeared several hundred thousand years earlier. However, it made an appearance several hundred thousand years *after* evidence for visual perception.

Spatial visualization and spatio-temporal ability are harder to recognize in the archaeological record. Spatial visualization is the ability to detect patterns hidden in complex backgrounds. It is tempting to argue that prepared core techniques required this conceptual ability, as with simple Levallois cores, but this would be a serious overinterpretation (Fig. 7). As Schlanger has argued (1996, see above), Levallois results from the application of a set of knapping conventions. These are very interesting in their own right, but do not include "seeing" a flake in an unknapped core. Better evidence might come from selection of blocks or nodules of raw material, if it were clear that knappers perceived embedded patterns that indi-

The Role of Archaeology in Cognitive Science

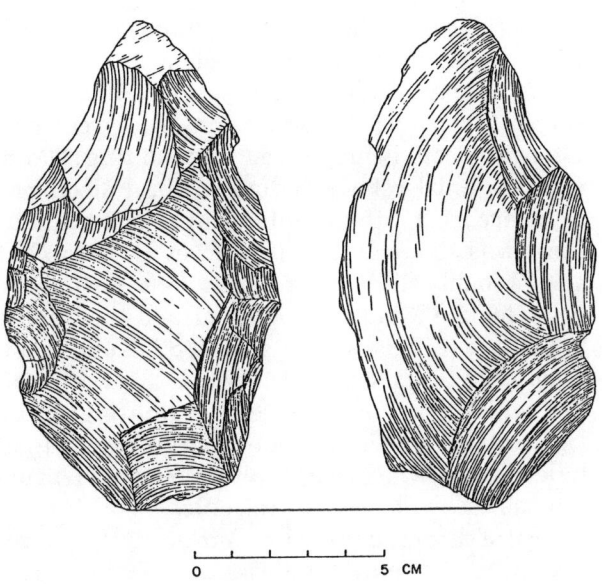

Fig. 5. Two-dimensional symmetry of this early handaxe (about 1.4 million years old) required developments in *spatial perception* beyond those known for apes, or for earlier hominids.

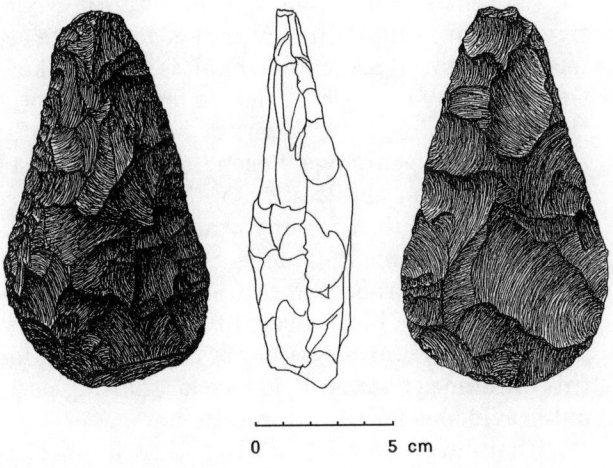

Fig. 6. The three-dimensional, congruent symmetries of late handaxes (after 500,000 years ago) required *mental rotation*.

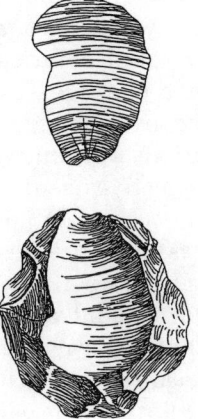

Fig. 7. A Levallois core and flake. Despite its significance in interpretations of culture history, Levallois reveals nothing significant about spatial abilities.

cated good knapping qualities. Similarly, selection of knappable edge angles might also require spatial visualization, but it would be necessary to eliminate trial and error knapping. It must be clear, for example, that a knapper selected the best knappable angle from a number of possible, but unused angles, and that he or she did this consistently. Raw material selection and platform selection are, then, possible lines of evidence for spatial visualization. Though there are obvious caveats (we rarely have the range of possibilities from which something was chosen, for example), the raw material and edge selectivity true of even the earliest stone tools suggests that some kind of visualization was in place.[2] However, truly convincing evidence of spatial visualization does not appear until very late, when early Upper Paleolithic artists recognized patterns in the walls of caves and expanded these natural patterns into images (Ucko and Rosenfeld 1967),

Archaeologists cannot document spatio-temporal ability directly because it incorporates an element of small scale timing. We must therefore rely on indirect evidence, and the only obvious candidate is the thrown projectile. Even here we must make a number of assumptions before it is possible to document an evolutionary sequence. Presumably, hunters would not invest a great deal of energy in the shape of a projectile unless they had some confidence in their ability to hit a target, so there should be a rough correlation between energy invested in projectiles and accuracy. I do not accept Calvin's (1993) and O'Brien's (1981) arguments that handaxes were projectiles (see Jerison, this volume), and in fact see no evidence for investment in *thrown* projectiles until very late. The Clacton spear (Oakey et al. 1977) and those from Schoeningen (Thieme 1997) document the existence of spears by late in the Lower Paleolithic, and Thieme has even argued that they were thrown. However, it is not until the Early Upper Palaeolithic that there is evidence for serious investment in projectile technology (Knecht 1993), and even then it seems unlikely that hunters tried to hit *moving* targets, which is the only circumstance in which spatio-temporal ability would have been of assistance.

In sum, evidence for spatial perception appeared early, evidence for mental rotation several hundred thousand years later, and convincing evidence for spatial visualization and spatio-temporal acuity appeared very, very late.

Although interpretations of hunting have been common in Paleolithic archaeology for over a century, evidence for the evolution of hunting is, somewhat ironically, harder to document than the evolution of spatial ability. Moreover, it is also the subject of considerable professional disagreement. There is no convincing evidence for hunting of large animals until relatively late in the Paleolithic. Evidence from late Acheulian sites indicates that there was some hunting (Singer et al. 1993), but also probably scavenging. Much the same picture holds for the Middle Palaeolithic and the Middle Stone Age (Singer and Wymer 1982; Kuhn 1993; Shea 1993). It is not until very late that archaeology paints a picture of large scale hunting on the logistical end of the forager-collector continuum. Indeed, this evidence appears at about the same time as the evidence for projectiles.

All of this argues against hunting as the selective agent for spatial abilities, and by extension against the hunting hypothesis for the sex difference in modern cognition as well. First, because the individual cognitive abilities apparently evolved at different times, it is unlikely that they were selected by the same specific activity. Selective pressures 1.5 million years ago *may* have selected for spatial perception, but it is very unlikely that the same pressures selected for mental rotation one million years later, or spatio-temporal acuity 50,000 years ago. Second, the evolution of mental rotation abilities was relatively late, and does not correlate with any other obvious evolutionary development. This is especially relevant for the sex difference literature because mental-rotation has the largest sex difference in modern populations. Hunting could not have selected for this difference because evidence for hunting postdates evidence for mental rotation!

There are several other hypotheses concerning the evolution of sex differences in spatial cognition, including the current most popular one that focuses on the female advantage in spatial visualization (Silverman and Eals 1992). All run into similar problems when set against the archaeological record. Spatial abilities do not appear to have evolved as a package; instead, they appear piecemeal in the archaeological record and have no clear correlations with known evolutionary developments or selective agents. This is in keeping with evidence for a female advantage in verbal memory, which is one of the most salient cognitive sex differences (Halpern 1992). It seems unlikely that selection for this advantage would have occurred prior to the appearance of modern language, which most paleoanthropologists believe was very late in human evolution. The paleoanthropological evidence just does not provide support for adaptationist hypotheses for the origin of the sex differences in cognition.

It is far more parsimonious to conclude that the modern sex difference in spatial cognition is an evolutionary side effect of selection acting on some other characteristic, or set of characteristics (Wynn et al. 1996). The most likely candidate is the timing of prenatal androgens. Male fetuses develop testes at about seven weeks, and the testes then flood the system with testosterone. This testosterone then inhibits cerebral development at a point when the right hemisphere is more developed than the left (Falk 1987; Geschwind 1987; Halpern 1992). As a consequence, male brains are more asymmetric, and function is more lateralized. Female brains develop more connectivity between hemispheres (they do not have the same androgen levels) and are less lateralized in function. This appears to be the physiological basis of the sex difference in cognition. While this could have been the proximate mechanism that was selected for in the adaptationist models, this seems unlikely. Selection would have had to favor males with the proper androgen timing by means of a slight spatial advantage in adulthood. It seems far more likely that selection operated directly on fetal viability. The minor sex difference in cognition was simply a side effect.

This scenario meshes better with the archaeological record than the adaptationist models. Selection may have acted on specific spatial abilities at several points in human evolution, but would have acted on males and females equally. The sex difference would appear at all stages because of its basis in fetal development. But the sex difference was itself never selected for. The archaeological record does not yet tell us what these selective pressures might have been, though the change in foraging pattern associated with early *Homo erectus* (Cachel and Harris 1996) presents the provocative possibility that better navigational skills might have been selected. Alternatively, selection might have acted on general intelligence, by any of a number of possible selective agencies. Spatial abilities would also have developed along with the general processing abilities, and, once again, the sex difference would have resulted from selection on the separate domain of fetal development.

In sum, the archaeological evidence suggests that the modern sex difference in spatial cognition is a side effect. It is not the result of selection. Given that spatial thinking provides the largest of all sex differences, this raises the possibility that many modern sex differences in cognition are simply evolutionary by-products. There may be no evolutionary reason for them at all.

General Conclusion

The archaeology of cognition adds important pieces to the puzzle of hominid evolution. It can describe specific cognitive abilities, and document the time they appeared in the evolutionary past. When added to the information supplied by other branches of paleoanthropology, the result is a richer picture of human evolution. It can even add to the resolution of traditional problems, such as the transition from Archaic *Homo sapiens* to modern humans. But to do this archaeology must employ established theories and concepts of cognition; implicit understanding simply will not do, except perhaps for popular story telling. The major task of an archaeology of cognition is methodological. It is necessary to translate the variables employed in cognitive theories into ones visible in the archaeological record. If this is done in an explicit and convincing way, then the underlying theory will supply powerful interpretive insights.

While the archaeology of cognition can, and probably should, address problems of traditional concern in paleoanthropology, it is not limited to these. Yes, insight into the cognitive components of the transition from *Homo erectus* to *Homo sapiens* would be important to our narrative story of human evolution, but this is largely a parochial issue grounded in the familiar categories of paleoanthropology. The archaeology of cognition can also broaden the scope of evolutionary studies by introducing the categories and concerns of cognitive science. It *is* possible to check the explicit and implicit evolutionary predictions of cognitive science against the archaeological record. Current evolutionary hypotheses for the sex difference in spatial cognition do not fit the archaeological evidence and must therefore be rejected or modified. Concerns like this one are not part of the traditional body of paleoanthropological problems, but they have direct relevance to our understanding of the modern human condition. Because of this, they can and should become concerns of Paleolithic archaeology.

Notes

[1] The following discussion has been influenced by Kate Robson Brown, who drew my attention to several of the relevant spatial abilities.
[2] Toth (Toth et al. 1993) has argued that the bonobo

Kanzi's stone knapping abilities are limited by his inability to recognize the best flaking angles, an ability clearly held by Oldowan knappers.

References Cited

Binford, L.
- 1981 *Bones: Ancient Men and Modern Myths.* Academic Press, New York.
- 1982 Comment on 'Rethinking the Middle-Upper Palaeolithic transition' by R. White. *Current Anthropology* 23:177-181.

Byers, A. Martin
- 1994 Symboling and the Middle-Upper Palaeolithic transition: A theoretical and methodological critique. *Current Anthropology* 35:369-400.
- n.d. Communication and material culture: Pleistocene tools as action cues. *Cambridge Archaeological Journal* (in press).

Cachel, S. and J. Harris
- 1995 Ranging patterns, land-use and subsistence. In *Homo erectus* from the perspective of evolutionary ecology. In *Evolution and Ecology of Homo erectus*, edited by J. Bower and S. Sartono, pp. 51-66 Leiden, Pithecanthrous Centennial Foundation.

Calvin, W.
- 1993 The unitary hypothesis: a common neural circuitry for novel manipulations, language, plan-ahead, and throwing? In *Tools, Language, and Cognition in Human Evolution*, edited by K. Gibson and T. Ingold, pp. 230-250. Cambridge University Press, Cambridge.

Dennell, R.
- 1997 The world's oldest spears. *Nature* 385:767-768.

Falk, D.
- 1987 Brain lateralization in primates and its evolution in hominids. *Yearbook of Physical Anthropology* 30:107-125.

Fodor, J.
- 1983 *The Modularity of Mind.*, MIT Press, Cambridge, MA.

Gardner, H.
- 1983 *Frames of Mind: The Theory of Multiple Intelligences.* Basic Books, New York.

Geary, D.
- 1996 Sexual selection and sex differences in mathematical ability. *Behavioral and Brain Sciences* 19:229-284.

Geschwind, N.
- 1987 *Cerebral Lateralization.* MIT Press, Boston.

Halpern, D.
- 1992 *Sex Differences in Cognitive Abilities,* 2nd ed. L. Erlbaum, New Jersey.

Jardine, R. and N. Martin
- 1983 Spatial ability and throwing accuracy. *Behavior Genetics* 13:331-340.

Knecht, H.
- 1993 Early Upper Paleolithic approaches to bone and antler projectile Technology. In *Hunting and Animal Exploitation in the Later Paleolithic and Mesolithic of Eurasia*, edited by G. Peterkin, H. Bricker and P. Mellars, pp. 33-48. Archaeological Papers of the American Anthropological Association no. 4, Washington, D. C.

Kuhn, S.
- 1993 Mousterian technology as adaptive response. In *Hunting and Animal Exploitation in the Later Paleolithic and Mesolithic of Eurasia*, edited by G. Peterkin, H. Bricker and P. Mellars, pp.25-32. Archaeological Papers of the American Anthropological Association no. 4, Washington, D. C.

Mithen, S.
- 1994a From domain specific to generalized intelligence: a cognitive interpretation of the Middle/Upper Paleolithic transition. In *The Ancient Mind: Elements of Cognitive Archaeology*, edited by C. Renfrew and E. Zubrow, pp. 29-39. Cambridge University Press, Cambridge.
- 1994b Technology and society during the Middle Pleistocene: Hominid group size, social learning and industrial variability. *Cambridge Archaeological Journal* 4:3-32.
- 1996 *The Prehistory of the Mind: The Cognitive Origins of Art, Religion, and Science.* Thames and Hudson, London and New York.

Oakley, K., P. Andrews, L. Keeley, and J. D. Clark
- 1977 A reappraisal of the Clacton spearpoint. *Proceedings of the Prehistoric Society* 43:13-30.

O'Brien, E.
- 1981 The projectile capabilities of an Acheulian handaxe from Olorgesailie. *Current Anthropology* 22:76-79.

Pelegrin, J.
   1990 Prehistoric lithic technology: Some aspects of research. *Archaeological Review from Cambridge* 9:116-125.

Piaget, J. and B. Inhelder
   1967 *The Child's Conception of Space*. (Trans. R. Langlon and J. Lunzer.) Norton, New York.

Renfrew, C. and E. Zubrow (eds.)
   1994 *The Ancient Mind: Elements of Cognitive Archaeology*. Cambridge University Press, Cambridge.

Robson Brown, K.
   1994 An alternative approach to cognition in the Lower Paleolithic: The modular view. *Cambridge Archeological Journal* 3:231-245.

Rozin, P. and J. Schull
   1988 The adaptive-evolutionary point of view in experimental psychology. In *Steven's Handbook of Experimental Psychology, vol. 1: Perception and Motivation*, edited by R. Atkinson, R. Herrnstein, and R. Luce, pp. 503-546 John Wiley and Sons, New York.

Schlanger, N.
   1994 Mindful technology: unleashing the *chaîne opératoire* for an archaeology of mind. In *The Ancient Mind: Elements of Cognitive Archaeology*, edited by C. Renfrew and E. Zubrow, pp. 143-151. Cambridge University Press, Cambridge,

Shea, J.
   1993 Lithic use-wear evidence for hunting by Neandertals and early modern humans from the Levantine Mousterian. In *Hunting and Animal Exploitation in the Later Paleolithic and Mesolithic of Eurasia*, edited by G. Peterkin, H. Bricker and P. Mellars, pp.189-198. Archaeological Papers of the American Anthropological Association no. 4, Washington, D. C.

Silverman, I. and M. Eals.
   1992 Sex differences in spatial abilities: Evolutionary theory and data. In *The Adapted Mind: Evolutionary Psychology and the Generation of Culture*, edited by J. Barkow, L. Cosmides, and J. Tooby, pp. 487-503. Oxford University Press, New York.

Sinclair, A.
   1995 The technique as a symbol in Late Glacial Europe. *World Archaeology* 27:50-62.

Singer, R. and J. Wymer.
   1982 *The Middle Stone Age at Klasies River Mouth in South Africa*. University of Chicago Press, Chicago.

Singer, R., B. Gladfelter, and J. Wymer.
   1993 *The Lower Paleolithic Site at Hoxne*. University of Chicago Press, Chicago.

Steele, J.
   1995 Stone tools and the linguistic capabilities of earlier hominids. *Cambridge Archaeological Journal* 5:245-256.

Thomas, H. and A. Lohaus.
   1993 Modeling growth and individual differences in spatial tasks. *Monographs of the Society for Research in Child Development* 58:9:1-192.

Thieme, H.
   1997 Lower Paleolithic hunting spears from Germany. *Nature* 385:807-810.

Toth, N., K. Schick, E. Savage-Rumbaugh, S. Sevcik and D. Rumbaugh.
   1993 Pan the tool-maker: Investigations into the stone tool-making and tool-using capabilities of a bonobo (*Pan paniscus*). *Journal of Archaeological Science* 20:81-91.

Voyer, D., S. Voyer, and M. Bryden
   1995 Magnitude of sex differences in spatial abilities: A meta-analysis and consideration of critical variables. *Psychological Bulletin* 117:250-270.

Wynn, T.
   1989 *The Evolution of Spatial Competence*. University of Illinois Press, Urbana.
   1993 Two developments in the mind of early *Homo*. *Journal of Anthropological Archaeology* 12:299-322.

Wynn, T., F. Tierson, and C. Palmer
   1996 The evolution of sex differences in spatial cognition. *Yearbook of Physical Anthropology* 39:11-42.

2. The Re-Emergence of Cognitive Archaeology

April Nowell

That the cognitive side of the archaeological record has been somewhat overlooked in favor of techno-environmental considerations, and that archaeologists may be unaccustomed to looking for it, then, is no argument in support of the putative difficulty of accessing the prehistoric cognitive systems in the material record of the past. (Whitley 1992)

Abstract

Once thought of as paleopsychology, cognitive archaeology has matured into a serious subject of study. Paleolithic archaeologists study human intelligence, language and symboling behavior through the material culture of prehistoric hominids. This study of the archaeological record is not performed in a vacuum and increasingly archaeologists are turning to psychology, primatology, and biology among other disciplines for additional theory and data. It is argued that with its emphasis on the scientific method, anthropological framework and multidisciplinary perspective cognitive archaeology is a natural development from processual archaeology in general and Paleolithic archaeology in particular.

Introduction

All archaeologists seek to understand human cultural variability in the past—this is one of the goals that defines our discipline. Archaeologists who study the Paleolithic period, however, face even more fundamental questions: At what point in time can we speak of "humans" in the archaeological record? When did hominids acquire language and symbolic thought? What were the capabilities and lifeways of our earliest ancestors? Uniting these issues is the question of the evolution of human intelligence.

Through the analysis of material culture, archaeologists have traditionally reconstructed subsistence strategies, technological innovations and even social structure and migrations. Delving into the human mind and tracing its evolution, however, is considered by some to be problematic at best and has been described as "paleopsychology" at worst (Binford 1965:203-210). This chapter will introduce cognitive archaeology, briefly trace three examples of archaeological schools that have approached studying human intelligence, and situate this area of research within a multidisciplinary approach to studying the evolution and development of the human mind.

Cognitive Archaeology and the Paleolithic Period

Since the 1980s human cognitive abilities has become an increasingly accepted area of research (see, for example, papers in Renfrew and Zubrow 1994; Mellars and Gibson 1996). This movement reflects a renewed interest in the evolution of the human mind that is taking place in other sciences as well (Mithen 1995). What is now called *cognitive archaeology* or the "archaeology of mind" can be divided into two types—cognitive archaeology of the Paleolithic and cognitive archaeology of the Neolithic period and onwards. The first type, as its name suggests has as its temporal dimension the Paleolithic period and focuses on the *evolution* and *development* of human cognition, language and symboling. Archaeologists within this area of study draw upon biology, primatology and psychology among other disciplines. They integrate these datasets and perspectives with archaeological data in order to coherently study and explain the evolution of the human mind. Examples of the kinds of studies which fall within the scope of this branch of cognitive archaeology include looking at what makes the human form of communication unique among the primates (Savage-Rumbaugh and Rumbaugh 1993; Noble and Davidson 1996), trying to understand the minimum cognitive abilities necessary to make stone tools (Gowlett 1984, 1986, 1996; Wynn 1979, 1981, 1991 and looking at how the similarities and differences between the minds of Neanderthals and modern humans might be expressed in the archaeological record (Mithen 1996). Much of the pioneering work in cognitive archaeology can be traced to papers by Thomas Wynn (e.g. 1979, 1989, 1998) who initially integrated archaeological evidence with developmental psychology and who more recently has begun to turn to other areas of cognitive science (see Wynn, this volume). Other important work has been accomplished by the archaeologist-psychologist team of Davidson and Noble (e.g. 1989, 1993a, b, this volume) and as well as by archaeologist Mithen (e.g. 1996) and psychologist, Donald (1991).

Cognitive archaeologists share many goals with other archaeologists and do not consider their specialization as anything more or less than a subdivision within archaeology (Flannery and Marcus 1993; Renfrew 1994). What makes cognitive archaeology of the Paleolithic unique is that at some point in time our ancestors were not "human" as we understand the term. Not knowing if the subjects we are studying had modern minds changes the rules of the game. Certain assumptions that we take for granted in later periods cannot be made here. For instance, in the search for the origins of art some scholars have made the argument that you cannot discount the possibility that hominids in the Lower and Middle Paleolithic in Europe engaged in complex rituals simply because we have no real evidence of them—that their absence may be due to taphonomy (Bednarik 1994, 1995; Chase and Dibble 1992; see also discussion by Byers, this volume). Employing ethnographic analogies, they point to the fact that many modern traditional societies engage in rich ceremonial rites that leave no trace for the archaeologist to discover (see for example Hewes 1989). In addition to the problems associated with arguing from negative evidence, the flaw in this type of argument is that with modern societies, they are just that—modern. We know to link certain behaviors with the modern mind. In the Paleolithic, we cannot assume what we are trying to prove (Chase and Dibble 1992; see also Gowlett 1986:243). In other words, when archaeologists study material remains as a means of documenting the emergence of the modern mind, they cannot assume that certain behaviors and objects must have existed in the past and then use their assumed presence to argue for modern symboling capacities among our early ancestors. At its extreme, this type of reasoning is circular and meaningless. Furthermore, as Chase (n.d.) notes, while differential preservation of materials is a serious problem "at a fine temporal and geographic scale...on a large scale it is less significant."

Cognitive archaeology extends the boundaries of traditional archaeological inquiry. New questions are asked of old material. Studying lithics now involves questions such as does the Oldowan represent a cognitive milestone relative to the material culture of the apes (Toth 1985b; Wynn and McGrew 1989; Toth and Schick 1993; Joulian 1996)?; can we discern handedness in the archaeological record (Toth 1985a; Pobiner 1999) and if so what are the implications for the development of cerebral asymmetries and language (Frost 1980; Wilkens and Wakefield 1995)?; and what can artifact manufacture, morphology and variation reveal about language (Dibble 1989; Wynn 1993; Graves 1994), mental templates (Gowlett 1984, 1996; Wynn and Tierson 1990, McPherron 1994) and decision making (Boëda et al. 1990; Schlanger 1994)?

We are also asking old questions of new material. Traditional questions concerning subsistence strategies and social structure are applied to primatology to provide models for human evolution. For instance, some primatologists emphasize the importance of what has been coined Machiavellian or social intelligence (see papers in Byrne and Whitten 1988a; Weaver et al., this volume; for further discussion). The social realm of primates is exceedingly complex (e.g. Cheney and Seyfarth 1990). It is here that access to food, mates, shelter and any other resource in limited supply is worked out. Some primates have been known to create alliances and to deceive one another (de Waal 1982; Byrne and Whitten 1988b, 1988c, 1990; Savage-Rumbaugh and Lewin 1994:272-76). By drawing on the work of primatologists and human paleontologists who look at the social structure, as well as the neo-cortex size of various primate groups (Aiello and Dunbar 1993), researchers make inferences regarding the evolution of human social intelligence and language. Similarly, another primatologist, Milton (1981, 1988) studies the relationship between body size, home-range, spatial and temporal distribution of plant foods in tropical forests and mental complexity in Howler and Spider monkeys. She suggests that the increase in brain and body size and the reduction in dentition size noted throughout hominid evolution reflect changes in diet and the participation of hominids in increasingly complex subsistence strategies such as cooperative hunting and a division of labor with food sharing. Therefore, conventional questions concerning subsistence strategies and social structure, when applied to primatology and linked with biology, can provide the cognitive archaeologist with valuable material with which to devise and test new hypotheses.

While it is obvious from this brief discussion that cognitive archaeologists are broadening the scope of traditional archaeological inquiry, this area of research does have its limitations. Wynn (n.d.) observes that there are three difficulties inherent to cognitive archaeology. First, is the problem of resolution. Because archaeological evidence is largely anecdotal it is difficult test hypotheses, quantify data and replicate experiments. Second is the problem of preservation. As noted, due to taphonomic reasons many aspects of the archaeological record will be lost. If early hominids practiced tattooing or ritual dance these behaviors will not be recoverable in the archaeological record. At the same time it should be possible to discern other correlates of symboling behavior if it exists. Finally there is the issue of the minimum necessary competence. As Wynn (n.d.:3) writes, "we cannot logically eliminate the possibility that our prehistoric subjects employed simple reasoning when producing and using stone tools, and much more complex reasoning in archaeologically invisible domains" but if all aspects of early hominid behavior are consistent with the lithic assemblages then the "interpretation of minimum competence becomes more reliable."

Cognitive Archaeology and the Neolithic Period

The second type of cognitive archaeology, while equally challenging, is not addressed in this volume and, therefore, will not be discussed in detail here. Essentially, this area of study focuses primarily on the religion, iconography and ideology of cultures from the Neolithic to the present, drawing on sources such as historical records and ethnographic accounts to supplement archaeological investigations. According to some researchers this approach is "an opportunity to make mainstream archaeology more holistic whenever possible (Flannery and Marcus 1996:351)." Examples of this type of research include studying petrogylphs in Britain (Bradley 1994) the symbolism of ancient Greek statues (Schnapp 1994) and ancient writing systems in order to document changes in information processing (Juteson 1994). Flannery and Marcus (1996:351) provide one of the first formal definitions of cognitive archaeology and it is most appropriate for the Neolithic period onward. For these researchers cognitive archaeology is:

> ...the study of all of those aspects of ancient culture that are the product of the human mind: the perception, description, and classification of the universe (cosmology); the nature of the supernatural (religion); the principals, philosophies, ethics and values by which human societies are governed (ideology); the ways in which aspects of the world, the supernatural, or human values are conveyed in art (iconography); and all other forms of human intellectual and symbolic behavior that survive in the archaeological record.

While these two forms of cognitive archaeology differ in their goals, methodology, theoretical frameworks, data and temporal focus, what they do share is an interest in the human mind and a belief that it can be studied through the archaeological record. With many notable exceptions what they also sometimes share is a lack of sound methodology. As Flannery and Marcus (1996:3511) note some researchers have, "seen [cognitive archaeology] as the shortcut to a kind of 'armchair archaeology' that requires no fieldwork or rigorous analysis of any kind. ... any fanciful mentalist speculation is allowed, so long as it is called 'cognitive archaeology'."

Many archaeologists cringe when they hear the term *cognitive archaeology* and it is precisely the "fanciful" approaches referred to by Flannery and Marcus that incur such a reaction. Cognitive archaeology is *not* about negotiating the archaeologist's relationship with the past and it is certainly *not* about trying to get *into the mind* of a Neanderthal in the sense of being able to reconstruct their actual thoughts. Cognitive archaeology *is* about documenting the evolution of human language, intelligence and symboling and understanding the wealth and diversity of human symbolic culture *when and where appropriate*. Most of the work done in cognitive archaeology is sound but if these less rigorous approaches are not weeded out then cognitive archaeology will remain stigmatized as unscientific and the human mind will remain epiphenomenal and beyond the realm of archaeological investigation.

Specifically, there are two reasons cognitive archaeology of the last decade has developed such a poor reputation. The first derives from an impression that this kind of research represents a short-lived trend in archaeology whose popularity will soon wane. In reality, there is a long tradition of archaeologists studying the human mind and a great history of debate concerning the best methods for doing so. Thus, how to study the mind is hardly a new question. The second reason underlying the reluctance to accept cognitive archaeology is the fact that its goals and methods are thought to be poorly defined. Ironically, it can be argued that cognitive archaeology has its roots in the scientific method and anthropological investigation. Each of these objections to cognitive archaeology will be addressed in turn.

Examples of Archaeologists Studying the Mind

It is beyond the scope of this chapter to present the work of all of those who have contributed to the question of studying the human mind through the archaeological record, such as the Marxists (see, for example, McGuire 1992, 1993) and a variety of social anthropologists. Instead, the processualists, the structuralists, and the post-processualists (as a unified group) will be discussed as three examples of archaeological schools that have wrestled with this problem.

Hawkes' Hierarchy

While, there are a number of ways of approaching a history of archaeologists studying the mind, the issue is most effectively thought of as a question of what behavioral inferences researchers believed could be made from archaeological data. Arguably, the starting point in this debate was the publication of Christopher Hawkes' 1954 paper in which he articulated his notion of a *'ladder of inference'*—an idea that became known as *Hawkes' Hierarchy*. Although the idea of an "ascending scale of difficulty in reconstructing a culture's technology, economics, sociopolitical organization and religious beliefs" (Trigger 1989:266) has played a major role in archaeology in Britain since the 1930s it was Hawkes who really elaborated the concept. He argued that without recourse to textual material and oral traditions, archaeologists could more easily reconstruct what is "generically animal in man [than what is] more specifically human (Hawkes 1954:162)." In other words, drawing on the natural sciences archaeologists could more easily reconstruct subsistence strategies and technology than religion because the former conform to relatively straightforward rules of physics and biology (see also Childe 1956). Accordingly, he proposed four levels of data: (1) *techniques*, (2) *subsistence-economics*, (3) *socio/political institutions* and (4) *religious institutions and spiritual life*. The hierarchical ladder of inference based on these categories means that it is it is easier to reconstruct a culture's technology than its social structure, which in turn is still easier to reconstruct than its ideological realm. Hawkes found it ironic that the ideational realm would, for the most part, remain unknowable because it is this realm that renders humans unique. But it is precisely its uniqueness that makes this area unpredictable and thus less knowable. He summarizes the situation as "the more human, the less intelligible (ibid.)." Since then "archaeologists have debated whether this hierarchy is inherent in the nature of archaeological data or results from the failure of archaeologists to address relevant interpretive problems (Trigger 1989:392)."

Binford and the Processualists

In the 1960s Lewis Binford and the processualists argued that archaeologists, using the appropriate methodology, could study *all* aspects of sociocultural systems (Trigger 1989:392). The processualists took Hawkes' assertion that it was difficult to reconstruct the past because it was not directly observable and tried to overcome it (Leone 1982:743). This confidence was a direct result of one of the basic tenets of processualism, which was a belief in the scientific method. Processualists emphasized scientific rigor, empiricism, and hypothesis testing. They made use of statistics and other methods of quantitative analysis as well as computer-generated random sampling (Steibing 1993). The processualists searched for "Laws" of cultural development and change and sought to understand human behavior.

Binford (1962, 1965, 1972) and the processualists following Leslie White (1959) viewed cultures as composed of three interrelated subsystems (similar to Hawkes' levels of inference): technology, social organization and ideology. Thus, in 1962, Binford suggested that an artifact had its primary role in only one of these three subsystems but by 1965 he came to believe that every artifact had technomic, sociotechnic and ideotechnic aspects and thus reflected all three subsystems simultaneously (Trigger 1989). This position allowed him to argue that it was possible to reconstruct a past culture through a systematic study of artifact assemblages and their contexts. In fact, he explicitly "repudiated the idea that it was inherently more difficult to reconstruct social organization or religious beliefs than it was to infer economic behavior" (Trigger 1989:298).

Ironically, Binford nevertheless reiterated Hawkes' stance by asserting that archaeologists were better able to reconstruct a culture's technological and economic systems than its ideational realm. Like Hawkes, he argued that whereas the physical and biological sciences provided ways of reconstructing a culture's subsistence strategy, the relationship between material culture and social organization and beliefs was not *yet* well understood. Binford (1962) maintained that in order to construct these kinds of correlations archaeologists must be trained as ethnologists. Only by studying modern groups where "behavior and ideas can be observed in conjunction with material culture was it possible to establish correlations that could be used to infer social behavior and ideology reliably from the archaeological record (Trigger 1989:300)." Therefore, while affirming that all levels of inference were possible, Binford was forced to admit that archaeology was not yet at the stage where it could reliably reconstruct the ideational realm.

Binford's confidence in the processualist approach was further based on the neo-evolutionist belief that there is a great deal of regularity to human behavior. Like 19th century unilinearists, neo-evolutionists argue that if one aspect of a culture can be reconstructed, especially if it is its subsistence pattern, then all aspects of a culture can be recovered. Binford cautioned that inferences could only be made from material culture if it could be demonstrated that a particular behavioral trait and an object were *always* paired—the discovery of one would then imply the existence of the other in the archaeological record. He further asserted that "shortcomings were not the responsibility of archaeologists but rather resulted from the failure of ethnologists to make good enough theories that related technology and environment to the rest of culture" (Trigger 1989:294).

Finally, Binford adhered to White's 1959 definition of culture as a people's extrasomatic means of adaptation to the environment around them. This led the processualists to describe all cultural change as adaptive responses to changes in the environment. In an idealized scenario any group of people living in a static environment would experience little cultural growth. Because of this reliance on external environmental factors to explain culture change, Binford discounted the importance of *psychological* factors for understanding prehistoric human behavior. Trigger writes, "within an ecological framework specific psychological factors can be viewed as an epiphenomenal aspect of human behavior that arose as a *consequence* of ecological adaptation (1989:303, my emphasis)." In other words, because of the many regularities inherent in human behavior, the particular thoughts and beliefs of individuals in the past were not only unknowable (archaeologists were *not* "paleopsychologists" [Binford 1972:98]) but irrelevant.

Structuralism

In contrast to the processualists, traditional structural archaeologists (e.g. Leroi-Gourhan 1967a, 1967b, 1968; Conkey 1978; Hodder 1984, Deetz 1977), drawing on the work of Claude Lévi-Strauss (1966) argued that reconstructing the mind, far from being epiphenomenal, was in reality of central importance to archaeologists (Leone 1982:743). They argue that the mind functions in a logical and predictable fashion analogous to mathematics or grammar. The underlying assumption of structuralism is that "the human mind categorizes and divides; creates contrasts and oppositions; that it reverses, displaces, and distinguishes between inside and outside, culture and nature, male and female; furthermore, that the mind uses a limited repertoire of contrastive categories like these to think about virtually all reality (Leone 1982:742)." According to structuralist theory, the past is knowable. Since the modern mind emerged, our brain has continued to categorize and divide the world in much the same way although the details of this organization vary slightly from culture to culture.

These regularities in the manner in which the brain structures reality allow archaeologists to study human cognition through *all* artifacts. For the structuralists the artifact categories envisioned by the early processualists (i.e. technomic, sociotechnic and ideotechnic artifacts) represented an artificial and meaningless division: all artifacts are equally capable of shedding light on past human behavior because the human mind created them. This is one of the main tenants of structuralism (Leone 1982). Indeed, similar to the processualists, the structuralists assumed that the past could be reconstructed if any of a culture's artifacts remained. "While the details of a culture may be lost, the components of the underlying logical organization cannot be (Leone 1982:743)." This view prompted some archaeologists to devise rules or a "grammar" of culture to explain how these categories related to the world (e.g. Deetz 1967; Glassie 1975).

A classic example of structuralism in Paleolithic archaeology is Leroi-Gourhan's study of Upper Paleolithic cave paintings (1967a, 1967b, 1968:58-

70). In his analysis of parietal art Leroi-Gourhan considered all paintings as forming part of one single tradition which could then be studied in terms of fundamental structuralist principles—inside/outside, nature/culture, male/female and life/death (Leone 1982). For example, weapons were considered to be "male" symbols while wounded animals were thought to be "female" symbols. Leone (1982:744) writes:

> Two fundamental oppositions, one inflicting pain and death and the other suffering pain and death, are essential to each other in providing life. Thus Magdalenian logic linked women and wounded bison, horses and ibex and counterpoised them to men, lions and bears while seeing all within a single paradigm, not as scattered unrelated items of sympathetic magic.

His study was praised because he was able to tease out commonalties between various painted caves irrespective of the time, place or subject matter of the individual paintings and to use this unity as a tool for getting at cognition. In a sense, it was a sophisticated treatment of the topic that went beyond the simplistic functionalist "hunting magic" hypothesis. Conkey writes (1996:301; see also Conkey 1989) that Leroi-Gourhan argued that:

> ...the images had been deliberately selected by species or type of geometric form, and placed in deliberate locales within any given cave, and in specific relations with respect to each other. That is he assumed there was an underlying structure or set of structural principals that had generated the imagery and it topographic arrangements.... The specific images had not been selected for depiction because of some sort of magical powers or needs; rather, they were selected because of the role they could play within a certain syntax that governed the arrangement as symbols. A well-preserved cave was itself a message (cave-as-text) with elements (frames and figures). The underlying 'formula' for the arrangement of images came to be known as a 'mythogram'.

By intentionally ignoring temporal and geographic variation, however, he dismissed the importance of real differences in style. For example, Altamira and Lascaux caves, both known to Leroi-Gourhan, exhibit tremendously different art styles and are separated by hundreds of kilometers and thousands of years. Furthermore, as Conkey (1996:297) notes, if one considers all of Paleolithic cave imagery, then horses and bison really are the two most frequently depicted species as Leroi-Gourhan argues, but in reality they do not always dominate *individual* caves—for example, Lascaux is characterized by aurochs and horses while Font-de-Gaume by bison and mammoth.

It is reasonable to collapse all of these paintings into one unit of study *only* if genetics are viewed as the driving factor behind what is seen as superficial variation. In order to conduct such an analysis one *must* assume that the modern mind structures reality in a particular way and always has since its emergence. If, on the other hand, differences in Upper Paleolithic parietal art are considered in the context of cultural change and ethnicity, which is how most researchers today believe they should be studied (see a related discussion in White 1992), then the genetic common denominator of Pleistocene hominids is irrelevant. Essentially, the structuralist approach conceals more than it reveals about hominid behavior and risks projecting modern symbols into the past. As Trigger (1989) notes, "how can it ever be proved that Leroi-Gourhan was correct or even moving in the right direction, in associating bison with female principals and horses with male ones in European Paleolithic cave art?"

Post-Processualism and the Contextual Approach

While the processualists were confident in their ability to reconstruct the ideational realm, Trigger (1989) notes that over the past three or four decades they have continued mainly to study the lower rungs of Hawkes' ladder. In surveying papers published in Schiffer's (1978-1986) first eight volumes of *Archaeological Method and Theory* he found that while 39% of them dealt with data recovery and chronology; and 47% with ecology, demography and economic behavior, only 8% of papers focused on social behavior, and 6% on ideology, religion and scientific knowledge. A more recent survey shows that these percentages have improved somewhat with the number of papers devoted to social organization and/or the ideational realm rising from 14% in 1986 to 30% in 1999 (pers. obs.).

Trigger argues that while the processualist approach has resulted in the identification of a number of external constraints on human behavior, in the vast majority of cases it is biological, ecological and technological factors that limit the range of forms that the economy, social organization and ideology can take. However, these constraining factors do not *strictly dictate* the forms taken. If it is true that there is no absolute correlation between these levels, then this observation represents a blow to archaeologists attempting to reconstruct the ideational realm because it means that the higher levels of Hawkes' ladder cannot be inferred from the lower ones.

Post-processualists or '*contextual*' archaeologists have attempted to overcome this conclusion by identifying constraints that apply specifically to these higher levels (Trigger 1989). As such, post-processualists seek regularities in human cognitive or psychological processes rather than ecological ones (ibid.). According to Hodder (1996) there are three basic tenets of the contextual approach. First, material culture is meaningfully constituted. In other words, there are socially constructed ideas and concepts that influence the way material culture is used. It seems from this statement that meanings are at least partially arbitrary and cannot be studied using cross-cultural scientific generalizations. However, "while material culture meanings may be historically arbitrary in this sense, they are not arbitrary in another sense. Any use of an artifact depends on previous uses and meanings of that artifact or of similar artifacts within a particular historical context (p.14)."

This brings us to the second tenet of post-processualism. Material culture has to be studied contextually. This contextualization depends on generalities and involves Geertz's (1973) notion of "*thick description*" of placing an object fully into its various contexts. The generalizations used to study archaeological data must be related to specific contexts and as a result contextual archaeology is not relativist (Hodder 1996). Third, material culture is active rather than passive. According to the post-processualists this means that material culture is not a by-product of human activity but rather it plays an active role in defining and shaping that behavior.

Hodder (1996:16) summarizes the problems that these points pose for the reconstruction of hominid behavior by writing:

> The three points discussed above raise a host of difficult questions. For example, if material culture is meaningfully constituted, how can archaeologists reconstruct the different meanings given to objects by long-dead people? If meanings are contextual, how do we know what the relevant context in the past was? If material culture is active and the meanings constructed, how can we use generalizations?

He addresses these difficulties by stating that not only is it possible to construct an understanding of "their" meanings using a contextual approach but that all archaeologists do so whether they admit it or not. He reasons, for example, that the mere act of calling an archaeological structure a house or a quarry or a temple is laying down a level of interpretation that could not be made without some assumption of what these structures "meant" to people in the past. He asserts that even the reconstruction of a subsistence system involves deciding what animals "they" thought were edible or useful for making clothes or tools. For Hodder, "the idea that archaeologists can get away without reconstructing ideas in the heads of prehistoric peoples is pure false consciousness and self delusion (1996:18)."

There is a great deal of difference between identifying a dwelling or describing the subsistence pattern of a group of people and surmising what a Neanderthal felt the first time he or she encountered a modern human or what modern humans thought about when looking at Upper Paleolithic cave art. As Hodder admits the ideas archaeologists reconstruct are not necessarily the conscious thoughts that would have been expressed by people if archaeologists could travel back in time and speak with them (ibid.). If this is the case then I do not believe that the processualists and post-processualists are as polarized on this issue as perhaps each would like to appear to be. It has been argued above that specific meanings and even conscious thoughts may be reconstructed where appropriate (for example, where textual evidence exists) but that in the majority of cases archaeologists must be content with more general meanings.

Processualism and the Roots of Cognitive Archaeology

While it is clear that there is a long history of archaeologists studying the human mind, knowing where to situate cognitive archaeology within this framework is less obvious. It is argued here that it is within processualism that the roots of cognitive archaeology lie. Some have mistakenly linked cognitive archaeology with the post-processualist movement of the 1980s and 90s described above (see for example, Whitley 1992). This confusion arises from a misunderstanding of the goals of cognitive archaeology, (e.g. Bender 1993). Essentially, post-processualism can be seen as a backlash against the perceived tyranny of the processualists' positivist and functionalist viewpoint (Renfrew 1994; see also a discussion in Patterson 1990). Post-processualists (e.g. Hodder 1982a, b;

Shanks and Tilly 1987a, b) underscore the need for the archaeologist to take into account his/her own biases in interpreting the past. Some have argued that there are no objective facts nor is there "one past" (see Hodder 1996:16). It is further argued that many voices are needed to produce a richer portrait of the past. This means including feminist and Marxist critiques and looking for ethnicity and the contributions of the individual in the archaeological record.

It should be clear that while the post-processualists fulfill a very important role in the history of archaeological theory, their goals have little to do with those of cognitive archaeology as they relate to the Paleolithic period. Arguably, cognitive archaeology *cannot* derive its approach or goals from post-processualism as this movement, in this specific context, functions best as a critique of the processualist school rather than a source of new methodology. Household archaeology and landscape archaeology represent examples of methodology derived from post-processualism, but they are less applicable to the Paleolithic.

Renfrew (1994, see also Renfrew and Bahn 1991) suggests that far from withering under the post-processualist gaze, processualism has entered a new intellectual phase in which its proponents increasingly focus on cognition. In fact, he uses the term *cognitive-processual archaeology* to acknowledge "both the principal field of study and the intention of working within a tradition that is broadly scientific...(1994:3)." While the term itself may be unwieldy, he does make a valid point. Studying cognition within this framework represents a development away from the romantic approach to archaeology. There is no place for the individual who wants to "dabble" in art or symboling behavior. The hypotheses that cognitive archaeologists propose must be based on concrete data and these data must somehow be reconciled with data from linguistics, neurology, psychology and so on. This is in great distinction to the post-processual or post-modern stance held by philosophers from many disciplines on the impossibility of objectivity. While we cannot achieve objectivity in an "ultimate" sense (Renfrew 1994:10), not all reconstructions of the past are equally valid as some post-processualists would claim and we must judge them based on the degree to which they explain the "facts" as we currently know them. As Renfrew (1982:11) writes: "The crux of the matter is to make explicit the assumptions and inferences which sustain the argument. To do this for archaeological reasoning in general was the principal goal of the New Archaeology. It remains that of processual archaeology, of which cognitive-processual archaeology is the logical and natural development and extension into the symbol-using fields of thought and communication"

While cognitive archaeology can be considered a "logical and natural" progression from processualism as Renfrew notes, it should be emphasized that while both emphasize behavior, processualism, as discussed above, ignores the thought processes behind this behavior. It is for this reason that Marcus and Flannery (1993) describe cognitive archaeology as a way of fleshing out the past— of rendering archaeology more holistic. Nonetheless, both processualism and cognitive archaeology use a scientific model to explain human behavior and in this way they differ from post-processualism.

A second aspect of processualism that laid the foundation for cognitive archaeology is its adherence to the aims and perspectives of cultural anthropology (Willey 1958; Binford 1962, 1965; Longacre 1970). It is the adoption of the anthropological perspective that allowed archaeology to move beyond the culture-historical approach. Cultural anthropology seeks to document and *explain* human cultural variation in the present while archaeology projects this study into the past. While this is a highly simplified definition of archaeology, it is this emphasis on human behavior and the explanation of culture change that acted as a fundamental precursor to archaeologists studying cognition. Without this intellectual framework we are geologists or natural historians but we are not anthropologists. These other disciplines (geology, natural history and the like) are not interested in cognition, language, art or symboling behavior nor are they equipped to study these questions. Since archaeology's inception as a discipline archaeologists have speculated about the minds, thoughts, beliefs, religion and ideology of past cultures, but it is only the joining of scientific method with anthropological concerns which provided archaeologists with a framework within which to study cognition. We should perhaps think of this union then as a *re-emergence* of cognitive archaeology.

Multidisciplinary Perspectives

Cognitive archaeology can also be seen as a natural development from Paleolithic archaeology because they both are very multidisciplinary in their approaches. Paleolithic archaeology has a long history of incorporating the data, theory and methodology of other disciplines. In the eighteenth century most scientists relied on the Bible to estimate the age of life on the planet. It was the work of nineteenth century geologists such as Lyell, Cuvier and Smith (see Grayson 1983) that established the antiquity of the earth. It was part-time geologist John Frere that first uncovered human remains with the bones of extinct animals, and his evidence coupled with stone tools collected by de Perthes, gave credence to the idea that humans were of considerable age as well (Grayson 1983). Darwin (1859), a natural historian, lays the foundation for all of the biological sciences when he developed the idea that natural (environmental) selection and differential reproduction led to the transmutation of species—a true milestone in what would become paleoanthropology. Neo-Darwinism or the "modern synthesis", the addition of genetics to natural selection, is now the basis for our understanding of human evolution. In a similar vein, the anatomist Huxley brought the comparative approach to Paleolithic studies when he published a detailed analysis of the anatomical structure of a modern human, apes and a Neanderthal (Huxley 1863; Grayson 1983).

In the 1960s Louis Leakey encouraged Jane Goodall, Diane Fossey and Birute Galdikas to study the chimpanzee, gorilla and orangutan respectively. The data they collected on social structure, sexual behavior, and tool use among other aspects of primate behavior are routinely employed to generate primate models for human evolution. Also in the 1950s and 1960s, contributions from chemistry and physics revolutionized the field through potassium-argon dating which doubled the age of the Paleolithic and for the first time provided chronometric dates for fossil hominids (Johanson and Edey 1981) and continue to do so today (e.g. Swisher 1994). Finally, while paleobotany and paleoclimatology as well as other disciplines also have a long relationship with Paleolithic studies, more recently we have also begun to incorporate mitochondrial DNA, behavioral data on bonobo chimpanzees, argon-argon dating, computer modeling and many other types of information into our reconstructions of early hominid lifeways.

These examples represent just a few instances of Paleolithic archaeology drawing data from other fields. In fact, we usually do not even think of these examples as "borrowing" because these ideas and techniques are integral to Paleolithic studies. It is the combination of radiometric dating techniques, a knowledge of the anatomy and behavior of primates, ethology of social carnivores, geological laws, excavation techniques, paleoclimatology, ethnographic analogies and artifact typologies placed within a evolutionary framework, that forms the essence of this discipline. In this context, cognitive archaeology can be seen as a natural development from processualism and from Paleolithic archaeology with its scientific method, anthropological framework and multidisciplinary perspective.

The Role of Archaeology

One question that remains is what role archaeologists can play within the larger realm of the cognitive sciences. While there may be a more obvious relationship between neuroanatomy or cognitive psychology and the study of intelligence, Paleolithic archaeologists contribute knowledge to cognitive studies in three very significant ways. First, they add a chronological dimension to this question. We have a database of 2.5 million years with which to work. When we speak of the evolution of human intelligence, the evolution of language and the evolution of symboling behavior we are emphasizing the need to study these questions from a diachronic perspective. What other discipline is better able to study change over time than archaeology? Second, Paleolithic archaeologists study the actual material culture of early hominids in the form of stone tools and faunal remains, and in later periods, the traces of dwellings, pits, hearths, and symbolic endeavors. This material culture is the direct result of hominid behavior and there exists a relationship between this behavior and hominid intelligence. One important goal of Paleolithic archeology is to explore the nature of this relationship and to understand its implications for the study of the evolution of human cognitive abilities. Finally, because of its anthropological perspective, archaeology can study human cognition on a larger scale and within a more encompassing framework. While psychologists focus almost entirely on the mind of the individual, for Paleolithic archaeologists interested in the mind, studying cognition involves looking at populations and societies, and culture which is a supra-individual phenomenon (Chase, pers. com.).

Conclusion

Cognitive archaeology in its present form is the product of a long history of archaeologists studying the human mind. This chapter has argued that cognitive archaeology has matured into a serious subject of study. Far from being "paleopsychologists" cognitive archaeologists are committed to studying the evolution of human intelligence, language and symboling behavior through the material culture of prehistoric hominids. Cognitive archaeology should not been seen as a separate branch of archaeology but rather as a way of making our subdiscipline more holistic. With its emphasis on the scientific method, anthropological framework and multidisciplinary perspective it is a natural development from processual archaeology in general and Paleolithic archaeology in particular. It is suggested that the union of the scientific method with anthropological concerns signals a re-emergence of cognitive archaeology.

This chapter further argues that the study of the human mind through the archaeological record cannot be not performed in a vacuum. As a result, while archaeologists can make important contributions to cognitive science, they must turn to psychology, primatology, paleoneurology and other disciplines for additional theory and data. It was emphasized that as in any scientific pursuit "fanciful" speculations have no place in cognitive archaeology. The hypotheses archaeologists generate must be reconciled with data from these other disciplines. Archaeologists can either approach the study of intelligence alone or they can choose to work with researchers from other disciplines and develop a multidisciplinary team. What they cannot do is ignore the other disciplines because the evolution of language, culture and intelligence cannot be understood from lithics alone.

References Cited

Aiello, L., and R. I. M. Dunbar
   1993   Neocortex size, group size, and the evolution of language. *Current Anthropology* 34:184-193.

Bednarik, R. G.
   1994   A taphonomy of paleoart. *Antiquity* 68:68-74.
   1995   Concept-mediated marking in the Lower Paleolithic. *Current Anthropology* 36: 605-634.

Bender, B.
   1993   Cognitive archaeology & cultural materialism (viewpoint). *Cambridge Archaeological Journal* 3:257-260.

Binford, L. R.
   1962   Archaeology as anthropology. *American Antiquity* 28:217-25.
   1965   Archaeological systematics and the study of culture process. *American Antiquity* 31:203-210.
   1972   *An Archaeological Perspective*. Seminar Press, New York.

Boëda, E., J.-M. Geneste, and L. Meignen
   1990   Identification de chaînes opératoires lithiques du Paléolithique ancien et moyen. *Paléo* 2:43-80.

Bradley, R.
   1994   Symbols and signposts—understanding the prehistoric petroglyphs of the British Isles. In *The Ancient Mind*, edited by C. Renfrew and E. B. W. Zubrow, pp. 95-106. Cambridge University Press, Cambridge.

Byrne, R., and A. Whitten (eds.)
   1988a   *Machiavellian Intelligence: Social Expertise and the Evolution of Intellect in Monkeys, Apes and Humans*. Oxford University Press, Oxford.
   1988b   Tactical deception of familiar individuals in baboons. In *Machiavellian Intelligence: Social Expertise and the Evolution of Intellect in Monkeys, Apes and Humans*, pp. 205-210. Oxford University Press, Oxford.
   1988c   Towards the next generation in data quality: a new survey of primate tactical deception. *Behavioral and Brain Sciences* 11:267-73.
   1990   Computation and mindreading in primate tactical deception. In *Natural Theories of Mind*. Edited by A. Whitten. Blackwell Scientific, Oxford.

Chase, P. G., and H. L. Dibble
   1992   Scientific archaeology and the origins of symbolism: a reply to Bednarik. *Cambridge Archaeological Journal* 2:27-57.

Cheney, D. L., and R. M. Seyfarth
   1990   *How Monkeys See the World: Inside the Mind of Another Species*. University of Chicago Press, Chicago.

Childe, V. G.
   1956   *Piecing Together the Past: The Interpretation of Archaeological Data*. Routledge & Kegan Paul, London.

Conkey, M.
   1978   An analysis of design structure variability among Magdalenian engraved bones from Northcoastal Spain. Ph. D., University of Chicago.
   1989   Structural studies of Paleolithic art. In *Archaeological Thought in America*, edited by C. C. Lamberg-Karlovski, pp. 135-54. Cambridge University Press, Cambridge.

1996 A history of the interpretation of European Paleolithic art: magic, mythogram, and metaphors for modernity. In *Handbook of Human Symbolic Evolution*, edited by A. Lock and C. Peters, pp. 288-350. Clarendon, Oxford.

Darwin, C.
1859 *On the Origin of Species by Means of Natural Selection*. Murray, London.

Davidson, I., and W. Noble
1989 The archaeology of perception: traces of depiction and language. *Current Anthropology* 30:125-155.
1993a On the evolution of language. *Current Anthropology* 34:165-166.
1993b Tools and language in human evolution. In *Tools, language and cognition in human evolution*, edited by K. Gibson and T. Ingold, pp. 363-388. Cambridge University Press, Cambridge.

Deacon, T.
1997 *Symbolic Species: The Co-evolution of Language and the Brain*. Norton, New York.

Deetz, J. F.
1967 *Invitation to Archaeology*. Natural History Press, New York.
1977 *In Small Things Forgotten*. Doubleday, Garden City.

Dibble, H. L.
1989 The implications of stone tool types for the presence of language during the Lower and Middle Paleolithic. In *The Human Revolution: Behavioral and Biological Perspectives on the Origins of Modern Humans*, edited by P. Mellars and C. Stringer, pp. 415-432. Edinburgh University Press, Edinburgh.

Donald, M.
1991 *Origins of the Modern Mind: Three Stages in the Evolution of Culture and Cognition*. Harvard University Press, Cambridge, MA.

Ember, C. R., and M. Ember
1999 *Anthropology*, 9th edition. Prentice-Hall, Upper Saddle River, NJ.

Flannery, R., and J. Marcus
1996 Cognitive archaeology. In *Contemporary archaeology in theory: A reader*, edited by R. Preucel and I. Hodder, pp. 350-363. Blackwell, Oxford.

Frost, M.
1980 Tool behavior and the origins of laterality. *Journal of Human Evolution* 9:447-459.

Geertz, C.
1973 *The Interpretation of Cultures*. Basic Books, New York.

Glassie, H.
1975 *Folk Housing of Middle Virginia*. University of Tennessee Press, Knoxville.

Gowlett, J. A. J.
1984 Mental abilities of early man: a look at some hard evidence. In *Hominid Evolution and Community Ecology*, edited by R. Foley, pp. 167-92. Academic Press, London.
1986 Culture and conceptualisation: the Oldowan-Acheulian gradient. In *Stone Age Prehistory*, edited by G. N. Bailey and P. Callow, pp. 243-60. Cambridge University Press, Cambridge.
1996 Mental abilities of early Homo: elements of constraint and choice in rule systems. In *Modeling the Early Human Mind*, edited by P. Mellars and K. Gibson, pp. 191-216. The McDonald Institute for Archaeological Research, Cambridge.

Graves, P.
1994 Flakes and ladders: what the archaeological record cannot tell us about the origins of language. *World Archaeology* 26:158-171.

Grayson, D. K.
1983 *The Establishment of Human Antiquity*. Academic Press, New York.

Harris, M.
1978 *Cows, Pigs, Wars and Witches*. Vintage Books, New York.

Hawkes, C. F.
1954 Archaeological theory and method: some suggestions from the Old World. *American Anthropologist* 56:155-68.

Hewes, G. W.
1989 Comment on 'The archaeology of perception: traces of depiction and language' by I. Davidson and W. Noble. *Current Anthropology* 30:145-46.

Hodder, I.
1982a *Reading the Past*. Cambridge University Press, Cambridge.
1982b *Symbols in Action*. Cambridge University Press, Cambridge.
1984 Burials, houses, women and men in the European Neolithic. In *Ideology, Power and Prehistory*, edited by D. Miller and C. Tilly, pp. 51-68. Cambridge University Press, Cambridge.
1996 *Theory and Practice in Archaeology*. Routledge, London.

Huxley, T. H.
1863 *Evidence as to Man's Place in Nature.* Williams and Norgate, London.

Johanson, D., and M. Edey
1981 *Lucy: the Beginnings of Humankind.* Simon and Schuster, New York.

Joulian, F.
1996 Comparing chimpanzee and early hominid techniques: some contributions to cultural and cognitive questions. In *Modeling the Early Human Mind*, edited by P. Mellars and K. Gibson, pp. 173-190. McDonald Institute Monographs, Cambridge.

Juteson, J. S., and L. D. Stephens
1994 Variation and change in symbol systems: case studies in Elamite cuneiform. In *The Ancient Mind*, edited by C. Renfrew and E. B. W. Zubrow, pp. 167-175. Cambridge University Press, Cambridge.

Leone, M. P.
1982 Some opinions about recovering mind. *American Antiquity* 47:742-760.

Leroi-Gourhan, A.
1967a *The Art of Prehistoric Man in Western Europe.* Thames and Hudson, London.
1967b *Treasures of Prehistoric Art.* H. N. Abrams, New York.
1968 The evolution of Paleolithic art. *Scientific American* 218:58-70.

Lévi-Strauss, C.
1966 *The Savage Mind.* University of Chicago Press, Chicago.

Longacre, W. A.
1970 *Archaeology as Anthropology: a Case Study.* University of Arizona Press, Tucson.

McDaniel, S. H.
1995 Collaboration between psychologists and family physicians: implementing the biopsychosocial model. *Professional Psychology: Research and Practice* 26:117-122.

McGovern, P.
1995 Science in archaeology: A review. *American Journal of Archaeology* 99:49-142.

McGuire, R.
1992 *A Marxist Archaeology.* Academic Press, San Diego.
1993 Archaeology and Marxism. *Archaeological Method and Theory* 5:101-157.

McPherron, S.
1994 A Reduction Model for Variability in Acheulian Biface Morphology. Ph.D. dissertation, University of Pennsylvania.

Mellars, P. and K. Gibson
1996 *Modeling the early human mind.* Cambridge: McDonald Institute Monographs.

Milton, K.
1981 Distribution patterns of tropical plant food as an evolutionary stimulus to primate mental development. *American Anthropologist* 83:534-48.
1988 Foraging behavior and the evolution of primate intelligence. In *Machiavellian Intelligence: Social Expertise and the Evolution of Intellect in Monkeys, Apes and Humans*, edited by R. W. Byrne and A. Whitten, pp. 285-306. Oxford University Press, Oxford.

Mithen, S.
1995 Paleolithic archaeology and the evolution of mind. *Journal of archaeological Research* 3:305-332.
1996 *The Prehistory of the Mind.* Thames and Hudson Ltd., London

Noble, W. and I. Davidson
1996 *Human Evolution, Language and Mind.* Cambridge University Press, Cambridge.

Patterson, T. C.
1990 Some theoretical tensions within and between processual and post-processual archaeologies. *Journal of Anthropological Archaeology* 9:189-200.

Pobiner, B.
1999 The use of stone tools to determine handedness in hominids. *Current Anthropology* 40:90-91.

Renfrew, C.
1982 *Towards an Archaeology of Mind.* Cambridge University Press, Cambridge.
1994 Towards a cognitive archaeology. In *The Ancient Mind: Elements of cognitive archaeology*, edited by C. Renfrew and E. Zubrow, pp. 3-12. Cambridge University Press, Cambridge.

Renfrew, C., and P. Bahn
1991 *Archaeology, Theories, Methods, and Practice.* Thames and Hudson, London.

Renfrew, C., and E. B. W. Zubrow. Editors
1994 *The Ancient Mind: Elements of Cognitive Archaeology.* Cambridge University Press, Cambridge.

Savage-Rumbaugh, E. S., and D. M. Rumbaugh
1993 The emergence of language. In *Tools, language and cognition in human evolution*, edited by K. Gibson and T. Ingold, pp. 86-108. Cambridge University Press, Cambridge.

Savage-Rumbaugh, S., and R. Lewin
  1994 *Kanzi*. John Wiley & Sons, New York.
Schiffer, M. Editor
  1978-86 *Advances in Archaeological Method and Theory*. Vol. 1-9. New York: Academic Press.
Schlanger, N.
  1994 Mindful technology: unleashing the chaîne opératoire for an archaeology of mind. In *The Ancient Mind*, edited by C. Renfrew and E. B. W. Zubrow, pp. 143-151. Cambridge University Press, Cambridge.
Schnapp, A.
  1994 Are images animated? The psychology of statues in Ancient Greece. In *The Ancient Mind*, edited by C. Renfrew and E. B. W. Zubrow, pp. 40-44. Cambridge University Press, Cambridge.
Scupin, R., and C. R. DeCorse
  1998 *Anthropology: A Global Perspective*, 3rd edition. Prentice-Hall, Upper Saddle River, NJ.
Shanks, M., and C. Tilley
  1987a *Re-Constructing Archaeology*. Cambridge University Press, Cambridge.
  1987b *Social Theory and Archaeology*. Cambridge Polity Press, Cambridge.
Steibing, W. H.
  1993 *Uncovering the Past*. Prometheus Books, Buffalo.
Swisher, C. C., G. H. Curtis, T. Jacob, and A. G. Getty
  1994 Age of the earliest known hominids in Java, Indonesia. *Science* 263:1118-1121.
Tilley, C.
  1984 Ideology and the legitimation of power in the Middle Neolithic of southern Sweden. In *Ideology, Power and Prehistory*, edited by D. Miller and C. Tilley, pp. 111-146. Cambridge University Press, Cambridge.
Toth, N.
  1985a Archaeological evidence for preferential right-handedness in the Lower and Middle Pleistocene, and its possible implications. *Journal of Human Evolution* 14:607-614.
  1985b The Oldowan reassessed: a close look at early stone artifacts. *Journal of Archaeological Science* 12:101-120.
Toth, N., and K. Schick
  1993 Early stone industries and inferences regarding language and cognition. In *Tools, language and cognition in human evolution*, edited by K. R. Gibson and T. Ingold, pp. 346-362. Cambridge University Press, Cambridge.
Trigger, B.
  1989 *A History of Archaeological Thought*. Cambridge University Press, Cambridge.
de Waal, F.
  1982 *Chimpanzee Politics: Power and Sex among the Apes*. The Johns Hopkins University Press, Baltimore.
White, L.
  1959 *The Evolution of Culture*. McGraw-Hill, New York.
White, R.
  1992 Beyond art: towards an understanding of the origins of material representation in Europe. *Annual Review of Anthropology* 21:537-564.
Whitley, D. S.
  1992 Prehistory and positivist science: a prolegomenon to cognitive archaeology. *Journal of Archaeological Method and Theory* 4:57-100.
Wilkens, W. K., and J. Wakefield
  1995 Brain evolution and neurolinguistic preconditions. *Behavioral and Brain Sciences* 18:161-226.
Willey, G. R., and P. Philips
  1958 *Method and Theory in Archaeology*. University of Chicago Press, Chicago.
Wynn, T.
  1979 The intelligence of later Acheulian hominids. *Man* 14:371-91.
  1981 The intelligence of Oldowan hominids. *Journal of Human Evolution* 10:529-41.
  1989 *The Evolution of Spatial Competence*. University of Illinois Press, Urbana.
  1991 Tools, grammar and the archaeology of cognition. *Cambridge Archaeological Journal* 1:191-206.
  1993 Layers of thinking in tool behavior in *Tools, Language and Cognition in Human Evolution*, edited by K. Gibson and T. Ingold, pp. 389-406. Cambridge University Press, Cambridge.
Wynn, T., and W. C. McGrew
  1989 An ape's view of the Oldowan. *Man* 24:383-98.
Wynn, T., and F. Tierson
  1990 Regional Differences in later Acheulean handaxes. *American Anthropologist* 92:73-84.

3. Memories Out of Mind: The Archaeology of the Oldest Artificial Memory Systems

Francesco d'Errico

Abstract

Humans are the only species capable of creating artificial memory systems, that is to say, devices specifically conceived to record, store and recover information outside the physical body. Archaeologists suggest that the first manifestations of this behavior may be found in sequentially marked objects from the Upper Paleolithic caves of Europe. These interpretations, however, remain controversial. The ethnographic record is used here as a reference to establish the general principles which make it possible to reduce information to a form that can be stored and later recovered by humans. It is then discussed how analytical methods, based on experimental replication of the Paleolithic marks, can help us to identify such devices in the archaeological record.

Introduction

A fundamental turning point in the evolution of human cognitive abilities was when humans were able to store thought in material symbols and to locate memory outside the individual. The next, and equally important step was when humans designed physical devices specifically conceived to record, store and recover information. Goody (1977, 1987) has convincingly argued that of all of these devices, writing has exerted the greatest influence upon the development of human societies. In recent years, however, the ability of humans to store coded information outside the physical body has been examined from a wider perspective.

The concept of writing as being a visual substitute for speech (and especially so if alphabetical), is still the dominate paradigm for many historians of writing systems (Gelb 1952; Pulgram 1976; Bottero 1980, Havelock 1982, Fevrier 1984, De Francis 1989, Postgate et al. 1995). Nevertheless, this paradigm has waned in the face of Derrida's philosophical argument against the logocentrism of Western culture (1967) and, in more recent times, as a consequence of Harris' semiological approach (Harris 1986, 1990, see also Hill Boone and Mignolo 1994). Partly in response to these new viewpoints, scholars from many disciplines study the full range of systems created by humans to store information in an attempt to develop integrated theories of communication (Sperber 1985; Harris 1994, 1995) or models for the evolution of human cognition (Festinger 1983; Goonatilake 1991; Donald 1991, Mithen 1996; Hutchins 1995; Thierry et al. 1996). These authors refer to these systems as "exosomatic systems", "external memory media" or "artificial memory systems".

It is commonly held that these systems played an important role in the evolution of human cognitive abilities. They mark for Leroi-Gourhan (1964) the final expansion of memory, when individual brains became incapable of storing and handling all of the information required for the functioning of a particular society. Changeux (1983) sees in the initial stages of the development of writing an appropriate metaphor for the evolution of the human brain and considers it to be a turning point in the evolution of mankind—the moment in which images and concepts became more stable than neurons and synapses. For Goonatilake (1991) the use of "external information systems" represents a major step in the sophistication of the information flow which characterizes the entire process of the evolution of life. In Donald's view (1991), what he terms "external symbolic storage" represents, from the very beginning, a major factor in human intellectual endeavor upon which the present level of conceptual development—referred to as "theoretic" development, is based. For Mithen (1996:245) the conscious use of material culture to store information is a fundamental feature of any specialized (domain specific) intelligence. External memory devices demonstrate the emergence of a "cognitively fluid mind" able to develop powerful metaphors and analogies which, following Kuhn (1979), form the basis of modern scientific thought. These are but a few examples of the central role commonly attributed to these devices in current models on the origin of modern cognitive abilities.

It is also widely accepted that many of these devices were used before the advent of writing as well as after it (see discussion in Harris 1986, 1995). Relatively little is known, however, about their origins and early stages of development or about the role played by different types of these retrieving systems in human biological and cultural evolution. The oldest possible examples of artificial memory systems consist of objects in bone, antler, and ivory engraved with sequences of marks produced by different techniques. Though most of these objects come from Upper Paleolithic (35,000-10,000 BP) sites in Europe, recently this interpretation has also been proposed for objects found in Middle Stone Age sites of South Africa (Knight et al. 1995; Watt 1999).

Sequentially marked bones from the French Upper Paleolithic were first discovered over a hundred years ago (Lartet & Christy 1865-75). Since then archaeologists have proposed a number of hypotheses to explain these markings. They have been interpreted as *"marques de chasse"* (marks recording the number of prey killed), devices to keep track of songs or the number of people attending a ceremony or as the result of some other notational/calculation system (see Robinson 1992 and d'Errico 1995a for references). Another hypothesis is supplied by Alexander Marshack. Marshack, who over the past three decades has examined a number of Paleolithic marked objects, has interpreted many of these marks as notation systems based on lunar phases (e.g. 1964, 1970, 1972a-c, 1988, 1991a-c). These interpretations, however, have faced repeated challenges since their publication (e.g. Rosenfeld 1972; see also comments in Marshack 1972a). For example, Marshack has repeatedly argued that morphological differences between incisions produced by burins were the product of the engraver using different tools. This suggested to Marshack that the marks were produced at different times. White (1982), however, demonstrated that these morphological differences could alternatively be the result of changes in tool orientation. Therefore, sequences of marks, interpreted as notations done by adding marks in different sessions, could all have been created at one time and not have been made for notational purposes.

I have stressed (D'Errico 1989; 1995a-b) that Marshack's analytical approach suffers from a lack of objectivity in that it is exclusively based on personal observations of the markings. No attempt is made to experimentally demonstrate that the morphological characteristics he interprets are, in fact, due to the causes he ascribes to them. Thus, when he describes the ways in which marks were produced on an object or when he identifies tool changes we do not know on what basis he is making his claims. Moreover, comparative microscopic analysis of experimental and archaeological marks, carried out with a range of microscopic techniques, has demonstrated that the criteria used by Marshack to study the objects are incomplete and, in some cases, erroneous (D'Errico 1992, 1995a,b). For example, a re-analysis of one of the most famous of Marshack's (1972) "lunar calendars", the La Marche antler, has shown that some basic technological data on which Marshack based his reading of the marks (e.g. number of marks, changes of tool, marking procedures), should be interpreted otherwise (D'Errico 1995a, 1996a).

I have (D'Errico 1995a, 1996b) also questioned the epistemological value of Marshack's method of testing his interpretations. Marshack claims that he has identified systems of notation but he does not explain what is meant by this term. How many different types of notation can humans conceive of and of these, how many of them will we be able to identify in the archaeological record? These questions remain unanswered. Counting marks is offered as a reliable way to "test" the calendrical hypothesis without first discussing what explanatory value a particular number has in archaeological reasoning. The question is cogent since no explicit (i.e. depictional) element exists which supports the hypothesis that these marks are related to lunar phases. Should we interpret any prehistoric object as a calendar if it bears a number of items totaling the number of days in a lunar month?

D'Errico (1989, 1995a, 1996a) and Robinson (1992) have noticed that Marshack's method of relating marks to lunar phases is done in such a flexible manner, i.e. changing the rules for each object or group of marks in a single object that, in the final analysis, it is more difficult to identify what does not fit the model rather than what does. Further, Robinson (1992) has rightly argued that the systematic interpretation of marked objects as observational calendars, i.e. produced by adding new marks without prior knowledge of the lunar month being recorded, is in contradiction to the hypothesis postulated by Marshack that these devices were the result of a continuous tradition throughout the Upper Paleolithic. Why would Paleolithic people have continued to make calendars for 20,000 years by adding single marks or groups of marks corresponding to the days that have passed between each moon observation as if they were unaware of how long a lunar phase would be? At least after a pioneering stage, a tradition would be more plausibly expressed by notations created in advance, for which there would have been no need to add marks at different times. Elkins (1996) also found a number of theoretical discrepancies in Marshack's work, and used his criticisms against this author to claim the impossibility of any "close reading" of Paleolithic markings. This is the extreme thesis that all attempts to interpret "evidence" through a macro or microscopic analysis are doomed to fail.

In sum, despite pioneering and stimulating research carried out in the past, we still do not seem to have a testable theoretical framework nor an explicit methodology with which to approach the archaeological record. Thus, more than a century after their initial discovery, the interpretation of Paleolithic sequentially marked objects, possibly the material expression of "artificial memory systems" remains controversial. In my view, to carry out research in this field, we need a theoretical model capable of defining the general principles that make it possible to reduce information to a form that can be stored and later recovered by humans. In creating this model, we must draw on advances made in linguistics, semiotics and the other sciences interested in human cognition. This should enable us to classify these devices, explain their variability, and evaluate their efficiency. We also need to discuss the applicability of this model to the study of the archaeological record, develop a suitable methodology to analyze this record and reduce, in this way, the range of possible interpretations.

In two recent papers (1995a, 1998), I have proposed a way to create this model by studying a large number of artificial memory systems known ethnographically. This was not done to establish a direct analogy between the ethnographic and Paleolithic records, as has been done in the past (Marshack 1974, 1985), but to understand the ways in which these devices work and the general rules that govern their codes. I have also suggested that a technological analysis of Paleolithic marked objects is essential for discussing their possible interpretation as devices specifically conceived to store information. I have further argued that this analysis should rely on diagnostic criteria established experimentally. In searching for this model I take a different epistemological path than that recently taken by Buissac (1994, 1997). This author argues, for instance, that in order to demonstrate that some examples of rock art should be interpreted as "paleographic writing systems" it must first be shown statistically that there is a systematic patterning to the organization and combination of signs.

Initial attempts to apply this composite framework to the study of a selected sample of Upper Paleolithic sequentially marked objects led me to the conclusion that the interpretation of some of them as artificial memory systems was indeed the more plausible one (D'Errico 1991, 1995a, 1998; D'Errico and Cacho 1994). The significance, however, of the interpretative guidelines used to put forward this explanation have not been fully discussed or generally understood. For instance, Elkins asserts that (1996:198) "analytical criteria for judging changes of tools or the passage of time..." is of no use in identifying Paleolithic "notations". The aim of the present work is to delve deeper into a discussion of the applicability of my theoretical model to the identification of possible Paleolithic artificial memory systems. To achieve this goal, I will first describe the main points of my theoretical and analytical approach. I will then focus on a particular type of memory storage device that is based on the accumulation of information over time. Finally, I will discuss how one might recognize such a device in the archaeological record.

From Ethnography to Semiology

AMS's held by modern or historically known human cultures provide no direct indication of what types of AMS's were used by humans in the remote past nor do they give an indication of the possible evolution of these devices. I assume, however, that there is a limited number of principles available to humans through which to reduce information to a form that they can manipulate and recover at a later point in time. Looking for these principles, I have created a database of 50 ethnographically known AMS's used for different purposes all over the world (D'Errico 1998) and I have studied the functioning of each system, i.e. the way in which the information is recorded, processed and recovered. This survey (Figure 1) has revealed that four major factors can govern an AMS code: 1) morphology of elements, 2) spatial distribution of these elements, 3) their accumulation over time, and 4) their number.

Different types of rosaries, used by Hindus, Buddhists, Muslims, Catholics and followers of the

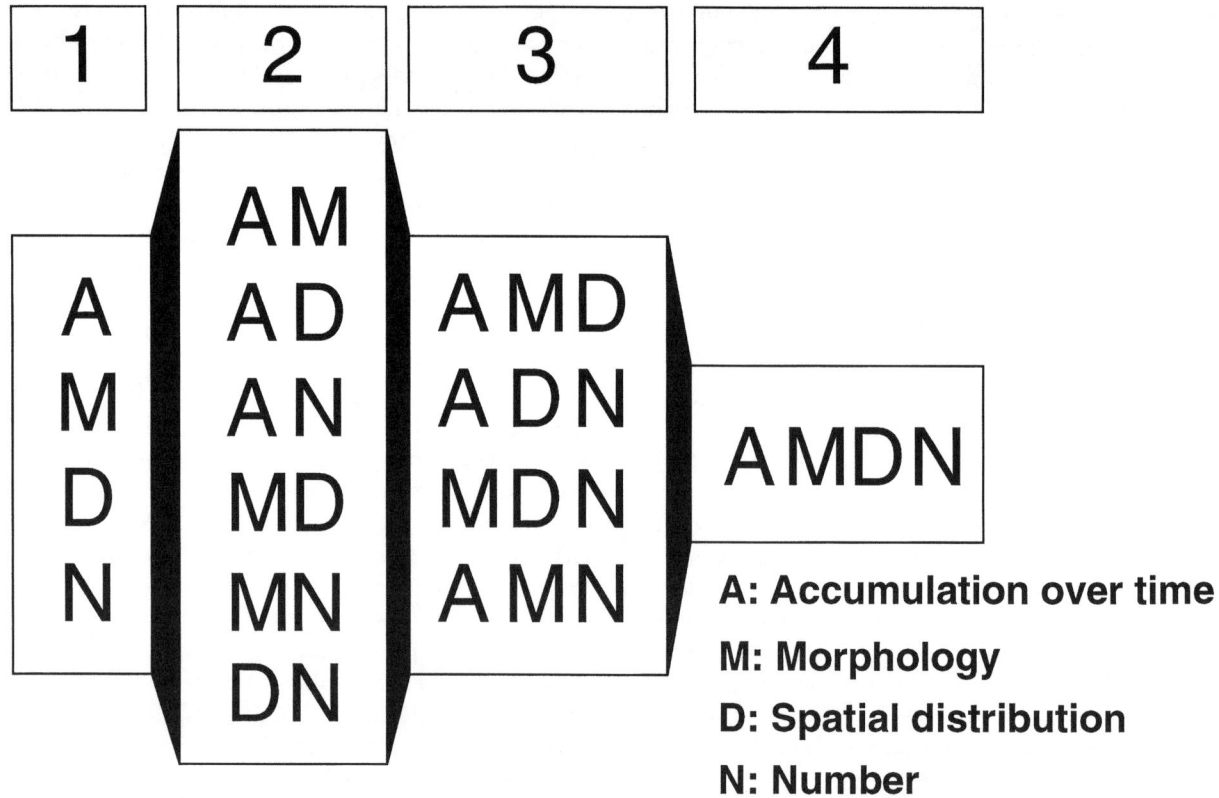

Fig. 1. Possible associations between factors organising an AMS code (see text).

Greek Orthodox faith are excellent examples of codes where information is transmitted through the spatial distribution of elements (Dubin 1987). For example, the Buddist monk's rosary, consisting of 108 beads, corresponds to the number of emotions that can be overcome through the recitation of prayers using the beads. A type of Muslim rosary, composed of ninety-nine identical beads, is used to recite the attributes of "God the Holy", "the Mighty", "the Forgiver", and so on. Another type of rosary, with one hundred beads, is reserved for saying the name of God and to recite short sentences honoring God. The same sentence can be recited a number of times or alternated with others. In no way does the morphology of the beads suggest the choice of the sentence or the rhythm adopted to alternate them—these choices are left entirely up to the user. In contrast to these rosaries, the code of the classic Catholic rosary, composed of 150 small beads divided into "decades" by 15 larger beads, is based on both the spatial distribution of the elements bearing information as well as on the morphology of these elements. The small beads are used for reciting the "Hail Mary" prayer while the larger ones are reserved for "The Lord's Prayer". Thus the order of the beads indicates the order of the prayers, while the bead dimensions or, in smaller more portable versions, the chain length between similar beads, indicates the type of prayer.

There are also codes based on the accumulation of elements over time. Imagine a situation where a notch is carved on a tally-stick every time an event occurs or an object must be recorded. The code of this type of AMS would be based on the accumulation over time of elements bearing information. Since notches cannot be distinguished by the naked-eye their morphology cannot play a role in the code. Spatial distribution does not play a role either. In fact you can carve a new notch in the space still remaining on the stick or between existing notches without changing the information that you are storing.

Two types of codes based solely on the accumulation of marks over time can be distinguished. The first type of code is when the number of events or objects one is recording is unpredictable, such as the number of animals or enemies killed. The second type of code is when there is a certain periodicity to the recorded objects or events. In the first case, each mark or group of marks represents one or more similar events that can occur simultaneously or over a period of time. A single mark can represent a number of events that are more or less contemporary or, alternatively, a mark can indicate each event separately, even when these events took place simultaneously (e.g. the number of animals killed as a result of a day's worth of hunting).

The second type of code ("periodic") records natural events (days, months, seasons, years etc.) or cultural events linked to these natural occurrences (festivals, ceremonies, seasonal migrations etc.). Within this second code, it is possible to distinguish between systems that record natural or cultural events that occur at regular intervals (every month, for example) and those that, while predictable, are separated by variable lengths of time (for example, certain seasonal migrations). For both of these categories, since marks represent culturally created breaks in a ongoing temporal process, meaning can be attached not only to each mark but also to the space between them. The periodic code can be adopted by users who have a knowledge of the precise number and timing of the events to be recorded as well as by users who partially or totally ignore this information. This system permits the former to know exactly where they are in the cycle, and for the latter to evaluate the number of events that have occurred since the beginning of the recording.

What characterizes all systems based on the accumulation of marks over time is not only the fact that information is updated at different times but also that this information is used in-between marking episodes. To clarify this point, imagine that identical marks are accumulated periodically with the goal of creating a calendar to be used once the recording is finished (at the end of the month, for example). In this case, updating the information is used just as a means to create a device that, once ready, will be governed by a code based on the spatial distribution of the signs and where no further modification or updating will take place.

When a code is based on numeric values, each different figure or each numerically different group of identical signs takes on a qualitatively different meaning. In other words, each numerical sign or accumulation of signs in not significant in its own right. Instead, they are used together to communicate a meaning which has no logical or iconic link with the smaller or larger values. For this reason, codes based on number generally present associations of elements which cover a relatively large range of numerical variation. This is an essential means of storing different types of information. An example of an AMS based on the number and spatial distribution of information-bearing elements is the Aroko, used by the Jebu of West Africa (Gollmer 1885; Adedo Ogunde 1997). This AMS consists of chains of shells which communicate different messages based on the number of shells and their positions on the string.

Although it is not employed as an AMS, another example of a code based exclusively on numeric information, is the *La Smorfia,* which is used by National Lottery players in Naples. Among these players there exists a widely held belief that if you can accurately interpret dreams, you can know in advance which numbers are going to be pulled at the next lottery. A traditional code assigns a number to each person or event appearing in a dream. However, often the dreams unfold in a very complex manner where the relationship between the event dreamed of and its corresponding number is not always obvious. Furthermore, the richer prizes come from the drawing of several numbers at the same time, which forces the players to identify in their premonitory dreams *several* events or people that can be then "translated" into numbers.

In sum, the ethnographic evidence shows that each code can depend on one, two, three or even all four of the factors cited above. From fifteen basic codes, several variants can be conceived of depending on the hierarchical organization of the factors within the code. An example of a code that is the result of hierarchical organization is the Inca AMS—the quipus (Hill Boone and Mignolo 1994). The quipus consists of a number of cords of different lengths and colors suspended from a topband. The position and type of knot on each cord communicates information about objects, humans and animals. This is a code based on the spatial distribution and morphology of elements. Spatial distribution is organized hierarchically in two stages—the order of the cords and the position of knots. The morphology of the elements is also organized hierarchically through the color of the cords and the types of knots.

My model also takes into account the ways in which information is recovered. Humans obtain information through visual, tactile, auditory, olfactory and taste senses. In a rosary, for example, information is processed only through tactile reception while the quipus is interpreted both tangibly and visually. In a tally-stick, information can also be processed by tactile or visual reception, or by a combination of both. The choice of a particular means is determined by neurophysiological constraints but also by cultural traditions as in the example of the Muslim and Catholic rosaries.

Auditory, olfactory and taste senses can be reasonably left out of our discussion of AMS's. Though rather complex musical instruments were used from the very beginning of the Upper Paleolithic (Buisson 1991; Hahn 1996), the ability to record sounds is a relatively recent invention. Olfaction and taste are well suited to retrieving information from the environment but problems arise when they are used to create specific codes and to store information in such a manner that it can be later recovered. There is a great deal of individual variation (information well known to wine testers!) in distinguishing scents and flavors. These differences can only be slightly modified by culture. Therefore, the difficulty in attributing specific meanings to odors or tastes reduces the possibility of using these methods to retrieve coded information. In addition, it is problematic to create hierarchically organized codes using these senses (see Leroi-Gourhan 1964-5:107-119 for precursory statements). In any event, even if codes based on scent or taste existed in the Paleolithic our chances of finding any evidence of them is minimal. Therefore, because of the problems associated with these senses and the difficulty in recovering any evidence of them, it seems logical to concentrate on the visual and tactile perception. The ethnographic examples I have had the opportunity to study so far suggest that vision is the most widely used sense to obtain information. Tactile and visual perception are often associated but, in these cases, the former often serves to help the latter.

I have also considered the efficiency of each system in terms of the quantity and quality of stored information, the possibility of updating information, and the effort required to learn the code. The importance of this approach is that it classifies AMS's on the basis of the formal, and probably invariable, elements that play a role in elaborating any type of AMS code and not on features specific to each particular AMS, such as its function or the meaning attributed to the signs. In other words, I use this model to ask questions about how these artifacts were used, not about what specific type of information was stored and recovered as I believe these kinds of questions are difficult to answer at this stage of our investigation.

Applying Ethnographic Data to the Archaeological Record

The model described above seems capable of accounting for the variability of AMS's created by different cultural systems. It remains, however, to ascertain exactly how such a model can help us to identify prehistoric AMS's. Analysis of ethnographic AMS's reveals that most of them are made of materials which would not survive in the archaeological record or would generate highly ambiguous evidence (for example, the beads of a rosary). The Upper Paleolithic, however, is rich in bone objects displaying sets of marks. Repeated changes of tools and variation in the marking techniques, arrangement and morphology of the marks constitute a more suitable ground upon which to discuss the AMS hypothesis. I will now focus on one particular type of AMS, one based exclusively on the accumulation of information through time and discuss the criteria for identifying this type of AMS in the archaeological record.

In Search of Analytical Criteria

Criteria for identifying Paleolithic AMS's with codes based on the accumulation of information over time derive from a technological analysis of the marks. When marks are created by stone tools, a tool change will probably take place between each new stage of marking, if the periods separating the stages are relatively long. Thus, isolating diagnostic criteria for identifying when a new tool is employed is crucial for identifying systems based on these types of codes. In sequences of parallel marks, repeated inversions in the position of the marking tools and in the marking direction are additional ways of identifying this type of AMS. These changes confirm that marks were added over different sessions and suggest that the engraver had no interest in positioning the marked object (or the tool) in the same way at each new marking stage.

Paleolithic marks (Figure 2) include sequences of single or multiple stroke lines (made by a single or repeated movement of a point), notches (produced by a single or repeated movement of a cutting edge), and microincisions (produced by the pressure, percussion or rotation of a point) carved on different types of material (bone, ivory, antler, stone). Analysis of experimental and archeological marks, conducted over the last 15 years with optical and scanning electron microscopes, computer stations for profile visualization and measurement, and image analysis systems, have provided diagnostic criteria with which to identify the marking techniques used by prehistoric engravers and to establish whether morphological changes between marks are due to a change of tool or to other causes (d'Errico 1991, 1995a, b, 1996b, 1998; Fritz et al. 1993).

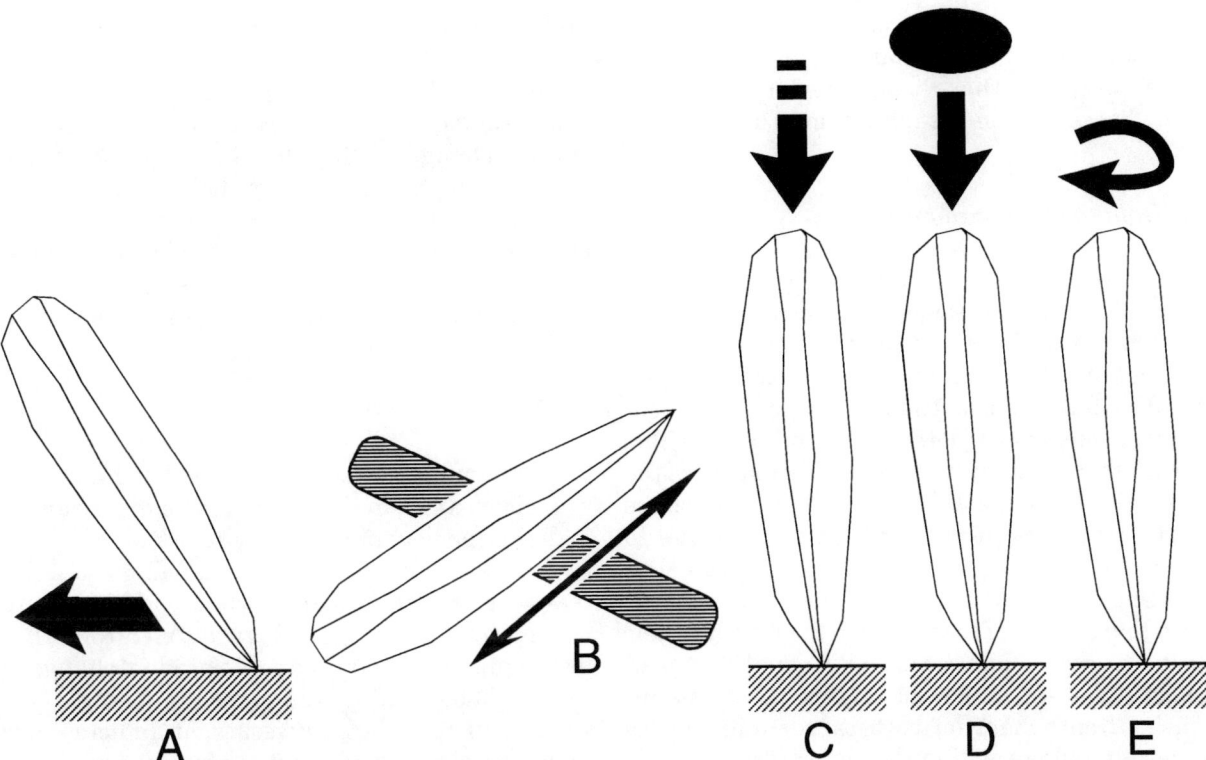

Fig. 2. Main marking techniques used in the Upper Paleolithic: single or multiple stroke lines, made by a single or a repeated movement of a point (A), notches, produced by a single or repeated movement of a cutting edge (B); microincisions, produced by the pressure (C), indirect percussion (D) or rotation of a point (E).

We know experimentally that sequential marks produced in a single session using the same technique and the same tool, are not exactly identical. Their morphology will change as a result of (1) slight variations in hand motion (i.e. changes in the tilt and orientation of the tool), (2) tool wear, (3) possible breakage of the tool, (4) resharpening or (5) deliberate changing of the tool's "active zone" (i.e. the region of the tool that is actually in contact with the surface of the material to be engraved). By "wear" I mean dulling of the active zone and the possible development of micro removals. "Breakage", by contrast, consists of the accidental removal of the all or a large part of the tool's active zone. To distinguish morphological variations introduced by these factors in sequences of marks produced in a single session from those created by a tool change is indeed crucial for identifying possible Paleolithic AMS's with a code based on the accumulation of marks over time.

Of the five phenomena cited above, the first three develop independently of the wishes of the artisan while the last two result from a deliberate choice. Wear will occur throughout the entire marking process, while the other phenomena will not necessarily happen and, thus, may not effect the marks' morphology. The extent of the morphological changes produced by each of these phenomena is a function of the marking technique used, the type of support on which the marks are created and the technical competence of the artisan. These factors are fundamentally interconnected. A particularly fragile active zone or one that is not well adapted to the technique selected for creating marks will wear rapidly and fracture easily, and possibly require either resharpening or replacement by another zone of the same tool. The more variable the hand motion of the artisan, the more rapidly these phenomena will develop. Similarly, the technical processes used will effect the rapidity of wear of the active zone and the probability of its fracture and/or replacement by another active zone of the same tool. Finally, it is evident that the hardness and morphology of the marked object will condition the rate of wear development, the appearance of fractures, and the need to change the active zone. I will now consider possible criteria for distinguishing the above five phenomena from the situation where a change of tool has occurred. I will then discuss if tool changes can be produced for reasons other than the creation of AMS's with codes based on the accumulation of marks over time.

Hand motion variability

Experimental work demonstrates that with marks made in succession, slight variations in the orientation and tilt of the tool rarely produce changes in a mark's morphology which can be confused with the artisan changing tools. When neuromotor constraints, determined by a range of factors (posture, type of movement, morphology of the tool and of the marked surface etc.), introduce gradual changes in hand motion, this results in progressive changes in the marks' morphology, thus allowing recognition that the same tool is being used to create all of the marks. If marks vary their shape as a consequence of an occasional change in the craftsperson's movement, this can be distinguished from a change of tool because similar marks generally reoccur within the series.

Differentiating hand motion variability from tool changes, however, is easier for some marking techniques than for others. Microincisions produced by pressure or indirect percussion, corresponding to the negative imprint of the active zone of the tool, generally show peculiar features which remain recognizable in spite of variations in the orientation of the tool (D'Errico 1995a). In a series of notches, the repeated to-and-fro movement of the tool generally compensates for occasional variations in hand motion, affecting only one or a few of the artisan's movements. Thus, consecutive marks are recognizable as the result of the same cutting-edge (D'Errico 1991). In marks created by single or multiple rotations of a point, variations in hand movement produce marks of different shapes. They can still be recognized as having been produced by the same tool, however, through an examination of the mark's internal morphology and the use of particular features to reconstruct the volume of the tool's active zone (D'Errico in prep.).

Marks produced by the unique displacement of a point (i.e. engraved lines) represent cases where variation in hand motion can occasionally produce puzzling changes in mark morphology. This occurs because of the limited contact of the tool's active zone with the marked surface. Each change in the orientation of the tool changes the area which comes into contact with the marked surface creating engraved lines with different sections of varying width and shape. If such a change occurs suddenly and remains stable in all of the subsequent marks it becomes difficult to differentiate it from a true change of tool. It is very rare, however, to find two of these changes within the same sequence. Thus, even if this phenomenon can be confused with a change of tool, the corresponding marks cannot be easily interpreted as an AMS based on accumulation of marks over time because a single change of tool would indicate only two stages in their accumulation.

Tool wear

Only marking techniques involving attrition wear down a tool and lead to corresponding changes in the morphology of consecutive marks which could be confused with an artisan changing tools. No real wear occurs in marks produced by pressure and indirect percussion (Figure 3). In the other techniques, the wear produces a gradual change in the morphology of marks carved in a single session. Each new mark possesses, at a microscopic scale, certain characteristics in common with preceding marks and some features of those which follow it. By identifying this trend it is possible to distinguish between changes resulting from wear and those produced by a change of tool (Figures 4 and 5). Experimental marking, however, demonstrates that wear affects the working edge of a tool in a different manner and at a different rate according to the technique used. Certain parts of the edge, subjected to more constant contact with the marked material, may be subject to more rapid modifications, while others remain relatively unaltered throughout the duration of the marking process. Those zones less affected by wear provide the most reliable morphological characteristics for identifying the use of the same tool. Experimentation also shows that tools used to cut notches or points used to engrave single or multiple stroke lines can wear out very quickly thus producing rapid changes in the first marks followed by a stabilization of the working edge morphology with a corresponding similarity of the marks.

Breakage

Differentiating changes in mark morphology produced by a change of tool from those generated by breakage depends on the marking technique used and the extent of the breakage. Breakage occurring during the production of a sequence of

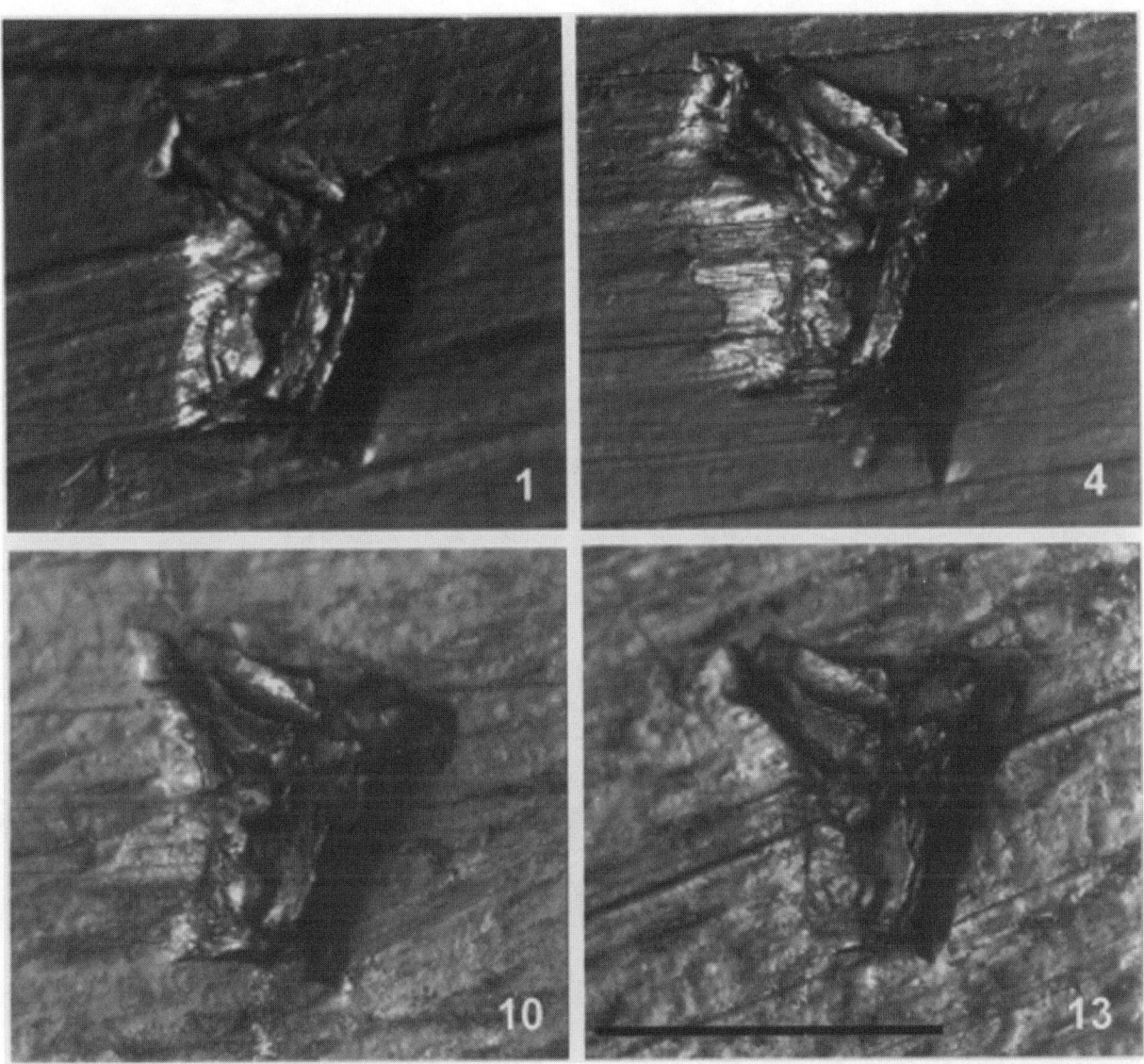

Fig. 3. Four experimental marks produced by indirect percussion selected from a sequence of 15 made in succession with the same burin tip. Notice that marks exhibit stable internal features allowing researchers to recognize that the same tool has been used throughout the sequence. Micrographs are video images obtained by examining transparent resin replicas in transmitted light. Scale = 1 mm.

notches will generally not prevent the recognition of the fact that the same tool has been used throughout the sequence. The removals will possibly engender changes in the position and orientation of the microsteps created on the notch walls by the cutting-edge movement. The general profile of the notch, however, and the angle formed by the notch walls, will remain the same, indicating that the same tool was used (D'Errico 1991).

In single stroke-lines and microincisions produced by pressure or indirect percussion, a partial breakage of the working edge results in marks which are often still recognizable as having been carved by the same tool. By contrast, complete removal of the working edge results in a radical change of the marks' morphology. Microscopic analysis of experimental single stroke lines on which such complete breakage has occurred, however, has revealed that, in many cases, clues exist to establish that a breakage and not a change of tool was the cause of the morphological change (D'Errico 1995b:24-5). In effect, the breakage produces a sudden interruption of the engraved line followed by the start of a new, slightly shifted and often wider line. The subsequent marks retain the new morphology confirming that the working edge,

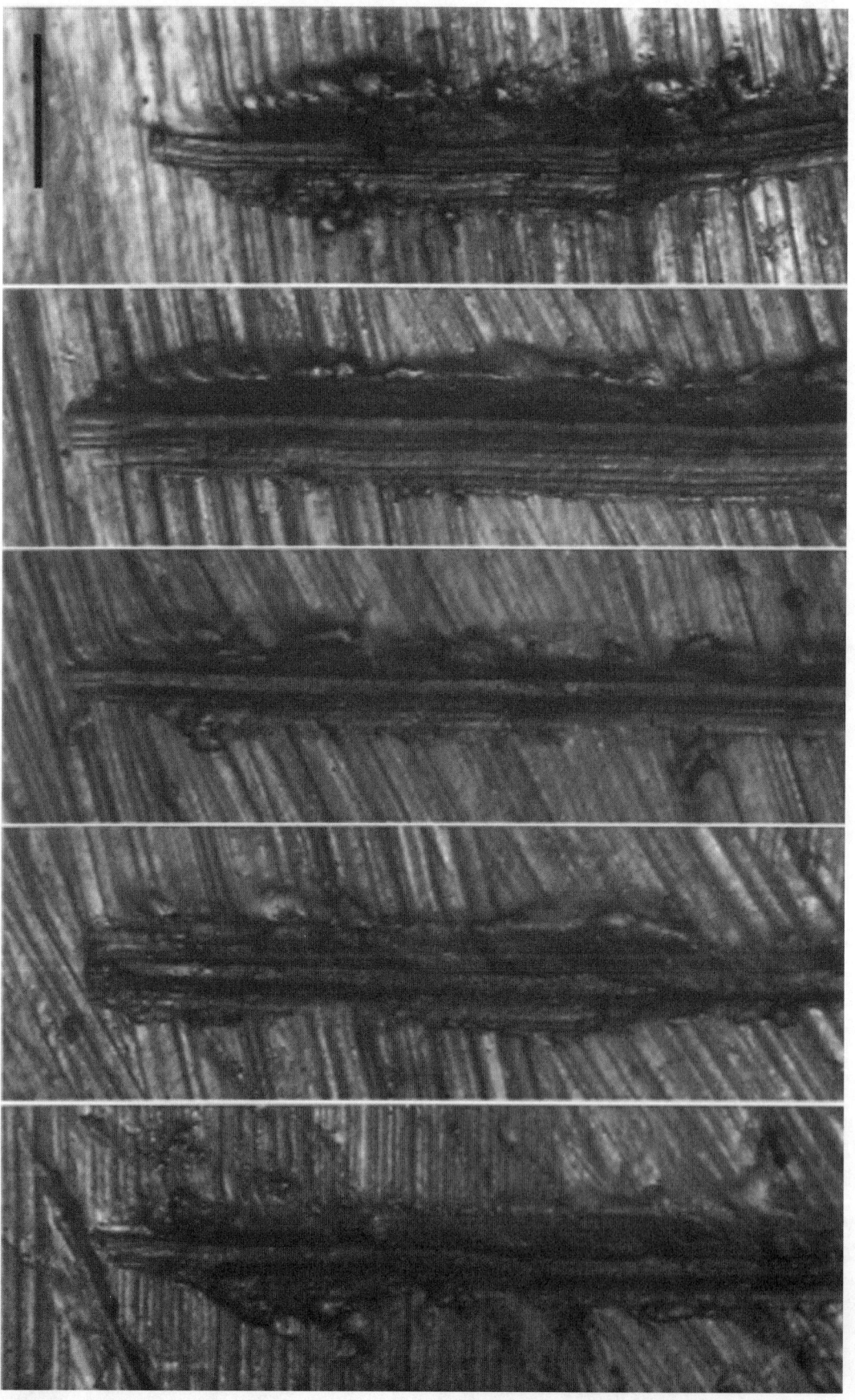

Fig. 4. Five experimental single stroke lines selected from a sequence of 27 engraved in succession with the same cutting edge on a red deer antler. From left to right are marks n. 1, 5, 12, 18, 26. Although the wear of the tool produces a gradual widening of the left side of the engraving with appearance of parallel striations, the use of the same tool is still recognizable. Horizontal striations are due to scraping of the antler, made to prepare the surface. Scale = 500 µm.

The Archaeology of the Oldest Artificial Memory Systems

Fig. 5. The two sides of the Tossal de la Roca broken pendant. Each side shows four sets of tiny marks arranged in parallel rows. Microscopic and morphometric analysis of these marks suggest that each set was made by a different point and often by a different marking technique. In two sets it is even possible that two different points were used. Notice the variation in the mark's spacing and orientation. Scale = 1 mm.

as a result of breakage, was responsible for their production. It is also been observed that the breakage generally leads to a stabilization of the working point morphology and a corresponding stability in the engraved line sections, with no further new breakages. These clues survive post-depositional processes as they have also been recognized on archaeological engravings (D'Errico 1995b: fig. 62c).

Finally, it has been observed experimentally that breakages which could be confused with tool changes do not occur in sequences of marks produced by pressure or indirect percussion on relatively soft support such as wet deer antler (D'Errico 1995a). They are, instead, present on 20% of those produced on dry antler. As with single stroke lines, however, these breakages very rarely occurred more than once in each series. Moreover, about 80% of them happened while engraving one of the first three marks. As with changes produced by hand motion variability (see above) breakages occurring in single stroke lines and microincisions due to pressure or indirect percussion can be confused with changes of tool. The corresponding marks, however, cannot be readily interpreted as an AMS based on accumulation in time since a single change of tool would indicate only two stages of mark accumulation.

Resharpening

A stone tool can be resharpened when it becomes ineffective for the task for which it was made or chosen. In the case of marks intended for visual display, resharpening is not necessarily caused by the impossibility of using the tool to make additional marks. The tool can also be deliberately modified or changed when the marks it produces no longer correspond to a particular pattern the engraver is trying to create. Whatever the reason, the need to resharpen a tool varies according to the morphology of its working edge, the marking technique, the hardness of the marked object, and the dimensions of the marks (length of single or multiple stroke lines, depth of notches and micro incisions). Certain techniques do not require a tool to be resharpened. Others techniques necessitate resharpening only when used on some types of material.

Experimental marking, for instance, has shown that a large number of short single stroke lines can be carved on a flat bone surface without needing to resharpen the point used to make the marks (D'Errico and Cacho 1994). When longer lines are engraved on stone, the wear of the point can produce a gradual widening of the line. This, however, is often visible only at microscopic scale and does not necessarily require a resharpening of the point nor does such resharpening guarantee that a point will produce a thinner line. Similarly, in the absence of particularly annoying breakages, there is no need to resharpen a tool used to make microincisions produced by pressure or indirect percussion. This is because it has been observed experimentally that dozen of marks of this type can easily be made in succession without producing any visible changes in mark morphology.

Resharpening is more useful when carving notches because a tool, used in a sawing motion on bone or stone, wears out relatively quickly, particularly if the cutting edge is narrow and unretouched. It has been observed experimentally that blades used to make notches on bone sticks generally become worn out after 10-20 notches, depending on the length of the cutting edge and the depth of the notch. Resharpening of blanks used to cut notches, however, can in many cases be distinguished from changes of tool. If marginal, retouch along the edge does not prevent the tool's identification because it will not affect the main profile of the notch. If extensive, the retouch can be resharpened only a limited number of times (2-3). Microscopic clues exist that allow the researcher to establish whether notches were cut by a retouched or unretouched tool and to know if the tool was worn to such an extent that it needed to be resharpened (D'Errico 1991, 1998). The section of the cutting edge and the orientation of the tool (position of the ventral and dorsal face of the blank) can also be determined. Thus only a transition from notches made by an unretouched cutting edge (indicating an advanced stage of tool wear) to notches made by a retouched edge (while still exhibiting a profile similar to that of the tool used for the first group) can be easily interpreted as the result of resharpening. Clear morphological changes, for instance, between notches produced by unretouched cutting-edges, involving changes in the notch shape and profile should probably be attributed to a change of tool.

Change of working edge

Instead of resharpening a worn edge, an engraver can either turn the blank around and use the opposite edge of the tool or use a new tool. Both behaviors produce changes in mark morphology which cannot be distinguished from tool changes in a situation where the marks were made over a considerable period of time. It should be noted, however, that knapped tools have at most two or three edges that are suitable for fashioning marks. Thus, repeated changes in mark morphology, similar to those produced experimentally by changing tools, cannot be interpreted as the result of a simple change of working edge.

Several issues already discussed in relation to resharpening also apply to some extent, to tool changes. Independently of the technique used, many marks can generally be engraved without requiring a tool to be resharpened or changed. The use of worn tools, requiring a resharpening or a change, can be identified through a microscopic analysis of the marks. Changing tools requires extra time and it is uneconomical since, for example, cutting edges used to mark hard material cannot be effectively used for other tasks (e.g. cutting or scraping soft material). Therefore, repeated changes of tool can hardly be explained simply by the need for sharper tools.

Predictive criteria

The question arising from these results is whether or not other elements, independent of the technical analysis of marks, can facilitate the identification of this type of AMS in the archaeological record. Repeated isomorphic marking on ethnographic objects not used as AMS's is often intended to communicate a message by the visual outcome of the engraving as a whole. This is achieved by carefully creating a symmetrical or asymmetrical pattern in relation to the object's shape and through the placement and composition of different groups of marks. Consistency or variation in the spacing and orientation of the marks can also contribute to creating a meaningful pattern. This is generally obtained by using the same marking technique and the same tool in a unique session or adopting techniques that make all marks look the same even if they were made at different times. For technical reasons, their productions required several work sessions.

Since AMS's based on the accumulation of marks over time are not intended to produce such a pattern, their marks should vary in distribution, spacing, orientation and dimension more than if they were created to communicate information as a coherent pattern.

If the number of items to be recorded by an AMS is not known or is only roughly known at the beginning of the accumulation, spacing cannot be established in order to precisely fill the available space with marks. Thus it is possible that at the end of the process, marking will occupy only a part of this space, that marking will be continued on a space originally not intended for this purpose or even that the marks will be crowded in one area of the object in order that all of the marks can be included in the available space. If the accumulation of marks is carried out over a long period, it is also possible that the object may be lost or damaged before the work is completed. In both of these last cases the sequences of marks will be notable by their lack of uniformity and symmetry in relation to the object's overall shape.

Special cases

The identification of repeated changes of tool and of changes in their position and directionality do not guarantee, however, by themselves, the identification of a AMS based on the accumulation of marks over time without further analysis. In the first place, it is unlikely that the use of these criteria will lead to the identification of all the AMSs of this type. Marks can be accumulated over time with the same tool. We do not know the length of time a Paleolithic craftsman was accustomed to keeping a tool used to mark objects. Moreover, the tool's use-life probably varied as function of raw material availability, the type of tool, the "social life" it was intended for since it was knapped within a particular cultural tradition. Many apparently unused blanks and retouched tools are found in Paleolithic sites, indicating that knapped stones were discarded relatively easily. Knapped stone tools, however, are associated with many Paleolithic and Mesolithic burials, suggesting that some tools, at least, were permanently and symbolically associated with some individuals—indicating a prolonged utilization by their owners. In sum, it is difficult to fix a minimum length of time beyond which the temporal accumulation of marks would have required the use of a new tool.

Clues exist which can help to identify marks carved at different times with the same tool. New marks may have been made with a different part of the tool, producing marks of different morphology in situations where no technical need is detected for such a change (absence of breakages, wear etc). Experimental marking indicates that even if using the same tool, the passage of time results in the marked object and the tool being handled differently at each new stage of marking. This situation often produces marks that are different in their morphology as well as in their orientation suggesting that something has interrupted the marking process.

Another problem arises when interpreting repeated changes of tool as the result of an accumulation of marks over time. In principle, repetitive tool changes constitute a valid criterion only when the resulting marks are not distinguishable by the naked eye. If by changing a tool clear morphological changes are produced, it becomes difficult to ascertain whether such tool changes are the consequence of an accumulation of marks over time or the result of a deliberate behavior, carried out in just one session, with the objective of attributing a particular meaning to marks of different morphology. In the latter case there is a desire to produce different signs that are part of a code based on the morphology of information-bearing elements and not on accumulation of marks over time. The reason for producing morphologically different marks by changing the tool used can also be due to the desire to produce "decoration" which is defined as a pattern in which the engraver does not attribute a distinct meaning to each mark or group of marks but rather to the visual effect of the entire engraving.

In these last two cases, changing the tool used to create the marks would simply be a technique for producing repetitive, visible variation in mark morphology. Experimental replication of different types of marking techniques found on Paleolithic objects, however, reveals that, in most of the cases, a simple change of tool within a sequence of marks done with the same technique, does not produce noticeable changes in mark morphology. It therefore constitutes by itself an unreliable means to create unambiguous systems of signs or recognizable visual patterns. Thus, it is probable that even if slight morphological changes appear within sequences of marks produced by different tools, these should more reasonably be considered as epiphenomenal—the result of the accumulation of marks over time rather than the desire to introduce changes in the marks' morphology.

In theory, a situation can exist where the identification of repeated tool changes might lead to an erroneous interpretation. This is the case in which a group of individuals marks an object in succession. If each of them uses a different stone tool to produce one or more marks on the object, the resulting sequence will look like marks that were made over a period of time where a different tool may have been used at each new marking stage. Although the situation of a number of marks being created at one time but by numerous people represents a particularly unlikely case, it remains nevertheless a possibility which cannot be eliminated *a priori* from our interpretation of the archaeological record. It is worth noting that since such marking could have been done for reasons other than to record information, the technical analysis of this object could suggest its use as a AMS based on the accumulation of marks over time when, in reality, the object might not be a AMS at all.

The sole means of identifying the above case would be to establish criteria that are independent of the marks themselves in order to ascertain if a lapse of time has passed between them. If, for example, the marks have accumulated over a long period of time, it is possible that the earliest marks may have undergone more alteration (functional use, transport, manipulation, accidental damage etc.) than marks created later in time. Whether such differences can be documented in a convincing manner several thousand years after deposition remains to be demonstrated. Microscopic analysis of bone, antler and ivory objects experimentally transported and manipulated has shown, however, that criteria do exist to distinguish these modifications and differentiate them from damage produced by other causes such as deliberate polishing, post-depositional damage etc. (Bromage 1984; d'Errico 1993). Microscopic analysis of a selected sample of Paleolithic objects has shown that some of them exibit wear similar to that found on experimentally worn objects.

Conclusion

The relevance of my approach for identifying the oldest artificial memory systems is not fully demonstrated by the above presentation as I have addressed only one of the four factors which can govern an AMS code—probably the one which leaves the least ambiguous traces in the archaeological record. Codes based on factors such as morphology or spatial distribution of signs as well as those governed by a combination of factors are more difficult to identify and, in some cases, they are impossible to distinguish from or are inextricably mixed with other forms of communication means. These questions need to be explored in the future in order to make explicit the limits and potential ambiguity of archaeological interpretation.

It has been argued that the identification of AMS's with codes based on the accumulation through time of isomorphic signs should rely on an technical analysis of the marks. The combined use of experimental marking and microscopic analysis demonstrates that diagnostic criteria do exist to distinguish tool changes, possibly indicating an accumulation over time of marks, from changes in mark morphology due to technical reasons. Other clues indicate changes in the orientation of the tool and whether or not an object has been subjected to long periods of transportation and/or manipulation. The reliability of these criteria for the study of the sequences of notches has been confirmed by two blind tests (D'Errico 1991). It is clear, however, that these clues cannot give definite answers in all cases and they should be supplemented in the future by new experimental data and the application of new analytical techniques.

Acknowledgements

I would like to particularly thank April Nowell for organizing the workshop on the *Archaeology of Intelligence*, for inviting me to prepare this paper, and for her stimulating critical reading of the manuscript. Helpful suggestions and criticisms were also provided by Joao Zilhão, Roy Harris and Yau Shun-Chiu. The CNRS/Royal Society Exchange program supported my 1996 stay in Cambridge, during which the first draft of this paper was prepared. I am also grateful to Jean Auel and the Aquitaine/Euskadi Navarre Joint Program for financial support.

References Cited

Adedo Ogunde, Ph.
  1997 The communicative and semiotic contexts of aroko among the Yoruba symbol-communication systems. *African languages and cultures* 10:145-56.
Bottero, J.
  1980 De l'aide mémoire à l'écriture. In *Ecritures-Systèmes Idéographiques et Pratiques Expressives* (Actes du Colloque International de l'Université Paris VII.) edited by A. M. Christin, pp. 13-35. Le Sycomore, Paris.
Bromage, T.G.
  1984 Interpretation of scanning electron microscopic images of abraded forming bone surfaces. *American Journal of Physical Anthropology* 64:161-78.
Buissac, P.A.
  1994 Art or script? A falsifiable semiotic hypothesis. *Semiotica* 100(3-4):349-67.
Buissac, P. A.
  1997 New epistemological perspectives for the archaeology of writing. In *Archaeology and Language 1. Theoretical and methodological orientations*, edited by R. Blench and S. Matthew, pp. 54-62. Routledge, London.
Buisson, D.
  1991 Les flûtes paléolithiques d'Isturitz (Pyrénées atlantiques). *Bulletin de la Société Préhistorique Française* 87, 420-33.
Changeux, J.-P.
  1983 *L'Homme neuronal*. Fayard, Paris.
De Francis, J.
  1989 *Visible speech: the diverse oneness of writing systems*. University of Hawai Press, Honolulu.
Derrida, J.
  1967 *De la Grammatologie*. Les Editions de Minuit, Paris.
D'Errico, F.
  1989 A reply to Alexander Marshack. *Current Anthropology* 30:495-500.
D'Errico, F.
  1991 Microscopic and statistical criteria for the identification of prehistoric systems of notation. *Rock Art Research* 8:83-93.
D'Errico, F.
  1992 A reply to Alexander Marshack. *Rock Art Research* 9:59-64.
D'Errico, F.
  1993 La vie sociale de l'art mobilier paléolithique. Manipulation, transport, suspension des objets en os, bois de cervidé, ivoire. *Oxford Journal of Archaeology* 12:145-74
D'Errico, F.
  1995a A new model and its implications for the origins of writing: the La Marche antler revisited. *Cambridge Archaeological Journal* 5:163-206.
D'Errico, F.
  1995b *L'art gravé azilien. De la technique à la signification*. XXXIe supplément à Gallia Préhistoire. CNRS, Paris.
D'Errico, F.
  1996a Marshack's approach: poor technology, biased sciences. *Cambridge Archaeological Journal* 6:111-17.
D'Errico, F.
  1996b Image analysis and 3-D optical surface profiling of Upper Palaeolithic mobiliary art. *Microscopy and Analysis* 39:27-29.

D'Errico, F.
    1998 Palaeolithic origins of artificial memory systems: an evolutionary perspective. In *Cognition and Material Culture: The Archaeology of Symbolic Storage*, edited by C. Renfrew and C. Scarre, pp. 19-50. McDonald Institute Monographs, Cambridge.

D'Errico, F. and C. Cacho
    1994 Notation versus decoration in the Upper Palaeolithic: a case-study from Tossal de la Roca, Alicante, Spain. *Journal of Archaeological Science* 21:185-200.

Donald, M.
    1991 *Origins of modern mind*. Harvard University Press, Cambridge and London.

Dubin, L. S.
    1987 *The history of beads*. Thames and Hudson, London.

Elkins, J.
    1996 On the impossibility of close reading. The case of Alexander Marshack. *Current Anthropology* 37:185-226.

Festinger, L.
    1983 *The human legacy*. Columbia University Press, New York.

Fevrier, J.
    1984 *Histoire de l'écriture*. Payot, Paris.

Fritz, C.; M. Menu; G. Tosello; Ph. Walter
    1993 La gravure sur os au Magdalénien: étude microscopique d'une cote de la grotte de la Vache (commune d'Alliat, Ariège). *Bulletin de la Société Préhistorique Française* 90:411-25.

Gelb., I. J.
    1952 *A study of writing*. University of Chicago Press, Chicago.

Goolmer, C. A.
    1885 On African symbolic messages. *The Journal of the Royal Anthropological Institute of Great Britain and Ireland* 14:169-81.

Goody, J.
    1977 *The Domestication of the Savage Mind*. Cambridge University Press, Cambridge.

Goody, J.
    1987 *The interface between the written and the oral*. Cambridge University Press, Cambridge.

Goonatilake, S.
    1991 *The Evolution of Information. Lineages in Gene, Culture and Artefact*. Pinter, London.

Hahn, J.
    1996 Le Paléolithique supérieur en Allemagne méridionale (1991-1995). In M. Otte (ed.), *Le Paléolithique supérieur européen. Bilan quinquennal (1991-1996)*, edited by M. Otte, pp. 181-86. ERAUL 76, Liège.

Harris, R.
    1986 *The origin of writing*. Duckworth, London.

Harris, R.
    1990 Quelques réflexions sur la tyrannie de l'alphabet. In *L'Ecriture: le cerveau, l'oeil et la main*, edited by C. Sirat, J. Irigoin and E. Poulle, pp. 195-200. Brepols, Turnhout.

Harris, R.
    1994 *La sémiologie de l'écriture*. CNRS, Paris.

Harris. R.
    1995 *Signs of writing*. Routledge, London.

Havelock, E. A.
    1982 *The Literate Revolution in Greece and Its Cultural Consequences*. Princeton University Press, Princeton.

Hill Boone, E. and W. D. Mignolo (eds.)
    1994 *Writing without words*. Duke University Press, Durham.

Hutchins, E.
    1995 *Cognition in the Wild*. MIT Press, Cambridge, MA.

Knight, C., C. Power, and I. Watts
    1995 The human symbolic revolution: a Darwinian account. *Cambridge Archaeological Journal* 5:75-114.

Kuhn, T.
    1979 Metaphor in science. In *Metaphor and Thought*, edited by A. Orton, pp. 409-19. Cambridge University Press, Cambridge.

Lartet, E. and H. Christy
    1865-75 *Reliquae Aquitanicae, being Contributions to Archaeology and Palaeontology of Perigord and the Adjoining Provinces of Southern France*. Williams and Norgate, London.

Leroi-Gourhan, A.
    1964-65 *Le Geste et la Parole*. 2 Vols. Albin Michel, Paris.

Marshack, A.
    1964 Lunar notation on Upper Paleolithic remains. *Science* 184:28-46.

Marshack, A.
    1970 *Notation dans les gravures du Paléolithique superieur. Nouvelles méthodes d'analyse*. Publications de l'Institut de Préhistoire de l'Université de Bordeaux, Bordeaux.

Marshack, A.
  1972a Cognitive aspects of Upper Paleolithic engraving. *Current Anthropology* 13(3-4):445-77.

Marshack, A.
  1972b *The roots of Civilisation*. MacGraw Hill, New York.

Marshack, A.
  1972c Upper Paleolithic Notation and Symbol. *Science* 178:817-28.

Marshack, A.
  1974 The Chamula calendar board: an internal and comparative analysis. In *Mesoamerican Archaeology*, edited by N. Hammond, pp. 255-70. University of Texas Press, Austin.

Marshack, A.
  1985 A lunar solar year calendar stick from North America. *American Antiquity* 50:27-51.

Marshack, A.
  1988 Paleolithic calendar. In *Encyclopedia of Human Evolution and Prehistory*, edited by I. Tattersal, E. Delson and J. van Couvering, pp. 419-20. Garland, New York.

Marshack, A.
  1991a The Taï plaque and calendrical notation in the Upper Palaeolithic. *Cambridge Archaeological Journal* 1:25-61.

Marshack, A.
  1991b A reply to Davidson on Mania and Mania. *Rock Art Research* 8:47-58.

Marshack, A.
  1991c *The Roots of Civilisation*. Moyer Bell, New York.

Mithen, S. J.
  1996 *The Prehistory of the Mind: a Search for the Origins of Art, Religion and Science.* Thames and Hudson, London.

Postgate, N., W. Tao and T. Wilkinson
  1995 The evidence for early writing: utilitarian or ceremonial? *Antiquity* 69:459-80.

Pulgram, E.
  1976 The typologies of writing-systems. In *Writing Without Letters*, edited by G. Haas, pp. 1-28. Manchester University Press, Manchester.

Robinson, J.
  1992 Not counting on Marshack: a reassessment of the work of Alexander Marshack on notation in the Upper Palaeolithic. *Journal of Mediterranean Studies* 2:1-16.

Rosenfeld, A.
  1972 Review of "Notation dans les gravures du Paléolithique supérieur", *Antiquity* 36:317-9

Sperber, D.
  1985 Anthropology and psycology: towards an epidemiology of representation. *Man* 20:73-89.

Thierry, B. G., J. Y. Theraularz, P. Y. Gautier and B. Stiegler
  1996 Joint memory. *Behavioural Processes* 35:127-40.

Watt, I.
  1999 *The origin of symbolic culture: the Middle Stone Age of Southern Africa and Khoisan Ethnography.* University of East London, unpublished Ph.D. thesis.

White, R
  1982 The manipulation of burins in incision and notation. *Canadian Journal of Anthropology* 2:129-35.

4. A Pragmatic View of the Emergence of Paleolithic Symbol-Using

A. Martin Byers

Abstract

I address theoretical impediments to demonstrating the emergence of symbol-using in the Paleolithic and through elucidating the notions of material warrant, pragmatic symbol and effortless reflexivity, I justify the claim that the emergence of symbol-using will be visible in the Paleolithic archaeological record as a discontinuity that I call the Style-1/Style-2 rupture. I confirm this approach by showing isochrestic styles are conventional styles, thereby marking rather than masking the emergence of symbol-using. I then articulate basic axioms and corollaries on which to stipulate the criteria by which to recognize a Style-1/Style-2 rupture and I conclude that the European Middle-Upper Paleolithic transition conforms to these criteria.

Introduction

All modern human populations are masterful symbol-users. Are we unique this way or are we simply the latest in a line of more or less skilful symbol-using hominids? There are currently two alternative viewpoints addressing this question, the gradualist and the discontinuist. In general, gradualist views deny the uniqueness of our mastery in favor of emphasizing the continuity theme, claiming that current human symbol-using is not that different quantitatively or qualitatively from that of more recent archaic human populations, such as Neanderthals, and probably even hominid populations of the earlier Pleistocene were at least minimally capable symbol-users (Bednarik 1992, 1995; Chase 1994; Marshack 1989, 1990, among others). Discontinuist views opt for some breaking point, a threshold manifested as a rupture in the material patterning. These can be grouped into two variants, early and delayed discontinuism. Delayed discontinuism claims that initially anatomically modern human populations of, e.g., the early Upper Paleolithic of Europe, were merely moderately competent symbol-users, hardly distinct from the symbol-using Middle Paleolithic populations. Only in the final stage of the late Pleistocene did a sudden explosion of symbol-using occur, constituting the later Upper Paleolithic cultures (Clark and Lindly 1989, Lindly and Clark 1990; Duff et al. 1992; Mithen 1994, 1996a, 1996b). In contrast to delayed discontinuism, advocates of early discontinuism take the symbol-using revolution as *more or less* contemporaneous with the emergence of anatomical modernity in humans, although they do not claim that this rough correlation of the two processes was necessarily causally linked (Noble and Davidson 1996; Mellars 1989a, 1989b, 1991; White 1982, 1989, 1992).

While the gradualist and discontinuist views seem to be at loggerheads, in fact, they share the same basic theoretical perspective concerning the nature of signs, symbols, symbol-using and human action. A central, largely shared assumption is that human representations—whether mediated by words, gestures or material artifacts—are used primarily to convey objective information by means of referencing, usually about social or personal identity (Wiessner 1983; 1985). Furthermore, with respect to Paleolithic prehistory, it is largely assumed that the earliest artifactual modes of signing would have been organic-based expressive artifacts, e.g., leather straps, wooden staffs, and so on (Bednarik 1992, 1995). In this view, clearly very little material residue of early symbol-using would survive in the record and, hence, there would be no data by which to confirm or reject explanatory models of the initial emergence of symbol-using. The broad influence of this basic theoretical framework comes out in the claim made by Duff et al. (1992:225) when they write that early symbol-using:

> ...[Probably] developed in contexts distinct from those in which it is manifest today...(and)...we will only be able to detect it archaeologically

long after it had become a significant part of the human behavioral repertoire. (emphases added)

I will call this view the *Invisibility Thesis*. It explains the "fact" that there appears to be a lack of evidence for Early or Middle or even early Late Pleistocene material expression. It results in a methodological stalemate. After all, in its terms gradualists can claim that early hominids could have been competent symbol-users while never expressing this in durable form (Chase 1991; Lindly and Clark 1990).[1] Reciprocally, of course, at least early discontinuists can claim that despite what appears in the early Upper Paleolithic to be only a limited amount of material cultural symbolic expressivity, this really demonstrates the simultaneous existence of a rich but archaeologically invisible symbolic realm (For the gradualist view see particularly Bednarik 1992, 1995. Also Chase 1989, 1991; Chase and Dibble 1987, 1992; Dibble 1989, 1991; Duff et al. 1992; Marshack 1989, 1990. For the discontinuist view see: Byers 1994; Davidson and Noble 1989, 1993; Mellars 1989a, 1989b, 1991; Noble and Davidson 1996; White 1982, 1989, 1992).

To break this stalemate I present below theoretical grounds supporting what I will call the *Visibility Thesis*. This thesis takes a holistic stance by rejecting the radical separation of functional and expressive spheres and the concomitant division of material culture into utilitarian and symbolic style. Instead it claims that, for symbol-using hominid populations, *all* and not merely some of their material culture would have an irreducible expressive-symbolic moment. This symbolic moment would be realized by what archaeologists usually call conventional style (Byers 1994). This is a major claim that will be elucidated in this paper. But, to briefly anticipate this argument, what it means is that conventional style will be ubiquitous because material cultural symboling serves what I term the *warranting imperative*. As I will use the term, warranting is a symbolic-expressive process that constitutively transforms our practices so that they count as institutional conduct and action characteristic of a given cultural population (Searle 1995). As a symbolic transformation this warranting moment is routinely realized by the responsible populations using tools that bear symbolic styles and these, by convention, manifest the users' public action intentions in the very instance of their regular behavioral interventions. In virtue of routinely realizing the warranting imperative, then, all their practical tool-using will simultaneously be symbol-using practices.[2]

The Visibility Thesis does not deny that for much of the Pleistocene the patterning of the lithics will be exhaustively explicable in practical tool-using terms. But it insists that this patterning is the outcome of *non-symbol-using* hominid populations producing what I have termed Style-1 assemblages (Byers 1994). This thesis claims that out of this Style-1 background there will emerge what I term Style-2 assemblages. These are, at least initially, tool industries that are equally utilitarian but, nevertheless, displaying an irreducible, rule-governed expressive moment, constituting them as pragmatic symbols. As Style-2 assemblages they make up the industries of symbol-using populations. This is the basis for postulating that the Style-1/Style-2 rupture marks the emergence of symbol-using. If I can theoretically elucidate the Visibility Thesis, then the conclusion that follows is that even the utilitarian lithics of a symbol-using population will necessarily be action-constitutive or pragmatic symbols in their own right.[3]

The methodological point that follows is that, since utilitarian lithics are the densest and most durable aspect of the archaeological data, rather than masking and thereby making the emergence of symbol-using invisible, their patternings will have embedded in them formal ruptures that mark the emergence. In terms of the Visibility Thesis, then, wherever and whenever symbol-using emerged the lithics would have been a material part of this relatively abrupt process, thereby being independent witnesses and constituents of what I call the Style-1/Style-2 rupture. The Middle-Upper Paleolithic transition in Europe, I will postulate, is one such rupture. Of course, there probably have been others that have not as yet been perceived.

Theoretical Structure: The Warranting Model

Introduction

Before I can say much about the conditions promoting the emergence of symbol-using what has to be made clear is precisely what this phenomenon is. Since we are the only symbol-users about whom we can have confident knowledge, I think I can best establish the conditions of symbol-using by first examining theoretically the ways modern humans use and constitute symbols. I use the court warrant as an analogical model in order to highlight what I consider to be the generic dynamics of material cultural style. I call this the Warranting Model of material cultural style. I then go on to explore the theoretical undergirding of our usage of style, treating it as residing in a unique aspect of human cognition that I term effortless reflexivity. I call this the Reflexivity Model and I conclude this essay by using this theoretical structure to delineate the nature of the Style-1/Style-2 rupture and to sketch out central criteria by which to recognize it.

The Warranting Model

Warrants are court documents that constitute their bearers as occupying the social positions of bailiffs, sheriffs, or some other officer of the court, and simultaneously, their use transforms the behaviors these persons perform into the types of institutional practices and activities that they are, namely, legal acts of different types (Bhaskar 1979; Searle 1995). This transformative capacity is a virtual power (Giddens 1976, 1979, 1984) in the sense that it exists only in the collective understanding and intentionality of the particular community that holds these pragmatic meanings of the warrant. Therefore, the tangible properties of these simple pieces of paper can transform them into the powerful documents we identify as warrants only if they are formed according to the normative rules of the society and, of course, only those who know the code can recognize and interpret these attributes as mediating the virtual property we can term social authority. Indeed, it is by recognizing the warrants as warrants that their power is constituted. This code forms a complex background of collective understandings that I will call the action-constitutive protocols and that can be articulated as "how-to" rules and standards. These protocols - implicit and explicit, informal and formal - stipulate both the observable properties these paper artifacts must display to be warrants *and* the way that they are to be used (Byers 1991; Searle 1995). Furthermore, they implicate the very beliefs that make their existence and use possible, e.g., the belief that has the existence of court system as its content. Hence, in the moment of the full exercising of the warrant, the court is itself virtually presenced and constituted.

But this predicating of transformative powers is not limited to paper artifacts termed warrants. Rather, a literate society is characterized by a whole slew of protocol-based textual documents that constitute critical material constituents of that social reality, e.g., licenses, coupons, passports, visas, tickets, passes, deeds, money, contracts, academic credentials, and so on - somewhat indefinitely (Douglas 1982; Searle 1995). But the fact of literacy is not what determines this process. Rather, the opposite may be primary, i.e., the warranting imperative partly explains the emergence of written texts. Indeed, the core of the Warranting Model claims that warranting, i.e., the symbolic expression and endowing of collective powers, is universal to all symbol-using human societies - whether literate or non-literate. Since the absence of documentation is the hallmark of non-literate societies, two logical consequences follow: 1) That the material cultural assemblages of non-literate societies must bear this expressive transformative power; 2) That it is mediated by conventional style. Thought of in this way, the forms of even the fully utilitarian tools of non-literate symbol-using societies will be caught up in "how-to" protocol, making them the ever-present constituent symbol-elements of their social interaction.

To characterize the operation of the warranting imperative by the terms "caught up" is not to fudge the situation. Rather, it follows from the logical possibility that there are two possible ways that rules can come to govern the usage of the tools: 1) by being directly applied to the formation of the tool assemblage; or 2) by being indirectly applied via the conditions in which the different tools are used. These are not mutually exclusive possibilities. The direct case is straight forward. In a society of savanna foragers a spear of Style A is typically used to perform animal sacrifice while in the same society a spear of Style B is typically used to procure animal food. Both behaviors require animal slaughter. But style A is taken as a warrant of sacrifice while style B, as a warrant of hunting. In this case, the different styles are partly constitutive of the conditions in which they are used and, although this is not necessarily the case, the morphological differences that define the two styles will tend to be adjunct in nature - e.g., sacrificial spears displaying some properties that are surplus to utilitarian needs that any animal killing involves while the hunting spear may be fully utilitarian in its properties. The reverse case, in which the conditions determine the styles, is illustrated in the following. In the same society different stages of their animal procuring practices, while requiring the same form of material intervention, slicing and cutting, "demand" formally different tools. For example, when the game animal is being pursued, hunting spears of style A with blades of Style M are used. But when the animal is to be butchered, knives with blades of Style N are used. Objectively, the two blade styles could be used interchangeably. In this case, the stages of a single process "demand" different styles and, the Warranting Model would predict, the stylistic variation of the blades that marks these stages as different would be largely isochrestic, in Sackett's (1982) sense. This means that we could not analytically isolate any particular objective properties that would be surplus to the equivalent slicing-cutting tasks that the blades involved in both stages required. Hence, the stages are symbolically as well as practically constituted. Of course, these two dynamics of style are both based on the same source, the warranting imperative. The Warranting Model and the warranting imperative it presupposes, then, claim that these protocols are materially realized as Style-2 assemblages.

Discussion

The Warranting Model presupposes a theoretical stance concerning the nature of human meaning, action and intention and the latter is the basis for the claim that we can distinguish the material cultures of symboling from those of non-symboling tool-producing and tool-using populations. I call it the action-constitutive theory of material cultural style (Byers 1991). Earlier I pointed out that both gradualists and discontinuists largely share the perspective I termed the tool-using/symbol-using dualism. This dualism clearly leads to the view, largely ubiquitous in archaeology, that symbol-using populations use their material culture to perform two radically different types of material activities: physical actions directed to satisfying end-goals and information conveyance actions directed to satisfying communicative goals. The Warranting Model is the basis for rejecting this dualism and for postulating the tool-using/symbol-using *duality*. This simply claims that for symbol-using populations an irreducible part of social life as realized in their ongoing interactions entails expressive signing in the very instance of behaving, regardless of whether this behavior is functional or communicative. Hence, symbol-using is not something that is added to the action repertoire of a Style-1 assemblage when symbol-using emerges. Rather, symbol-using entails an emergent reconstruction of the forms of these assemblages into Style-2 types that serve the warranting imperative.

This radical claim requires justification. I will start by critiquing the Invisibility thesis and its tool-using/symbol-using dualism. This dualism rests on two (inadequate) assumptions about the nature of the sign meaning of an object: 1) that this meaning is an actual *property of the object*; 2) that this meaning can be largely equated with *information conveyance*. It is very easy to correct the first assumption. If it is accepted that both the function and the sign meaning of material things are contingent on the collective intentionality of the users, then it follows that meaning is a property of that intentionality and not of the objects being used (Searle 1983:28). Indeed, in these terms, to speak of the meaning of an object is to speak of the ways it is used.

Treating sign meaning in information conveyance terms is treating signs as exclusively designative rather than expressive in meaning (Taylor 1985). I consider this an error. As I articulate it, the Warranting Model rests on an expressive theory of meaning so that we do not use style to refer to something else; rather we use it to express our collective intentionality. But I find that treating signs in terms of their designative meanings is ubiquitous in archaeology. Therefore, Chase (1991) cannot be faulted for continuing the tradition when he uses Peirce's tripartite classification of signs in classifying artifacts according to *how* they refer. In this scheme, indexical signs refer to their referents by means of having a natural connection with them; iconic signs refer by means of resembling their referents; and symbols, in this scheme, refer to their referents by means of arbitrary convention.[4]

Treating material culture in terms of designative meaning makes the tool-using/symbol-using dualism almost inevitable. If signs are used to designate, and if straight forward utilitarian tools also bear conventional style, thereby being symbolic signs, then what else could they do but serve two contrasting and separate purposes, end-goal and information conveyance? Furthermore, as these two purposes are radically different they must be mutually independent. This makes it logically necessary that using conventional style-bearing tools must result from performing two different and separate activities. Furthermore, as its designative meaning capacity hardly approaches the flexibility of speech, the designative usage of style must be limited largely to referencing social and/or personal identity (Wiessner 1983).

I find all this very problematic. But much can be resolved by looking at style from the perspective of the alternative expressive approach to sign meaning that I mentioned earlier. Expressions, following Taylor (1985:219), can be defined as manifesting and something is manifest:

...when it is directly available for all to see.

It is not manifest when there are just signs of its presence, from which we can infer that it is there, such as when I "see" you are in your office because of your car being parked outside...If you have an expressive face, I can see your joy and sorrow in your face. There is no inference here...Expressions...make our feelings manifest; they put us in the presence of people's feelings.

How does this work? Take tears as natural expressive signs. Of course, because they have a physiological level of causality they can be brought on by purely objective conditions, such as one being exposed to excessive wind or cold or even pain. But as expressive phenomena natural tears are an intrinsic part of the intentional emotive state that they manifest. For this reason it is mistaken to speak of tearful behavior as "pointing to" or "designating" sadness. Rather, for the person crying the tears are part of the state of affairs that they are about, namely, sadness, despair or some complex compound of such emotions. This is not to deny that humans can be deceptive in their expressive behavior, thereby expressing what is, in effect, not there. But this is something that they must work at. For example, preventing tears in circumstances that call for them can be extremely effortful for emotional persons and for phlegmatic persons producing tears in the same circumstances can be equally effortful (Burling 1993).

But tears and smiles can be understood as natural signs, in the sense that they are not rule-based. That is, treated naturally smiles do not count as part of being happy. They simply *are* part of this emotive state. Nevertheless, if the subjects are symbol-users, and if we treat the sign meaning of gestures as the ways they are used, then the conditions that enable and/or constrain these tears or smiles are largely constituted by convention. A death calls for mortuary behaviors and these are performed by relatives and acquaintances of the dead. But constitutive rules define who occupies these positions and, of course, only if the participants know a whole slew of mortuary protocols are they able and thereby enabled to perform appropriately the natural expressive behaviors that these circumstances call for such that they will count as *and* be part of the mortuary practices of that society (Byers 1999).

Similarly, even though iconic signs are also "natural" - in that they bear some observable resemblance to what they are about - their usages must be treated in conventional expressive terms so that as signs they not only presence what they represent but also they are produced and caught up in "count as" constitutive protocols. In this sense, eagle feathers are "natural" iconic signs if they are experienced by their users as participating in and presencing the power of eagles. But the conditions enabling and making the wearing of these feathers possible will be largely conventional. For example, in some societies protocol may stipulate that only adult males who have achieved warrior status may wear them, and only when going into battle. Indeed, in warranting terms the wearing of feathers is partly constitutive of both the warrior status and the behavioral interventions they mediate so that they count as and are battle activities.

Finally, note that the expressive sign relates to what it is about differently than does the designative sign. While the latter is based on the premise that the sign and its referent, what it is about, are independent phenomena, the expressive sign is the tangible form of what it is about. Hence, it manifests its object and, when reified, it presences what it is about. The religious symbol, the icon, is part of the spiritual entity it represents and, therefore, is taken by its users as presencing the powers of that entity. Hence, when treating the material assemblage of a symbol-using people in expressive terms, *all* and not merely some of the items that they regularly use will be caught up in convention, thereby making them all symbols-in-use, i.e., action-constitutive or pragmatic symbols and, for our theoretical purposes, this means that the social systems of symbol-using populations are immanent in the expressive moment of material cultural usage. I now want to address the cognitive conditions that make warranting both possible and necessary. Clearly, these cognitive conditions are part of what brings about the symbol-using threshold.

The Reflexivity Model

Monitoring

All intelligent sentient beings have neuro-sensory systems by which to perceive their environments, assess them for information, and behave effectively in terms of their intentions and needs (Fay 1987). This complex perceptual/behavioral/world interaction can be termed monitoring. Some intelligent monitoring species can also attend to and assess their own monitoring, thereby having second-order awareness of their first-order perceptual/behavioral activities. Following an old tradition, I will call this reflexive monitoring or, simply, reflexivity (Bhaskar 1979; Giddens 1979). As a mental capacity, reflexivity involves the ability of the subject to shift focal awareness from the objects and states of affairs of her/his first-order monitoring to the first-order monitoring experience itself (Searle 1983).

Assessing means applying knowledge in guiding behavior so that all monitoring is knowledge-guided. But since intelligent non-reflexive monitoring species are, by definition, unable to attend to their own monitoring, they cannot think *about* their behavior or *about* their perceptions. Rather, they simply do and perceive and they do and perceive according to the knowledge contents and background of their non-reflexive monitoring. In contrast, since reflexivity entails second-order awareness of first-order monitoring, the reflexive subject can and often does think *about* the behaviors she/he performs and the perceptions she/he has and can and often does modify her/his behaviors or reinterpret her/his perceptions according to the assessing that the second-order monitoring affords. In short, thinking, contemplating, and reflecting are mental practices that are possible only because of reflexivity.

The wrinkle that I propose is to claim that in order for symbol-using to emerge and be sustained an intelligent species must have a special form of reflexivity, what I term *effortless reflexivity*. While effortlessly reflexive subjects attentively monitor the immediate world, they also sustain an implicit, practical ongoing second-order assessing of this monitoring itself and, independently of any external stimulus, they can shift their focal awareness from the objects of the first-order monitoring to the first-order monitoring itself. This self-instigating reflexivity sustains a chronic state of dual-level awareness, thereby constituting the cognitive condition that not only promotes effortless thinking, contemplating, and reflecting but also the emergence of self-awareness. And this combination constitutes the basis of symbol-using. This is because, as I made clear earlier, symbol-using, i.e., warranting, is coded behavior by which the active subject effortlessly communicates her/his action intentions to relevant others in the very instance of her/his material behavioral interventions, thereby transforming these into the types of material activities they are. Such ongoing code-based or protocol-governed behavior entails this effortless dual-level cognitive capacity (Bhaskar 1979).

Effortless reflexivity presupposes its contrasting condition. This is not simple or non-reflexive monitoring. Rather, it is effortful reflexivity, which I postulate is characteristic of most intelligent, reflexive species, including our hominid forbearers. The default condition of effortful reflexivity is a chronic practical first-order monitoring and only when externally instigated disruptions of everyday routines occur can subjects shift into second-order awareness by which they assess their first-order monitoring itself. Furthermore, the subject must exert cognitive *effort* to sustain this dual-level mental state. Therefore, as soon as accommodations to these disruptions are satisfactorily effected, routines are reasserted and first-order monitoring re-emerges. In short, dual-level monitoring for effortfully reflexive subjects is postulated here to be a part-time mental process. While having the intrinsic capacity to focus on their own awareness and its content, for much of their active lives they operate as if they were non-reflexive agents. This quasi-non-reflexivity does not promote thinking, contemplating, and reflecting nor, of course, full self-awareness and self-consciousness, thereby making the emergence of warranting highly improbable. Rather, their behavioral interventions and the material objects that they use as physical leveraging for these interventions will be shaped largely according to practical end-goal intentions and needs.

The "How-To/Know-How" Duality

I said earlier that the knowledge involved in monitoring is part of the monitoring awareness itself, making up its contents. I will call part of this content "know-how," and this is the total technical knowledge that subjects draw upon when carrying out their productive tasks. But subjects must also have natural knowledge, "know that," i.e., knowledge about the natural world, e.g., knowledge of the biological realm, the natural history realm, and so on, that acts as both the rest of the knowledgeable content of the monitoring and of the knowledgeable background within which the exercise of their technical "know-how" can occur.[5] Practical artifacts, e.g., tools, manifest this "know-how" and presuppose the "know-that" that makes the former possible. The "know-how" makes up the affordance or functional meanings that things hold for their users. Hence, to speak of the *affordance meaning* of an object is to speak of the practical tasks this thing physically enables its users to do. For example, fire *affords* cooking and heating; chert *affords* stone knapping, which *affords* slicing and cutting, and so on (Gibson 1979).[6] Since simple monitoring precludes the reflective posture, non-reflexive subjects are not aware of their own mental states nor their "know-how" contents. Rather they simply exercise this contextualized "know-how" in behaving intentionally. But as reflexive monitoring entails second-order awareness of first-order monitoring, even the effortfully reflexive subject can become aware of her/his own monitoring, at least under special conditions. Therefore, although rare she/he can also become aware of the "know-how" content of this monitoring and change it in the light of new information (Fay 1987).

Having said this, however, I am inclined to the view that effortful reflexivity has much less of an impact on the qualitative nature of the content of second-order awareness than does effortless reflexivity and that this difference is the reason that symboling is possible by populations having the latter capacity and, in general, impossible or highly unlikely for populations that lack it. This is because with only effortful reflexivity, the second-order and first-order "know-how" contents are largely identical in nature, both pretty well exhaustively characterizing the world of objects in terms of practical affordances. This follows logically from the theoretical claim that for the most part the monitoring of effortfully reflexive subjects is quasi-non-reflexive in nature and from this we could predict that when practical routines are externally disrupted, effortfully reflexive subjects will modify their behaviors accordingly using the qualitatively same type of practical "know-how" standards. This means that material modifications brought about under these circumstances could be exhaustively and objectively characterized in the same end-goal affordance terms as those produced by the routines.

In contrast, for effortlessly reflexive subjects the second-order and first-order contents are different in nature and this difference is the basis of warranting. Because of their chronic state of self-awareness, the practical "know-how" content of first-order monitoring is seamlessly transformed into evaluative (i.e., normative) "how-to" content of the second-order awareness, constituting what I term the "how-to/know-how" duality. This duality is "know-how" that is given a normative aspect, constituting ideal standards that stipulate the particular forms that artifacts *ought* to have so that in using them to mediate behavioral interventions their users satisfy the warranting imperative by manifesting the appropriate intentionality and, thereby, eliciting expressions of understanding and, hopefully, approval from all parties involved. Such public approval collectively constitutes these interventions as the types of social practices they are (Searle 1995). That is, effortlessly reflexive subjects chronically and seamlessly assess the affordance properties of the objects of their first-order monitoring, i.e., material artifacts, in terms of the normative second-order "how-to" aspect of their "know-how," modifying these artifacts and their use of them in order to satisfy both their end-goal and their warranting needs.

This difference between effortful and effortless reflexivity leads to quite different regional distributions of lithic patternings. For example, as Wiessner's (1983) investigation of San arrow styles illustrates, even if two neighboring and similarly adapted effortlessly reflexive populations regularly mix and interrelate, they will tend to generate and sustain functionally equivalent (i.e., isochrestic) tool assemblages that recognizably differ from each other, thereby constituting separate but equivalent normative protocol-based expressive Style-2 assemblages. In contrast, I postulate that, if two populations of the same effortfully reflexive species were spatially separate but adapted to similar environmental contexts, thereby not regularly interacting, because their second and first order monitorings are guided by the same range of affordance standards, they would tend to generate largely identical tool assemblages, thereby constituting two separate but formally similar Style-1 assemblages.

What a Style-1/Style-2 Rupture Might Look Like

Isochrestic Variation and Isochrestic Style

According to the Visibility Thesis and the theoretical view underwriting it as developed to this point, the emergence of symbol-using in a given region is marked in the archaeological record by a Style-1/Style-2 rupture. I am now going to argue that the latter is correlated with the emergence of local isochrestic styles from the undifferentiated background of utilitarian, nonconventional lithics manifesting the full range of potential isochrestic variation. But according to Chase (1991), the first isochrestic styles were nonconventional and, therefore, rather than marking, actually masks the emergence of symbol-using. His is probably the most comprehensive elucidation of the Invisibility Thesis. Therefore, if I am to sustain the Visibility Thesis, I must address what I consider the most relevant part of his argument with respect to the Visibility Thesis.

Now it is generally recognized that there is a difference between isochrestic variation and isochrestic style. According to Sackett (1982) isochrestic variation is the total *possible* range of functionally equivalent variation available to a tool-making population, given the practical material needs they have and the objective production constraints that they face. But he claims that this never becomes actualized since a tool-making population always focuses on a sector of the available variation and sticks to it, thereby largely unwittingly producing an assemblage manifesting spatio-temporally boundaried patterning. Despite being exhaustively accounted for by the archaeologist in functional terms, its patterning is the distinctive signature of that population and, according to Chase (1991), it forms an isochrestic style that is a useful objective indexical sign for the archaeologist.

Now this raises several important points. If isochrestic variation actually exists, it is possible only because end-use affordance function is insufficient to stabilize the instrumental forms of tools. That is, function *underdetermines* form. Different forms can serve the same function and the outcome of this underdetermination is *isochrestic variation - not isochrestic style*. This means that for the emergence of an isochrestic style some extra factor is required to provide regional stabilization. Following Sackett's (1982) initial lead in this regard, Chase (1991) has argued that the surplus element is the juncturing of arbitrary (in the sense of random) choice and social habituation. A community comes to share a sector of the isochrestic range as a result of an initial random choice becoming entrenched first by habit and then by practical learning, a form of practical socialization. This forms a social tradition which the archaeologist comes to identify as a distinctive isochrestic style.

If I have understood Chase correctly in this regard, then I have to disagree with him. Without some normative factor stabilizing form, the "derandomization" of choice cannot occur. According to the notion of random, i.e., no event controls or influences the subsequent event, a system of affordance production choices based on random selection from among equivalents ought to result in *all* the possible functionally equivalent forms being present in a given practical assemblage. Each generation will replicate this random selection. Differential proportion of the equivalents can only result from constraints stemming from environmental and demographic variations. In these terms, such assemblages will manifest the whole range of isochrestic variation. When isochrestic styles emerge, it is because random choices are no longer made because normative standards emerged to stabilize the choice around a sector of the variation. Therefore, isochrestic styles cannot mask symboling because they are the original warranting styles, i.e., their appearance in the archaeological record marks the emergence of Style-2, (Byers 1994). Furthermore, although it is always welcome, no independent evidence such as art or bodily ornament is needed in order to demonstrate that these patternings were produced by symbol-using populations. This follows from the Warranting Model since, as material warrants, Style-2 assemblages are pragmatic symbols in their own right, thereby constituting tool-using as symbol-using.[7]

In these terms, I can conclude that Chase (1991:207) is mistaken in claiming that the "meaning of Mousterian lithic typology ... is essentially irrelevant to the question of symboling." Instead, I take Dibble's (1989, 1991) view that the Mousterian lithics are end-use tools, the variations of which are largely the result of systematic reduction of the working edge. And this is simply another way of saying that this widespread tradition evidences the full range of isochrestic variation, given (1) the particular level of cognitive-based competencies of the hominid populations involved, (2) the objective constraints and (3) the local demographics. This is precisely what I would expect of a Style-1 assemblage.

Conclusion

I will use the above to ground the following general axioms.

Axiom 1: The material assemblage of a non-symbol-using population will characteristically display the whole isochrestic range, constrained only by objective environmental and demographic variations.

Corollary 1: The material assemblages of both neighboring and distant non-symbol using populations will be largely mirror images of each other, controlling for sources of objective variation, as above.

Axiom 2: The material assemblage of a symbol-using population will characteristically have mutually exclusive intra-assemblage isochrestic fields, manifested by some degree of utilitarian redundancy of end-product types (e.g., Type A and Type B spears, Type M and Type N blades, and so on).

Corollary 2: Neighboring but independent symbol-using populations (i.e., adapted to the same environment) having similar demographic dynamics will, in general, develop overall mutually exclusive inter-assemblage isochrestic styles, thereby indexing observable social boundaries. These may or may not be ethnic boundaries.

Using these axioms and corollaries I can now stipulate two methodological criteria by which to recognize a Style-1/Style-2 rupture in the utilitarian lithics, demarcating the emergence of symbol-using, since this simply involves identifying the emergence of isochrestic style from within a pre-existing Style-1 assemblage.

Criterion 1: Out of pre-existing Style-1 lithic assemblages there will be evidence of mutually exclusive intra-assemblage isochrestic fields formulated by mixed end-use (e.g., expedient tools) and end-product technologies.

Criterion 2: While the end-use aspect of these technologies will cut across and knit together assemblages within a region, distinctive regional isochrestic styles will emerge out of the end-product aspects of these mixed technologies, thereby demarcating a full Style-1/Style-2 rupture.

As symbol-using rests on the "how-to/know-how" duality, it will favor lithic production that enhances the range of possible formal variations without reducing the efficiency and variation in practical functions. There seem to me to be two lithic production strategies that can be used to pursue this end:

1. Retaining the macrolithic bias of Style-1 assemblages while favoring those techniques that increase the predictability of macro-blade over macro-flake production; or
2. Shifting from macrolithic production of Style-1 to a microlithic production. This should favor elaborating techniques for producing compound tools.

Both offer flexibility in form without diminishing utilitarian efficiency and they are not mutually exclusive strategies. For my immediate purposes, I will focus on the macrolithic strategy since this appears particularly appropriate for analyzing the Pre-Upper and Early Upper Paleolithic of Europe.

Criterion 1 is evidenced with the emergence of blade-type production from within a pre-existing background of Levalloisian flake industries in pre-Upper Paleolithic assemblages. I interpret this as local communities shifting from macrolithic end-use technologies to mixed macrolithic end-use and end-product technologies (Kozlowski 1990; Otte 1990). Largely simultaneously Criterion 2 is evidenced with the emergence of mutually exclusive regional Style-2 assemblages as end-product blade industries of these mixed technologies take on distinctive intra-regional patternings (Mellars 1989a, 1989b, 1991). This means that such early Upper Paleolithic assemblages as the Szeletian, Chatelperronian, Uluzian, and the early Aurignacian, among others, mark the Style-1/Style-2 rupture. These conclusions echo the more detailed descriptive characterizations of the trends as given by Mellars (1989a:354) when he characterizes the Middle-Upper Paleolithic Transition in terms of the emergence of the relative rapid changing of tool morphology, when compared to the Middle Paleolithic, along with the increasing complexity of regional patterning, and the imposing of greater "morphological standardization."

Notes

[1] Chase (1991) arrives at this methodological conclusion. Bednarik (1992, 1995) also applies the same point as expressed in the Invisibility Thesis to argue vigorously that as early referencing by means of material symbols was necessarily ephemeral in nature, we should not expect to find an abundance of hard evidence of symbol-using in the early and middle Pleistocene record. Since this is precisely the case, he concludes that symbol-using is a pre-modern phenomenon on the very questionable grounds that there are some vestiges of apparent "concept-mediated" markings!

[2] But the perceptive reader might immediately point to ethnological instances in which the symbolic nature of tools does not seem to be universal and, therefore, no warranting imperative is being served, e.g., the rough and ready appearance of Australian Aboriginal stone tools. However, is it the case that these stone tools lack style? This assessment may arise from the assumption that stylistic properties must be non-utilitarian or adjunct in nature. That this is not the case has been cogently argued by Sackett (1982) and Wiessner (1983). Although all Australian tool kits may display strong utilitarianism, when looked at comparatively they may also bear distinctive formal variation that makes them observably mutually exclusive with this variation not being attributable to objective conditions. I will terminate this essay by specifying the empirical criteria for identifying symbolic style in material cultural assemblages that are exhaustively utilitarian and the

Australian data could be a test case. Therefore, this is an empirical question that can be resolved with further research and, in terms of the warranting imperative, it should become quickly clear that temporal and spatial formal variations are discerned that are not attributable to objective ecological or demographic conditions.

[3] I am using the term "pragmatic" in the linguistic sense of the way humans use their signs and representations to act and make a difference in the world. If we act by means of symbols, e.g., by making promises and giving orders, then such symbol-usage is not only pragmatic but, using the ordinary sense of the term, practical. By treating tools in symbolic pragmatic terms, I am claiming that even our "practical" utilitarian actions require a symbolic medium so as to be constituted as the types of actions we take them to be.

[4] See Parmentier (1997) for a different reading of Peircean signs. He particularly critiques the treatment of this triple classification as categorizing different types of signs rather than as different degrees of meaning that all signs have. For him, this classification is better treated as a hierarchy and a sign realizes all three (and possibly more) elements of index/icon/symbol. Certainly, in human culture, symbols have strong iconic and indexical aspects, e.g., the icons of the Eastern Christian Churches. Although in his essay he criticizes the Warranting Model, he does so on the basis of a serious misinterpretation of my work. "Byers argues that material objects can be usefully divided into those features that are interpretable according to end-goal functioning...and those features whose regular variation corresponds to no evident functional requirement (Ibid. p.47)." He construes my argument as based on treating style as largely non-utilitarian adjunct form, which is clearly the opposite point that I am making, both in this paper and in my 1994 paper. Of course, sometimes style is realized in adjunct form. But the thrust of this paper makes it clear that it can be realized in fully utilitarian form, as Sackett has also argued (1982, 1985, 1986a, 1986b). Recently Mithen (1994, 1996a) has argued that human symboling is based upon a mental property he terms "cognitive fluidity." Initially I considered his concept to be largely equivalent to my concept of effortless reflexivity. I am no longer sure of this. If I understand him correctly, he claims that reflexivity was constrained to a narrow range of cognition. Cognitive fluidity is the result of the demodularization of cognition. This "releases" reflexivity so that it broadens its scope but does not change its nature. In my scheme, reflexivity is unchanging in its scope, operating in across the background cognition of human knowledge and experience, but evolves in terms of ease of exercise from effortful to effortless. Clearly, I have much more work to do to clarify how my concept and Mithen's concept differ in interpreting the archaeological record. Treat this as a work in progress.

[5] In brief, I am claiming here that technical "know-how" is only exercisable when it is integrated with knowledge about the world and the way it works. This is one reason I find Mithen's (1996) view of cognitive fluidity suspect. According to him, with the exception of the cognition of modern human populations, biological, natural and technical "knowledges" are independently localized in specialized mental modules. Therefore, our ancestors were unable to think about the natural world while focusing their thoughts on technical concerns, and so on. In my view, this makes any thought impossible. Rather, my approach postulates that even effortfully reflexive hominid populations were "cognitively fluid" in the sense that they were naturally able to integrate their different knowledges. As I point out below it is just that they were not able to easily carry out second-order assessment of their first-order awareness and, therefore, unable to assess it in terms of socially constructed standards, i.e., norms. Instead they would use naturally constructed standards, i.e., those based pretty well only on considerations of efficiency. In short, standards based on practicality and efficiency would largely prevail.

[6] Although I am using Gibson's term "affordance," I do not fully agree with the way he defines it. For him affordances are properties of things relative to the utility they have to users. I have no problem with this. But he then goes on to argue that these affordance properties are meanings and these meanings, then, are a property of the objects themselves. The implication of this is that he accepts that we derive meaning - as a property - from the things themselves, as if just by perceiving things as having functions we absorb the "meanings" that they have. (See Pinker 1997 for a parallel application of this objectification of meaning when he argues that "symbols" are objective patternings that have meaning, i.e., information, and that are able to transfer this information to the human mind through the process of human monitoring.) This is absolutely contrary to my view, of course. But I find Gibson's term useful and his theory of affordance meaning as useful since it might be a good "as if" way of expressing how effortfully re-

flexive subjects actually experience the world. Since according to my theory they are not chronically reflective, they would identify meaning with function and experience both as being properties of the world and its objects. This would suggest that pre-symbolling hominids would largely see the world in the same meaningful terms, varied only according to the variation in material ecology.

[7]Of course, since for a symbol-using population all and not simply some of their regular material behaviours will be caught up in conventions constituting them as social practices, it is quite likely that non-utilitarian materials will show up as constituent media of an increasingly complex social system. White's (1982, 1989, 1992) notion of metaphorical thought to account for the rather sudden appearance and expansion of body ornament (primarily beads) in the Upper Palaeolithic is fully compatible, in my view, to the position I have developed here. The only change I would make is to stress that when treated in expressive terms this is "literal" figurative thought, i.e., beads are expressive icons that presence the powers that they represent.

References Cited

Bednarik, Robert G.
- 1992 Palaeoart and archaeological myths. *Cambridge Archaeological Journal* 2:27-43.
- 1995 Concept-mediated marking in the Lower Paleolithic. *Current Anthropology* 36:605-634.

Bhaskar, R.
- 1979 *The Possibility of Naturalism*. Harvester Press, Brighton.

Burling, Robbins
- 1993 Primate calls, human language, and nonverbal communication. *Current Anthropology* 34:25-53.

Byers, A. Martin
- 1991 Structure, meaning, action and things: the duality of material cultural meaning. *Journal for the Theory of Social Behavior* 21:1-30.
- 1994 Symboling and the Middle-Upper Paleolithic: a theoretical and methodological critique. *Current Anthropology* 35:369-399.
- 1999 Communication and material culture: Pleistocene tools as action cues. *Cambridge Archaeological Journal* 9:23-41.

Chase, Philip G.
- 1989 How different was the Middle Paleolithic subsistence? a zooarchaeological perspective on the Middle to Upper Paleolithic transition. In *The Human Revolution*, edited by P. Mellars and C. Stringer, pp. 321-37. Edinburgh University Press, Edinburgh.
- 1991 Symbols and Paleolithic artifacts: style, standardization, and the imposition of arbitrary form. *Journal of Anthropological Archaeology* 10:193-214.
- 1994 On symbols and the Paleolithic. *Current Anthropology* 35:627-629.

Chase, P. G. and H. L. Dibble
- 1987 Middle Paleolithic symbolism: a review of current evidence and interpretation. *Journal of Anthropological Archaeology* 6:263-296.
- 1992 Reply to Bednarik. *Cambridge Archaeological Journal* 2:43-51.

Clark, G. A. and J. M. Lindly
- 1989 The case for continuity: observations on the biocultural transition in Europe and Western Asia. In *The Origins and Dispersal of Modern Humans: Behavioral and Biological Perspectives*, edited by P. Mellars and C. S. Stringer, pp. 626-676. Edinburgh University Press, Edinburgh.

Davidson, Iain and William Noble
- 1989 The archaeology of perception: traces of depiction and language. *Current Anthropology* 30:125-158.
- 1993 Tools and languages in human evolution. In *Tools, Language and Cognition in Human Evolution*, edited by Kathleen R. Gibson and Tim Ingold, pp. 363-388. Cambridge University Press, Cambridge.

Dibble, Harold L.
- 1989 The implications of stone tool types for the presence of language during the Lower and Middle Paleolithic. In *The Origins and Dispersal of Modern Humans: Behavioral and Biological Perspectives*, edited by P. Mellars and C. S. Stringer, pp. 415-431. Edinburgh University Press, Edinburgh.
- 1991 Mousterian assemblage variability on an interregional scale. *Journal of Anthropological Research* 47:239-257.

Douglas, M
- 1982 Primitive rationing. In *In the Active Voice*, edited by M. Douglas, pp. 57-81. Routledge and Kegan Paul, London.

Duff, Andrew I., Geoffrey A. Clark & Thomas J. Chadderdon
   1992 Symbolism in the early Paleolithic: a conceptual odyssey. *Cambridge Archaeological Journal* 2:211-229.

Fay, Brian
   1987 *Critical Social Science: Liberation and its Limits*. Cornell University Press, Ithaca.

Gibson, J. J.
   1979 *The Ecological Approach to Perception*. Houghton-Mifflin, Boston.

Giddens, Anthony
   1976 *New Rules of Sociological Method*. Hutchinson, London.
   1979 *Central Problems in Social Theory*. Macmillan Press, New York.
   1984 *The Constitution of Society*. University of California Press, Berkeley.

Kozlowski, Janusz K
   1990 A multiaspectual approach to the origins of the Upper Palaeolithic in Europe. In *The Emergence of Modern Humans: An Archaeological Perspective*, edited by P. Mellars, pp. 419-437. Cornell University Press, Ithaca.

Lindly, J. M. and G. A. Clark
   1990 Symbolism and modern human origins. *Current Anthropology* 31:233-261.

Marshack, Alexander
   1989 On depiction and language, comment on Davidson and Noble (1989). *Current Anthropology* 30:332-335.
   1990 Early hominid symbol and evolution of the human capacity. In *The Emergence of Modern Humans: An Archaeological Perspective*, edited by P. Mellars, pp. 457-498. Cornell University Press, Ithaca.

Mellars, Paul
   1989a Technological changes across the Middle-Upper Paleolithic transition: economic, social and cognitive perspectives. In *The Origins and Dispersal of Modern Humans: Behavioral and Biological Perspectives*, edited by P. Mellars and C. S. Stringer, pp. 338-365. Edinburgh University Press, Edinburgh.
   1989b Major issues in the emergence of modern humans. *Current Anthropology* 30:349-384.
   1991 Cognitive changes and the emergence of modern humans. *Cambridge Archaeological Journal* 1:63-76.

Mithen, Steven
   1994 From domain specific to generalized intelligence: a cognitive interpretation of the Middle/Upper Paleolithic Transition. In *The Ancient Mind: Elements of Cognitive Archaeology*, edited by Colin Renfrew and Ezra B. W. Zubrow, pp. 30-39. Cambridge University Press, Cambridge.

Mithen, Steven
   1996a *The Prehistory of the Mind: a Search for the Origins of Art, Religion and Science*. Thames and Hudson, London.
   1996b On early Paleolithic "concept mediated marks," mental modularity, and the origins of art. *Current Anthropology* 37:666-670.

Noble, William and Iain Davidson
   1996 *Human Evolution, Language and Mind: A Psychological and Archaeological Inquiry*. Cambridge University Press, Cambridge.

Otte, Marcel
   1990 From the Middle to the Upper Paleolithic: the nature of the transition. In *The Emergence of Modern Humans: An Archaeological Perspective*, edited by P. Mellars, pp. 438-456. Cornell University Press, Ithaca.

Parmentier, Richard J
   1997 The pragmatic semiotics of culture. *Semiotica* (Special Edition) 116(1) Part 3:42-61.

Pinker, Steven
   1997 *How the Mind Works*. W.W.Norton, New York.

Sackett, J. R.
   1982 Approaches to style in lithic archaeology. *Journal of Anthropological Archaeology* 1:59-112.
   1985 Style and ethnicity in the Kalahari: a reply to Wiessner. *American Antiquity* 50:154-159.
   1986a Style, function and assemblage variability: a reply to Binford. *American Antiquity* 51:628-634.
   1986b Isochrestism and style: a clarification. *Journal of Anthropological Archaeology* 5:266-277.

Searle, J. R
   1983 *Intentionality*. Cambridge University Press, Cambridge.
   1995 *The Construction of Social Reality*. The Free Press, New York.

Taylor, Charles
- 1985 Theories of meaning. In *Human Agency and Language*, edited by Charles Taylor, pp. 248-292. Cambridge: Cambridge University Press.

Wiessner, P.
- 1983 Style and social information in Kalahari San projectile points. *American Antiquity* 48:253-276.
- 1985 Style or isochrestic variation? a reply to Sackett. *American Antiquity* 50:160-166.

White, Randall
- 1982 Rethinking the Middle/Upper transition. *Current Anthropology* 23:169-192.
- 1989 Production complexity and standardization in Early Aurignacian bead and pendant manufacture: evolutionary implications. In *The Origins and Dispersal of Modern Humans: Behavioral and Biological Perspectives*, edited by P. Mellars and C. S. Stringer, pp. 366-390. Edinburgh University Press, Edinburgh.
- 1992 Beyond art: toward an understanding of the origins of material representations in Europe. *Annual Review of Anthropology* 21:407-431.

Wynn, Thomas
- 1991 Tools, grammar and the archaeology of cognition. *Cambridge Archaeological Journal* 1:191-206.

5. Nonmaterial Artifacts: Retelling the Natural History of Artifacts and Mind

Shirley C. Strum and Deborah Forster

Abstract

We have inherited assumptions and models about "mind" from past work on human cognition. These are currently being challenged in contemporary research in Cognitive Science. Among the "renewed" options are other ways of thinking about cognition which seem particularly germane to understanding non-human primates. These not only change the way we look at cognitive processes in real time but they suggest a radically different story about development that may be appropriated for evolutionary scenarios. The new approaches, distributed cognition and situated action, allow us to look at phenomena that were not previously considered "cognitive" because they were "in the world" rather than in the "head". They emphasize the importance of "coordination" as a central part of cognitive processes. We suggest that artifacts cannot be assessed without being seen as part of a larger system that began and accumulated before material culture and of which material culture is a part. We use two sets of behavioral data on baboons to illustrate deficiencies of past interpretations for discussions of cognition. We recast the data in the new cognitive framework and discuss implications from developmental and evolutionary perspective.

Introduction

Cognitive archeology investigates the evolution of the human mind using information from the archeological record (Nowell this volume, Wynn this volume). Since "mind" leaves only indirect traces this interpretative process forces cognitive archeology to turn to other disciplines such as paleoanthropology, the study of nonhuman primates, biology, psychology, and cognitive science for guidance. The primate perspective is essential; in mind as in behavior, to define what is human we need to know what is not. However, the recognition of primate mind is recent and the study of primate mind has many challenges (some of them similar to those of cognitive archeology). Furthermore, there is a striking discontinuity between humans and other species manifest in material culture and technology that challenges attempts to understand biological continuity between species. Gibson (1993:4-5) discusses past efforts addressing continuities vs. discontinuities between animal and human minds. Here we approach the issue by focusing on process rather than outcome in the search for possible precursors to material culture in nonhuman primate societies.

The chapter's standpoint has been forged by our (individual and joint) efforts to understand the minds of baboons in the context of recent research. We focus on changes in how science characterizes the *social*, *cognitive* and *technical / material* realms of life central to arguments about the nature, development and evolution of the primate mind. At the most abstract level these shifts are part of a larger change to an anti-reductionist and non-mechanistic world view (see Capra 1997 for a very accessible overview) which offers an opportunity to re-frame (or extend) our understanding of artifacts and their role in cognitive processes. Beginning with a basic premise that artifacts cannot be understood unless they are seen as part of a larger process that originates before material culture and of which material culture is a part, the question becomes "what is the nature of this larger process?" And given the interests of cognitive archeology "what are its cognitive entailments?"

The chapter begins with a consideration of the "social world of primates", both the nature of socialness and of society. The emphasis is on newly discovered primate "social complexity" and its constituent relationships, skills and competences. This social reality is very different from past descriptions and has important consequences for discussions of primate mind and for interpretations of the evolution of human mind. The next section explores "cognition". Starting with a bit of history (how social complexity and social strategies became cognitive), it considers a current controversy about different models of "mind". The "classic" or cognitivist model is contrasted with the alternatives of "situated action" and "distributed cognition". What this might mean to interpretations of primates behavior is briefly examined. The third section re-examines both artifacts and what happens to artifacts when shifts in the social and cognitive are taken into account. The focus is changed from that of artifact as a product (or result) to the *artifactual process* (and system) from which it emerges and in which it participates. The framework of artifactual activity makes it possible to explore artifacts "in action" both in the richness of human cultural settings and in nonhuman primate societies. The last section illustrates how these transformations (in the social, cognitive and technical) might affect evolutionary stories. It presents a model and a scenario for the interaction of society and technology that integrates the nonhuman and human primate data in a way more consonant with the previous discussions.

We conclude that shifts in the way we conceptualize the social, the cognitive, and the technical/material realms of life have profound consequences for cognitive archeology since they fundamentally reconfigure questions about the evolution of material culture and of cognition. Taken together, this perspective provides strong support for our claim that there is a dynamic continuity between the artifactual processes in nonhuman and human primates. We argue that what nonhuman primates do is homologous (historically related patterns of the same underlying process) rather than analogous to what humans do, despite the nonmateriality of their artifacts. Thus those interested in the evolution of human mind must first understand a process that began before material culture left its accumulated traces for archeologists to find.

The social world of primates

Society and socialness are not synonymous. Theoretically, aggregations of individuals can have cohesion without any social structure. This is Hamilton's (1971) "geometry of the selfish herd" where individuals, fleeing predators, congregate together to avoid being eaten. Togetherness offers some degree of protection, at least to those who are not on the edge. But individuals in *persistent* aggregation should eventually adapt to the presence of the others (Strum and Latour 1987). Thus a secondary social environment is created and with it, a secondary social adaptation.

During the last two decades, data from nonhuman primates illustrate just how this might work. The discovery of primate *social complexity*, particularly among baboons, macaques and chimpanzees, suggests a very different kind of social world than previously imagined. The diversity of social relationships which range from kinship ties to nonkin friendships (Strum 1975; Seyfarth 1978; Ransom 1979; Altmann 1980; Hinde 1983) vary in intensity, stability and duration. They come together in social networks. Relationships and networks rather than simply being noise generated by so many individuals living together, as originally thought, represent critical resources created and used by individuals to solve strategic problems of survival and reproduction (Kummer 1978; Strum 1982; 1983a,b; Dunbar 1984a; Stein 1984; Smuts 1985).

Baboons are the archetype of the new socially complex primate actor although chimpanzee social complexity is probably better known (de Waal 1982; Goodall 1990). In the baboons' matrilineal society kinship relationships can extend from great grandmother to great granddaughter and encompass aunts, great aunts, nieces and cousins into a large interacting network. Baboons negotiate and maintain a range of other relationships as well and create what might appropriately be called behavioral kin. The most important of these are friendships between adults (mainly males and females), between adult males and immatures, between immatures, and less intensely between adult females. These vary in the amount of work that is necessary and the durability of the bonds. What they share is the process, they are *negotiated* at every step from their inception to their demise. A social negotiation can be completed in one interaction or can span days, even weeks (see examples in Strum 1987); they can be simple or complex. Among baboons, few actors ever have the last word. In general, durability is limited. The most stable set of relationships, that between kin, relies on nearly constant proximity and continual interaction. Surprisingly, without these even kinship ceases to be a *social* category although it remains a biological and evolutionary one (Gouzoules 1984; Strum 1994). Since bonds disintegrate when contact and interaction decline, durability is sensitive to any factor that interferes with or disrupts behavioral reinforcement of relationships (in other words, lack of coordination). Social ties become vulnerable as the group gets larger and/or competition for social partners increases. Then the amount of work necessary to maintain (and monitor) relationships at their previous level increases, as does the degree of negotiation at each point of interference (see Strum 1994; Dunbar 1992, 1993, 1996). What accumulates in this social world is limited to the experiential constraints of the individuals and is ephemeral except for what can be embodied (i.e. marked in the physical body of the individual). Thus the temporal continuity of baboon society is a monumental task (Strum 1994a).

Individuals use the social network to their own advantage. What this means for baboons, is that it is impossible to directly and simply execute an action. Rather, there is constant interference from others who want the same thing or because the individual has become entangled in the social web of a potential competitor or protector or someone who has a score to settle (Strum and Latour 1987; Strum 1994a). Interference requires constant co-ordination between individuals (conflict interference requires just as much coordination as does cooperation or collaboration, see for example the description of ritualized fighting in wolves, [Moran et al.,1981]).

Relationships are good investments when seen as part of a system of social strategies that provide successful alternatives to aggression during competition and defense. *Agonistic buffering* (Strum 1983 a,b; Dunbar 1984b; and see papers in Taub 1983) shows how. Baboon males, for example, grab an infant or a female to use as a shield against the aggression of another male. However, the success of the tactic depends primarily on whether there is a pre-existing relationship between the male and his buffer. If the two are friends, the buffer trusts the user and cooperates. The result is that (for complex reasons embedded in the social workings of the group) the other male's aggression is deflected. If the buffer doesn't cooperate, however, the outcome is different. The user is then seen as the villain and becomes the target of the group's wrath rather than its beneficiary. Therefore, investing in a relationship, whether one with an infant or a female, can be thought of as the transformation of an unwilling and frightened actor into an effective, albeit passive, social *tool*. Kummer (1967) first suggested social relationships as investments and other conspecifics as *social tools* when trying to explain the dynamics of protected threat where a smaller individual uses the proximity of a larger or more dominant one to threaten an antagonist (see also Kummer 1995). The use of the term *social tool* suggests the extension of properties usually ascribed to inanimate objects to animate con-specifics.

But the idea of social tools did not really figure into scientific explanations until the late 1970's and early 1980's when a variety of similar social tactics were described from the wild and captivity (see references below). It is the sophistication and diversity of such social strategies in which social relationships are worthwhile "investments" that explains primate social complexity and clarifies the meaning of alliances and coalitions among primates (de Waal 1984; Datta 1986; Harcourt 1988; and see papers in Harcourt and de Waal 1992). The premium that some primate species place on "reconciliation" between actors (Cords 1988; de Waal 1989; Judge 1991; Aureli 1992) when the social fabric has been torn by aggression can only be understood in this context.

Socially complex baboons stand in dramatic contrast to the earlier view of a rigid, simple baboon society structured around males and their dominance hierarchy (DeVore and Hall 1965; DeVore and Washburn 1963). Baboons now become more active and more skillful. In order to navigate this complexity baboons need social knowledge, social sophistication and new competencies; they must be able to negotiate, test, assess situations and carefully manage relationships (for example Strum 1975, 1981, 1982, 1983 a, b). The new social reality changes options, even for a big and physically powerful male baboon. To maximize his reproductive success a male needs much more than size, strength or dominance rank (Strum 1994b).

The literature on nonhuman primates has exploded with evidence of social complexity in a number of primate species. All primates follow the same initial path in building complexity: from socialness to social relationships via attraction, proximity, interaction and attachment. What varies is the difficulty, the amount of social work and the extent and durability of the resulting network. The amount of work is itself influenced by the type of interactants (e.g. kin vs. nonkin, young vs. old, male vs. female, group member vs. outsider), and by the species (ultimately by the species' evolutionary adaptation to its environment). Baboons and common chimpanzees illustrate interesting variants. Baboon groups are relatively large and cohesive, probably as a defense against large predators in environments where it is not always possible to run to the safety of trees. By contrast, common chimpanzees live in very fluid and flexible groups which coalesce and disperse in response to the availability of food.

The basic social tasks change in emphasis between baboons and chimpanzees. Among baboons, individuals have great familiarity with each other due to their constant and close proximity. Overcoming strangeness and breaching the initial social distance is not a problem, except for immigrant males. The real social work is creating and maintaining relationships in the face of constant interference by others who are trying to do the same thing. As a result, baboons have become expert social negotiators trying to manage the situation to their advantage.

Chimpanzees spend long periods apart, dispersing to find food and then coming together again. This time apart creates a lack of familiarity between actors making them almost strangers. Without information about what has happened, the predictability essential to successful social encounters declines. Chimpanzee social work, then, involves re-establishing contact, overcoming recently developed fears, recreating social proximity and probing and improving predictability. To help with this work, chimpanzees have an elaborate and ritualized system of reunion and reassurance communication. Intermittent contact also means that individuals interfere less than among baboons but it also means there is less possibility of social management. In the end, chimpanzees have less diversified and less lasting social networks than baboons. However, shifts in circumstances can change all this. Baboons can become more chimplike and chimpanzees can become more baboonlike, at least temporarily. Captive chimpanzees in the famous Arnhem colony (de Waal 1982) can't practice the fission/fusion dynamics of wild chimpanzees. Instead they live in constant and close proximity – in essence they have become facultative baboons (Strum 1988) who no longer need to reestablish social contact in elaborate reunions. But now they have new baboon-like tasks: overcoming the constant interference of others through sophisticated social management. The Arnhem chimps perform elaborate social strategies reminiscent of baboons but rarely seen among wild chimpanzees. Similarly, when a wild baboon group was temporarily held captive in individual cages, they faced the chimp problem of reestablishing social contact at the time of release. These baboons displayed very uncharacteristic chimpanzee-like reunion and reassurance behaviors before resuming their normal pattern of interaction (Strum 1987).

As we might expect, negotiation and stability of a relationship are related. However, it is not a simple inverse function. It depends on the characteristics of the actors, some of which, like sex, size, age, reproductive condition and physical condition are not negotiable since they are encoded in the body and biologically *marked*. History is also important. The intensity of negotiation depends on the turnover rate of various characteristics. The higher the turnover rate, the greater the negotiation. The extensiveness of negotiation depends on the number and heterogeneity of potential social actors (Strum 1994a).

This new view of the social world forces us to think about process in addition to outcome, about context in addition to behavior, and about new competencies that social actors need. Beyond that, as we will discuss below, the demands of social complexity transform the social link into a process of acquiring knowledge about "what the society is" and determining what the society "is to be" with consequences for both cognition and material culture.

The cognitive world of primates

The change in the nature of "the social" has had major consequences for how we consider the "cognitive" abilities of nonhuman primates. During the last 15-20 years, social strategies quickly became *primate politics* (de Waal 1982; Byrne and Whiten 1988; Schubert and Masters 1991; Silverberg and Gray 1992). Social complexity generated new types of primate tactics and strategies which could not be explained by "genes" (see Strum and Fedigan 1999 for a historical review) – it was too variable, the variations happened too fast and in too many flexible combinations. Agonistic buffering or strategies of consort turnovers or the development of predatory behavior or the social politics of give and take required a new level of cognition. Just to manipulate or manage relationships, actors had to be able to perceive the social complexity that existed, predict multidimensional interactions, and control these for their own ends. In the late 1980's, the social manipulation of strategic partners became *Machiavellian Intelligence* (Byrne and Whiten 1988). This was a useful short hand which emphasized the most essential aspects of social intelligence. More than that, social complexity came to be viewed as one of the primary causes of the evolution of higher cognition (Kummer 1967; Humphrey 1978; Western and Strum 1983; Whiten and Byrne 1988; Cheney and Seyfarth 1990). Actors caught up in social complexity had to creatively respond to a constantly changing social game which placed a premium on behavioral flexibility, problem solving, declarative knowledge and perhaps even mindreading and self-reflection.

The cognitive revolution had finally come to nonhuman primates. The field of cognitive ethology (Griffin 1976, 1984, 1992; Ristau 1991) epitomized this change in attitude towards nonhuman minds. However, the revolution brought more than just an appreciation for the importance that mind might play in behavior; it also unwittingly imported a specific model of the mind popular at the time in the study of human cognition.

The cognitivist approach, rising as a reaction to classical behaviorism, relied on the centrality of representational accounts in describing cognitive phenomena. In other words, it argued for a separate level of description that is valid and necessary to explain human action and thought (Gardner 1985/87). For cognitivists, representations are the stuff that is found between input (perception) and output (action). They include internal mental entities such as symbols, rules, images, and the like. How these entities are joined, transformed and contrasted with one another, speaks directly to the nature of cognition. In this view, other levels of description (e.g., neuroscience) may be valid but are inadequate on their own. Moreover, the neural and social levels of description should ultimately map onto representational accounts in a straight forward reductionist fashion. But what exactly are representations and how can they be explored?

The early cognitive scientists made a number of choices about how to proceed. Computers offered a useful metaphor for constructing a representational account of human cognition, namely, the serial processing of physical symbols (Simon 1969). Cognition was accepted as synonymous with computation and information processing, with representations functionally identical to the digital one/zero bits manipulated by computer programs. Cognition could be described by lists of rules, with routines and sub-routines, goals and sub-goals, or as a *general problem solver* program (Newel and Simon 1972). Human action amounted to working systematically and serially from a cognitive blueprint or plan towards a goal. A controversy ensued about what sort of cognitive architecture could support such specifications. Some supported a 'central processing' model. These were the generalists (e.g., Anderson 1983) who stood in contrast to those who supported a 'modular' view (e.g., Fodor 1983; Gardner 1983). Gardner, for example, envisioned a marriage between cognitive psychology and artificial intelligence as the most promising future for cognitive science. Yet this focus, by extension, also defined what was to be overlooked or completely ignored in explaining human cognition, especially the role of murky concepts such as affect, context, history and culture (Gardner 1985/87; Hutchins 1995).

The interpretation of agonistic buffering just presented illustrates how the traditional model found its way into primate data. Males, for example, had to strategically plan in advance. Friendships (a possible sub-goal) were intentional, anticipating the need for "social tools" (as a sub-routine) essential to surviving conflicts without damage (and maximizing genetic fitness). Males figured out how to co-opt the troop's defense of threatened infants and females for themselves. To do this, individuals implementing social strategies must be manipulating mental representations which informed their behavior from internal programs or scripts. The cognitive abilities ascribed to nonhuman primates soon began to rival that of humans. What then was unique about the human mind? To find out, most recent work has focused on topics like intentionality, imitation, theory of mind, language, and self-awareness (see detailed review in Tomasello and Call 1997).

In the past two decades, while the cognitivist model of mind continued to influence much of cognitive ethology, it faced serious challenges in human cognitive sciences. The scrutiny and disillusionment was primarily due to the fact that the traditional models and their predictions did not fit the reality of human behavior very well. Computational (artificial intelligence) models easily did things that humans find pretty difficult (e.g., play expert chess, compute long numerical sequences without error) and yet were unable to perform tasks that people could do effortlessly (e.g., face and other pattern recognition, natural language processing). Thinking of cognition as serial processing of physical symbol systems made it very difficult to imagine how human cognitive skills could have evolved. In addition, new, albeit marginalized, research on humans offered alternatives which emphasized cognitive activity as part of social and physical environments.

A ready mapping between representational accounts and neuroscience was also not forthcoming (Gardner 1985/87). Neural pathways (e.g., vision) did not conform to what the classic model tried to impose on them and a uni-directional account from perception to action was nowhere to be found even in the simplest real brains. The most serious challenge and first major alternative to the physical symbol processing model came from what became known as PDP (Parallel Distributed Processing), or Connectionism (Rumelhart and McClelland 1986a; Elman et al. 1996). Neuroscience and 'neurally-inspired modeling' as McClelland and Rumelhart called their connectionist approach, suggested that both real brains and simulated networks did not fit into the neat depiction of serial symbolic processing. Moreover, even simple connectionist networks showed surprising and exciting similarities to known paradoxical behavioral patterns, such as the mis-application of grammatical rules in the course of language acquisition (Rumelhart and McClelland 1986b). Most importantly, a representational account of cognition could still be given but only if representations were allowed be depicted as *distributed patterns of activation* across many neural units, rather than the classic view of them as physically rigid symbolic structures.

An equally powerful (and overlapping) voice of dissent came in the form of a renewed emphasis on developmental perspectives on cognitive issues, be it in brain (e.g. Edelman 1987) or in behavior (e.g. Mandler 1992; Karmiloff-Smith 1992). This perspective was also successfully explored with connectionist models (see examples in Elman et al. 1996). Piaget was still the center-piece of cognitive developmental research but his position was more ambiguous now. Description of sensori-motor and conceptual development as a set of formal and sequential rigid stages did not fit the recent detailed behavioral data or with connectionist approaches to mind (Karmiloff-Smith 1992). And, so, when early 20th century Russian psychology was finally introduced into the United States (Vygotsky 1978), several already marginalized cultural psychologists welcomed the alternative (see Cole 1996 for a recent comprehensive review).

The early 20th century Russian cultural-historical psychology of Vygotsky (1978), Luria (1979), Leont'ev (1981) and others suggested that behavior should be studied as processes of change with a focus on social activity rather than on the individual. The Russians linked the social and the cognitive by suggesting that interpsychological processes enacted in social interaction could be and were appropriated by individual minds. This process happened largely through cultural mediation. Cultural tools (which Vygotsky and Luria took to include language) were the entities that organized activity in such a way to provide the conditions for appropriation from interpersonal space to intrapsychological space. The introduction of this perspective in the United States gave rise to the *situated action* approach (Suchman 1987; Lave 1988; Clancey 1993) which had a novel view of cognition. Behavior, rather than arising only from stored representations in the mind, is mostly the result of continuous negotiation with social and physical elements of the environment. In this way, structure in the environment plays an important role in cognition. Cognition is itself defined as systems of coordination among elements both inside and outside the individual (Schmidt et al. 1990). Situated action thus expands the unit for cognitive analysis beyond the individual to include its environment. The emphasis in this view is on 'communities of knowledge and of practice' through which social and cognitive processes unfold. Learning occurs mostly by informal apprenticeship, or guided participation, rather than by direct and intentional pedagogy (Rogoff 1990), and the transformation from novice to expert happens through situated learning in a process of legitimate peripheral participation, or LPP (Lave and Wenger 1991). This movement's most significant contribution was its attempt to reintroduce the historical and cultural context of human cognition.

An interesting variation on this theme came in the form of a more comprehensive attempt by Hutchins (1995) to apply a cultural perspective to the cognitive sciences which he formulated as "distributed cognition". Hutchins kept the central notions of the traditional model of cognition, namely the centrality of a representational account and the idea of cognition as computation (or information processing). However, he challenged the assumptions that were made about them by favoring research in culturally rich settings of human practice 'in the wild', in contrast to the controlled laboratory conditions favored by cognitive psychologists.

For Hutchins, cognitive phenomena (e.g. problem solving, memory, learning, and decision-making) were more likely to be distributed across a collection of individuals and cognitive artifacts than be the accomplishment of the isolated brains of his human subjects. In addition, it was possible for systems such as the cockpit of a jet plane, or the deck of a navy ship to have cognitive properties (e.g. computing a location for navigation). Moreover, these system properties were not necessarily identical to the cognitive properties of the individuals that participated in the systems. It is not necessary, for instance, for a deck-hand on a navy ship to understand the complex distributed computation in which he is participating reliably and effectively. Hutchins argued that we tend to overattribute to individuals cognitive abilities that belong to the system as a whole (Hutchins 1995:365)

Rather than having a goal the cognitive system regularly and reliably produces a recognizable *outcome*. Individual elements in the system have *functional abilities* to establish and maintain coordination with other elements in order to reliably produce the outcome. Most importantly, the system has cognitive properties (such as problem solving, memory, attention) that are emergent from the interactions of the elements but not necessarily reducible to them. Individuals in the system may have cognitive properties as well (memory inside the head, etc.) but these properties may or may not be similar or identical to the system level properties. Thus system outcomes may be reproduced reliably even if individual participants do not share the same intentions or awareness of the regularities on the larger system level.

Cognition was now transformed into a process by which representational states were propagated and transformed across the various media in the system. By making the boundary between the individual and the environment permeable Hutchins was able to use the same language (of coordination among bits of representational structure) to explain what happens both inside and outside of individual heads. Thus the way to analyze a cognitive system ala Hutchins is to follow the trajectories of the representational states as they move around the system and not assume that things happen only in one place (inside the head).

Situated action and distributed cognition make claims that are particularly relevant to thinking about nonhuman primate cognition (despite their origins in research on humans). They can be briefly summarized as follows:
1. social living does not represent a unitary cognitive challenge —it changes with the nature of the social interaction
2. cognitive processes occur both in social interaction and in the minds of individuals
3. the cognitive processes in social interactions are distributed among diverse actors, including social partners and the structure of the physical environment
4. actors participate in distributed cognitive processes, the outcome of which none could have produced alone
5. distributed cognition is a resource for individuals; it creates learning opportunities and the possibility of appropriation from the social world (from outside the head) to the "mind" (inside the head) of an individual.

Adopting these alternative notions of mind necessitates revising the cognitive requirements and skills of nonhuman primates. For example, the cognitive challenges of baboon agonistic buffering might have a different interpretation. First, more emphasis should be on elements like affect which were ignored by the cognitivist model. A major aspect of conflict for baboons is tension and anxiety. Initially, at least, a male might be motivated to simply seek contact comfort (Harlow and Harlow 1965) to reduce the stress created by an aggressive challenge rather than be executing an elaborate plan to mobilize the troop through the use of a buffer. But males have limited options for contact— only an infant or female friend can be "touched". The initiation of friendship could start the same way: wanting to be able to sit close, groom and have contact comfort from another baboon (rather than being a premeditated plan to transform a conspecific into a social ally). In the agonistic buffering sequence, the next step might then involve a reduction in anxiety produced by contact comfort. A less anxious opponent reacts differently, changing the costs and benefits for the aggressor. This new situation could produce a different reaction. If the aggressor persists, then the troop's defense is mobilized because of the nature of the society without regard to the male's initial intentions.

The situated action and distributed cognition perspective illustrates how the social setting constructs behavior, at least in part. The outcome would not necessarily have to be represented as an internal *goal* for it to unfold with repeated regularity. In other words, an interaction organizes its elements in ways that produce regular patterns. Its unfolding provides constraints for what follows—the kinds of coordinations that can be established, options can be explored and outcomes that can result. Repetition allows individuals to have multiple experiences of these patterns and thus offers an environment in which to learn to be more effective. The threatened male learns through interaction how to coordinate better and perhaps which youngster to choose as a buffer. The aggressor is usually a higher ranking younger male who may learn the inhibiting effect of confronting an opponent carrying and infant. A "hostage" buffer, not having joined willingly, may also gain information although is perhaps the least aware of the dimensions of the interaction. As infants grow they make more active choices about participating. Thus an infant may avoid a male who tries to pick it up, or even more striking, may initiate the appropriate contact with a male during conflicts, as if offering to be a buffer. In these repeated experiences, an infant may also learn about "buffering": how to contact males when challenged by another baboon, an excellent baboon case of situated learning by legitimate peripheral participation.

Reinterpreting agonistic buffering, from exceedingly "mindful" to socially structured does not imply the absence of mind. Instead it offers a more compelling explanation of how cognitive solutions might arise. Repeated social interactions with distributed cognitive properties accumulate into a coherent pattern of behavior that can organize even new actors into the pattern. Socially enacted solutions could become a major source of future in-the-head cognitive *schemas*.

Another important implication of the new cognitive research is that complexity at one level does not necessarily imply complexity at another level. Complexity can emerge from simple rules at lower levels and complex networks can exhibit overall simple and coherent behavior, as complexity theory has demonstrated for the physical sciences (see Capra 1997). This contradicts an underlying assumption of current cognitive interpretations of nonhuman primate behavior but it does not rob baboons of mind or of social complexity. Rather it highlights how the current critique of the classic model of mind in human cognitive sciences is relevant to the way that primate mind and the evolution of human mind might be interpreted. There are many ways to think about cognition, only one of which has been tried in cognitive ethology (and cognitive archeology?).

These new approaches hold special promise for research in nonhuman primate cognition and for those who want to explore the nonhuman/human evolutionary continuum (see Strum et al.1997). First, reframing cognition in these ways, changes our definitions and the unit of analysis. Second, it offers new principles and opens up research opportunities because it allows us to "see" cognition in the world (in interactions) and not just in the head. This solves a major problem for cognitive ethology whose subjects cannot otherwise be easily interrogated. Third, in an interesting reversal of the current stance, we can start by finding out what cognitive processes are already (distributed) in the world and then proceed to explore what needs to be "in the head" (Hutchins 1995). Finally, a core (sometimes hidden) issue in development and evolution is really about the origin of novel structure. In this case, how do new cognitive skills arise over time? Traditional explanations about the nature of mind (e.g., modularity, serial symbol processing) usually ignore questions of development and evolution—how things got to be that way, whether modules or symbols, in the lifetime of an individual and in the history of a species. Situated, distribution cognition provides some possible answers: interacting individuals can jointly produce new patterns that gain stability through repetition. Individuals may learn from the distributed results of a process in which they were participants. If Vygotsky is correct individual ontogenetic appropriation of solutions is plausible, making phylogenetic appropriation an intriguing possibility. It is easier to consider the co-evolution of cognition and material culture in this framework and it leads us to expect material artifacts to be a fundamental part of the cognitive system.

Retelling the natural history of artifacts

The shift to social complexity was a move towards a contextual and relational characterization of primate societies. The emphasis on alternative models of cognition has made primate cognition more situated and distributed. In this section we take artifacts as our focus and reexamine them in a similar way. The result is also an emphasis on process, context and complexity (Capra 1997).

Anthropologists generally ascribe to an "artifact as object" interpretation making the study of artifacts part of the study of material culture (Cole 1996). Ingold (1993) defines artifact as "an object shaped to some pre-existent conception of form" (see also Ingold 1986). Others similarly depict an artifact as a 'thing' that results from the shaping of naturally given raw material to a preconceived cultural standard (e.g. Goodenough 1981; Parker and Milbrath 1993).

Common usage is similar but less restrictive. The word originates in the act of producing something "with skill". Merriam-Webster's (10th ed.1993) first listing is "something created by humans usually for a practical purpose. Especially: an object remaining from a particular period." The common sense meaning is of a material product made by purposeful human skilled activity that occupies physical space and is stable over time. However, subsequent definitions diverge in interesting ways. An artifact can be "something characteristic of or resulting from a human institution or activity" (as in, "I contend that the division between social and technical skills is an artifact of the modern Western distinction between nature and society" – Ingold 1993:443). This definition does not make intended purpose or materiality of the product necessary conditions. Or an artifact can be "a product of artificial character (as in a scientific test) due usually to extraneous (as human) agency". Here the requirement that there be a purpose or a preconceived standard is possibly violated. Unintended artifacts of an experimental setting in scientific research are the things we try to factor *out* of our interpretations. In this sense, artifact links to "artificial" as something contrived by art (by humans) rather than by nature.

These less restrictive definitions have interesting implications raising the possibility that artifacts can be products of both intended and unintended (human) actions and may have defining nonmaterial features. If artifacts need not be restricted to prototypical physical objects but instead can occupy other "spaces" such as in the social and cognitive realms it is easier to extend the agency of artifactual activity to nonhuman as well as human primates.

What might be meant by the concept of a nonmaterial artifact? First, reverting back to the core meaning of artifact, it is something extrasomatic and skillfully *made*. But to qualify as an artifact, an entity must also be *stable* enough to warrant its own level of description and offer some explanatory power beyond the simple combination of elements from which it is created. Thus, an artifact (of any kind) is made through skilled activity and in turn constrains and organizes the very form of the elements from which it emerges.[1]

Putting nonmaterial and material products of artifactual activity as part of the same continuum focuses attention on the underlying process with major cognitive implications. To illustrate how, we examine two current views about humans and artifacts: artifacts as the end product of a long ago process and artifacts in action. Making cognitive inferences from the first vista requires re-construction or reverse-engineering. The other approach captures artifacts in the context of their manufacture and use. Anthropology of living societies, cultural psychology (and the alternative models we explored in the section about cognition), are examples of activity and practice theories in which such a cultural-historical view is prevalent. Each entry point provides a different characterization of the role of artifacts in human activity, in general, and cognitive processes, in particular. More specifically artifacts-in-action challenges the assumptions, and hence the inferences, made about cognition by artifacts-as-end-products. These challenges, in turn, provide a fresh perspective on historical or evolutionary connection between human and nonhuman primates and act as guides to plausible precursors of material culture.

Artifacts[2] as end-product objects

Artifacts are conceived of as objects shaped to pre-exiting concepts of form within a cultural standard, itself implying communication and social negotiation between humans (see Ingold 1993). They are often the result of an artisan whose solitary intentional actions derive from a mental representation of the plan for the product (Davidson and Noble 1993; Reynolds 1993). Artifactual remains evoke the cognitive skills of the past makers and users. Inferences range from linguistic abilities, decision making and planning, to representational abilities, sequencing of motor actions, eye-hand coordination, etc. (for a variety of examples see Gibson and Ingold 1993).

For example, Holloway (1969) describes the process of stone tool manufacture as parallel to the production of speech. Similarly Gowlett (1984) relates linguistic behavior to the necessary sequence of operations in production of stone cores (for choppers etc.). A different inference for linguistic abilities is suggested by Toth and Schick (1993) in their analysis of Oldowan flakes which shows a preferential right-handedness. Handedness, in turn, implies lateralization of the brain crucial to the evolution of language. Wynn (1993), by contrast, rejects the possibility that tool-behavior and linguistic behavior share homologous features. He argues that tool behavior is a system with at least three layers of thinking: biomechanical, action sequences, and problem solving.

A similar inferential process applies to artifacts and decision-making. Bradley and Sampson (1986) see evidence for master decision making in the products of Acheulean knappers – the ability to calculate which piece of raw material would provide a certain shape. Fagan (1989) concurs that the maker sees the shape of the artifact in a mere lump of stone.

Density and locational continuity of artifacts are an important source of insights about formal and cultural standards and learning processes (Toth and Schick 1993), and about mathematical (Gowlett 1984) and aesthetic sense (Edwards 1978), although other interpretations are possible (Davidson and Noble 1993).

Comparisons are sometimes between humans and nonhuman primates rather than early and later humans. These usually pay special attention to any evidence of material culture among nonhuman primates (e.g. chimpanzee tool behaviors both in the wild and in captivity, see McGrew 1992, 1993) and emphasize the striking discontinuity in material culture. Studies of skill, for example, focus on particular skills or features that seem critical to human artifact manufacture such as Piaget's sensorimotor development (Parker 1990), DiLisi developmental model (Parker and Milbrath 1993), questions of imitation (see review in Whiten and Ham 1992), of teaching (Boesch 1991), and of linguistic abilities (for apes see Savage-Rumbaugh and Rumbaugh 1993). Parker and Gibson (1979) offer nut-cracking Tai chimpanzees as the most likely representative of the protohominid pattern.

All these analyses, by virtue of several common assumptions, create problems for evolutionary explanations of artifacts (and of cognition). Artifact-as-object discussions revolve around whether specific behaviors are or are not present in earlier humans or in nonhuman primates however behavioral patterns are defined in homo-centric rather than species neutral terms (see Caro and Hauser 1992 and King 1994 for thoughtful discussions). Limiting the scope to physical objects largely ignores context (a stance reinforced by the notion of the lone artisan). And the classical cognitivist model of mind is asummed (see section 2 above). This model sees general problem solving in terms of mental subgoals and subroutines and asserts a tight correspondence between prepositional logic and cognition-in-the-head and between the formal analysis and real life production of artifacts.

Archeologists have questioned some of these assumptions. For example, Davidson and Nobel (1993) dispute whether artifacts-as-objects are the result of self-conscious prior intent which they call the *finished artifact fallacy*. Similarly, Toth and Schick (1993) interpret spheroids (part of Oldowan technology) to be well curated hammerstones whose globular shape results from use rather than ias intentionally shaped tools.

Ingold (1993) cautions about the *puzzle solving approach* to tool use where each episode of action is reduced to mental manipulation of a plan followed by its behavioral execution. He echoes Reynold's (1993) emphasis that planning and memory are activities carried out by people in a social context. According to Reynolds, archeology has created its own artifact, the *great tool-use fallacy* (that an artisan working alone produces tools) ignoring the social nature of tool behavior in humans:

> ...when archeologists and social historians describe technical processes, they often seem to forget everything they ever knew about social relations and go at it like authors of cookbooks, listing sequences of steps without ever actually describing who does what when in real-life situations. This literary approach to technical skills is very misleading, for even simple technical processes that could in theory be performed in sequence by a single group or even by a single person working alone may in fact be subdivided into a host of subtasks, each one performed by a different task group. (pp. 417)

Ingold (1993:340) concurs:

> ...the quality of 'being an artifact' no more inheres in the object itself than does the quality of 'being a tool'. As regards the latter... it entails the conjoining of the object to the technical skill of a user, whose presence is presupposed when we call the object a tool of a certain kind. Likewise, 'being an artifact' depends on its conjunction to an intended project. In short, both the instrumentality and the artifactuality of objects are conditional upon the situational contexts of their engagement in practical activity, and as objects endure over time, have histories of such engagement, so their status can change.

Artifacts as objects in action

These archeological critiques suggest the need for another way to think about artifacts, one that puts them in context, refrains from prejudging process or outcome and is agnostic about the nature of cognition. The second view of artifacts that we explore does this by emphasizing the *activity* from which artifacts emerge and which they mediate. Artifacts then acquire a dual nature; they partake in a process that is dialectical, cumulative and transformative and is social rather than solitary both in activity and in cognition.

The discussion that follows takes its inspiration from critics like Ingold and Reynolds as well as other research programs that converge on activity and practice theories (e.g., Bourdieu 1977; Cole 1996; Goodwin 1994; Hutchins 1995; Latour 1994; Lave and Wenger 1991; Nardi 1996; Norman 1991; Rogoff 1990; Suchman 1987; Vygotsky 1978). In this small space we can only present a brief summary of the points that are most useful to our project.

Reynolds (1993), for example, studying stone tool making by Australian Aborigines, argues that "The essence of human technical activity is anticipation of the action of the other person and performance of an action *complementary* to it, such that the two people together produce physical results that could not be produced by the two actions done in series by one person." (emphasis ours) He calls such a process *heterotechnic cooperation* to contrast it with *symmetric cooperation* in which all the participants do the same thing at the same time. Moreover, this process is manifested in all human societies by a form of social organization he calls the *face-to-face task group*, which he defines by the shared intention to transform matter and energy through the cooperative and complementary use of tools and tool-using skills by a group of people in face-to-face contact. It is not the end-products that define this activity as is clearly obvious in watching children play pretend games. What children practice for a long time before they practice *making* things are the kind of social interactions that take place in the face-to-face task groups.

Cole (1996), a cultural psychologist and an activity and practice theorist, argues for the dual nature of artifacts:

> ...an artifact is an aspect of the material world that has been modified over the history of its incorporation into goal-directed human action. By virtue of the changes wrought in the process of their creation and use, artifacts are simultaneously ideal (conceptual) and material. They are ideal in that their material form has been shaped by their participation in the interactions of which they were previously a part and which they mediate in the present. (pp. 117)

Activity and practice theories generally assert that the cumulative force of a history of interactions is one of several identifiable characteristics of what can be called *artifactual activity*. Artifactual activity invariably takes place in social groups of mixed age and mixed of levels of expertise. This heterogeneity supports the kind of complementary participation described by Reynolds. It also allows the transfer of skill through the very same interactions from which manufactured products emerge. The acquistion of skill often happens through apprenticeship (guided participation) and situated learning (legitimate peripheral participation) where individuals appropriate and internalize social solutions. Innovations can be the unintended results of opportunistically using physical and social structures in the environment (for modern industry examples see also Leonard 1995). The distributed nature of all these interactions supports both complex dialectical relations and the emergence of system level properties that are not reducible to the actions of individuals.

Activity and practice theories also share the idea that cognition is stretched across mind, body, activity and setting (Lave quoted in Cole 1996). In Hutchins' distributed cognition framework (1995), culture is a process and artifacts are the *residua* of the process. They accumulate through history in an adaptive process that saves partial solutions to frequently encountered problems. These 'solutions' are manifest in both the physical form of the artifacts and in the practices and activities (their ideational form) they mediate.

The residua of the artifactual process are *artifactual entities*. The circular causality of complex systems means that artifactual entities both organize and mediate the process from which they emerged.

We can now examine *artifactual activity* more closely in the search for evolutionary connections between species. To do this the identifiers of artifactual activity need to be made more species neutral (less human-centric) and more cautious about the intentionality and purposefulness of the participants, about whether the activity is truly collaborative, and about the representational status of structures. The presence of *material* residues, the traditional artifact-objects, need not be a criterion.[3]

In summary, *artifactual activity* is defined as the *skillful modification over time of an extra somatic aspect of a social unit*. This activity involves mixed age and mixed expertise, in complementary asymmetric and distributed interactions (i.e. not everyone doing the same thing at the same time) the outcomes of which are emergent and leave residues in social, cognitive and/or physical space. The residue of repeated interactions accumulates to a recognizable entity that, in turn, organizes and mediates the activity of the participants. Any artifactual entity/product is inseparable from the social structure and activity through which it was created.

The fundamental evolutionary question about the origins of artifactual activity now becomes whether there is *anything* that accumulate in nonhuman primates and that qualifies as an *artifactual entity*.

Activity and practice theories also offer some provocative corollaries of the presence of artifactual activity that should be studied in nonhuman primate social interactions:

1. Developing skill is more likely to happen through situated learning by legitimate peripheral participation than by active or formal pedagogy.

2. Individuals may appropriate patterns of behavior that are first enacted interpersonally.
3. Innovation may result from opportunistic (or even accidental) use of structure in the environment.
4. The existence of artifactual entities depends on their physical characteristics and on the continuity of agents that were active in their creation. This means that an event that leaves only an experiential residue in participants (and bystanders monitoring the interactions) can contribute to *artifactual activity* only in as much as that experience is subsequently available through the presence of at least one of the experiencing individuals.

Artifactual activity in baboons:

Baboons socialize incessantly. Although artifactual activity is inherently social it is not the case that social activity is necessarily artifactual or in any way skilled (think about social insects). Our search among baboons begins with a demonstration that at least some social *activity* qualifies as artifactual. Next, traces of artifactual activity must be identified. Finally, the residue we identify must be shown to accumulate over time into recognizable *artifactual entities*.

Is there artifactual activity among baboons? Social activity in baboons is indeed skilled, asymmetric and distributed. Most baboon social interactions happen in mixed-age and mixed-skills polyadic clusters and involve intricate and coordinated participation—monitoring, anticipation and continuous complementary adjustment to actions of others (often more apparent to observers in the slow motion of videotape than in real time). Some social activity results in the modification of extrasomatic aspects of the group.

If extrasomatic actions are the material manifestations of social activity, these communication gestures and behavioral elements do not endure long afterwards. Thus it is not easy to identify their traces or measure how their residues accumulate through repeated activity. Even when an aggressive encounter leaves a physical mark, the bite is relatively ephemeral.

We suggest, instead, that the traces of social interactions are to be found in the more intangible realm of experience and social and cognitive structure. For something to count as an artifactual entity, it must be stable enough to warrant its own level of description and offer some explanatory power beyond its mere manufacture. It must participate in a process of emergence and circular causality.

Hinde has most clearly described social relationships this way, in terms of the accumulated effect of interactions and association patterns. He argues (1987) that a number of distinct levels of complexity must be recognized in social behavior: "interactions, relationships and social structure, the latter leading on to yet further levels of complexity concerned with the interrelations between groups." Hinde is careful to point out the dialectics between levels of complexity where one level is not reducible to those beneath it:

> Each of these levels has properties that are simply not relevant to the levels below. Thus the behavior of two individuals interacting, but not that of a single individual, can be described as synchronous or well-meshed. Some properties of relationships...like commitment, or intimacy, are hardly if at all applicable to isolated interactions. Indeed...properties concerned with the temporal patterning of interactions, or with their relative frequency, can apply only to relationships...And within a group the relationships may be arranged hierarchically, centrifocally and in many more complex ways—issues not applicable to the individual relations ... It is equally important to remember the tow-way relations between (levels). The nature of an interaction or a relationship depends on both participants. At the same time, the behavior the participants show in each interaction depends on the nature of the relationship: what an individual does on each occasion depends on his assessment of and expectations about the interaction in which he is involved, or of the relationship of which it forms part...At the next level, the participants' view of the relationship affect the nature of interactions within it, and the nature of the relationships is determined by its constituent interactions...etc. (1987:25)

We offer social relationships as candidate artifactual entities. The friendships between female and male baboons may be archetypes. A friendship is identified by patterns of proximity, grooming, and support during conflicts. Grooming, for example, contributes to making the friendship in systematic ways. In this case, males initiate more grooming interactions at the beginning (and during active consorts) but later females do the majority of the grooming. The form of grooming interactions (who initiates or terminates, who is the groomer and who the groomee, duration and number of bouts in a session) is a telling sign of the status of the friendship. Thus friendship both emerges from, and mediates, grooming activity.

Thus friendships are the accumulated traces of artifactual activity, the product of a process that links grooming, proximity and support. These residues exist in social and cognitive spaces but as artifactual entities they organize and mediate the activity of the participants.

Similarly, male-male alliances and male-immature special relationships are likely to qualify as artifactual entities. They are traces of the same process although different behaviors might be involved, or the same behaviors combined in different ways (e.g. male allies rarely groom each other).

Besides specific social relationships, some patterns of polyadic interactions in baboons may also be artifactual entities which accumulate over time to organize and mediate the activity of participants. Take, for example, agonistic buffering. Agonistic buffering unfolds in predictable ways despite the variety of participants (perhaps due to biological and physical constraints). In this way it resembles human *schemas* classified by Cole and others as human artifactual entities (see discussion of schemas and artifacts in Cole 1996:128-130).

Many other baboon (and nonhuman primate) activity patterns seem likely to be artifactual activity that produces nonmaterial artifacts. Dominance hierarchy, types of consort turnovers (see *sleeping near the enemy* in Forster and Strum 1994), configurations of troop movement, even particular foraging practices come to mind.

For nonmaterial artifacts like these to accumulate *continuity* is crucial. Continuity ensures that experience is brought to bear on subsequent activity to form coherent regularities that can leave traces. As comparison between the social work of baboons and chimpanzees made clear, continuous physical presence is important for maintaining relationships within the social group. When social continuity is disrupted, compensatory behaviors are required (and evolve, e.g., the ritualized greetings in fission/fusion species such chimps, elephants, and social canids). While the process of artifactual activity may be the same in baboons and chimpanzees, differences in continuity could result in different artifactual entities in the two societies.

The new continuum

Looking at process as well as at outcomes provides another measure of the continuity or discontinuity in artifactual activity between humans and nonhumans. Doing so suggests that there is a continuum from baboon-like artifactual activity to human-like artifactual activity. One produces nonmaterial residues while the other generates nonmaterial entities as well as material products. Nonmaterial artifacts have problems of stability and durability. Material artifacts are more easily separated from the activity and the agents that produce them because they are more stable and durable.

In addition, nonhuman primate artifactual activity seems more biologically constrained, less arbitrary, and less intentional. Human artifactual activity appears more complex in its coordination, hierarchical structure and communicative aspects. Material artifacts can more easily support innovation and the kind of opportunistic use of structure described, for example, by Hutchins (1995) and Leonard (1995). Because they can exist "independently", material artifacts participate and mediate activity differently than nonmaterial entities, an argument that we take up in the next section.

The interaction of society and technology

The preceding discussions of the social, the cognitive, and the artifactual suggest the reexamination of a familiar question: what was the impact of material culture on primate behavior and society. Strum and Latour (Strum and Latour 1987, 1991; Strum 1994a; Latour 1991, 1994) propose a model and scenario that articulates with the new meanings of the social, cognitive and technical just presented. They first claim that society does not exist *a priori*. Instead society must be created by its members (Strum and Latour 1987). Actors do not enter into an already built structure but have to negotiate and maintain its meaning. The task, to build society, is the same for all primates, human and nonhuman. Strum and Latour argue that different resources build different types of society from the same process. In essence, humans and baboons differ in the *practical* means they have to enforce their version of society or to organize others on a larger scale. They then discuss how "technical" mediation or material culture has come to play a central role in the divergence of human from nonhuman society particularly in solutions to the problems of stability, durability, flexibility and size of society.

Baboon society is contrasted with human society. The first is complex, the second is complicated. This is a crucial distinction although the two terms are often conflated in common parlance. In a complex society, a profusion of variables impinges simultaneously on each social negotiation. Others incessantly interfere and actors have little power to enforce their version of society on others. Society is built and repaired with limited resources: baboon bodies, social skills and whatever social strategies they can create. This makes it difficult to isolate and negotiate any one thing at a time or to create long-lasting social stability.

Human society, by contrast with baboons, is complicated. Something is complicated when it is made of a succession of simple units, steps or operations, like in the construction of crystals or computers. These simple units can accumulate to become very large and stable structures. Complicated society results when actors can simplify social negotiations. The increase in the constancy of characteristics and relationships means that they can be effectively ignored or taken for granted, at least for a while. This also reduces the interference of many "others". Not everything is up for grabs at once and many aspects of society can be "black-boxed". In this way a larger, more durable and stable structure can emerge. The shift from complexity to complication requires a way to reduce the number of simultaneous factors impinging on the situation in order to simplify the negotiations and make them more stable and durable.

The most likely new resources for building society, for enforcing or reinforcing a particular view of "what society is" are material culture, artifacts, technique and technology. Strum and Latour imagine the progression from social complexity to social complication, from baboons to modern industrial societies (1987:792, Figure 1) in which more and more material and "extra-somatic" means are used to simplify social negotiations, until individuals can organize others on a large scale. Going from a situation where one baboon can't interact with another if she refuses to look, we arrive at a situation where social roles can be assigned to someone who is not even present. These additional material resources allow social actors to make weak and renegotiable associations (such as coalitions between males baboons or friendships between males and infants) into strong and hard to break units which do, in fact, become the social structure into which modern humans enter.

Strum (1994a) discusses a possible mechanism enabling a shift from social complexity towards the first stages of social complication. "Marking" relationships and behaviorally created characteristics would increase their durability and reduce the amount of social work normally needed to "represent" them. (It would also insure greater continuity and make them more powerful artifactual entities – in terms of the argument presented in the previous section.) This "marking" would have to be both flexible and durable because the outcomes of social negotiations in complex societies are variable. Baboons already make use of a type of material artifact in foraging when they use animal trails as guides for travelling. Trails become cognitive material artifacts for baboons. They hold knowledge "in the world" and organize behavior in a way that simplifies negotiations.

Latour and Strum (1991; Latour 1991, 1994) expand upon the notion of marking to create an imaginary scenario for the co-evolution of society and technology which has relevance to our basic premise about artifacts. They employ several novel principles to build this process including "symmetry between social and nonsocial partners" (treating both animate actors and inanimate objects as potential interactants in social negotiation), "discontinuities within the program of action" (the recognition that most goals are not reached directly but require detours from the original program of action because of interference from others), and the "crossing over of competencies between social and technical realms" (in line with the first symmetry principle which allows the swapping of characteristics and skills between animate and inanimate social actors). The story starts with artifacts being used to improve on a primitive marking system thereby reducing the amount of social work necessary to build society and ultimately the nature of the social task. Thus in the first stage, marking helps reduce complexity and enhances stability of characteristics, relationships and outcomes. In the next stage, artifacts help organize society by re-representing cognitive schemas or acting as "scaffolds" for joint action. Afterwards, artifacts begin to "stand for" social relationships and social actors. In the final stage, artifacts become "stand ins" replacing animate social actors and thereby modifying not only society but the nature of social negotiation. In each step, the cognitive process is more widely distributed and the cognitive characteristics of all participants are modified in the dynamic of social creation.

For Strum and Latour, the genealogy of society and technology cannot be separated and their approach makes it possible to tell an evolutionary story of continuity rather than discontinuity which, in this case, goes from baboon agonistic buffers to seat belts, door closers and speed bumps using the same set of principles. Throughout, the interaction of the social and technical changes both (and affects cognition) in unexpected ways. The underlying principle of symmetry between all actors, animate and inanimate, extends the unit of analysis beyond the individual to the interaction. Latour goes further, renaming society "the collective" (Latour 1994) in order to accommodate the variety of actors who have a say in building society. The model for the entry point of nonhumans (meaning *material artifacts)* into the "collective" (1994: Figure 6, pp. 46) includes *translation*, the means by which features of our social order are inscribed onto different "matter", *crossover* where properties are exchanged among members of the collective, *enrollment* of nonhumans into the collective, *mobilization* of nonhumans as unexpected resources for the social task, and the *displacement* in shape and extent of the collective once its composition has been altered through the inclusion of these new actors. Latour documents how properties of humans and nonhumans are exchanged and redistributed within this larger "corporate" body. Not all of this is cognitive, but what is, is situated in a much larger social context and distributed across all types of actors, animate and inanimate. In effect this applies the situated action and distributed cognition model to the evolution of society. It also situates the artifactual process in a larger context.

The process of creating, accumulating, and feeding back which is often seen as the hallmark of humanness because of its manifestation in material culture can only be understood if we start with socially complex baboons who have nonmaterial artifacts but of limited durability, stability and extensiveness and proceed to modern humans who face the same task but are able to build a very different society through technical mediation.

Conclusions

Ideas about the social, the cognitive, and the socio-technical domains, and about artifactual acitivity and its consequences are the building blocks of many interpretations of the evolution of mind. We contend that future interpretations of the archeological record need to take seriously these shifts in meaning and focus. Minimally:

The social must now begin with a complex primate society based on relationships that have to be established and maintained. Nonhuman primates already engage in artifactual activity that produces residues, nonmaterial artifacts made from social interactions and the characteristics of social partners. Both social complexity, which requires the skillful management of social partners and social information, and artifactual activity that results in the accumulation of nonmaterial traces have major cognitive corollaries.

The cognitive must now include situated, distributed cognition where cognitive solutions can be enacted before they are thought and where these social resources may be appropriated later by individuals. Situated, distributed cognition is particularly relevant to artifactual activity; material artifacts and technical mediation inevitably become part of the evolution of cognition in humans.

The social and technical cannot be segregated; they have converged in the shift from complex nonhuman primate to complicated human society. Now the "collective" includes both animate and inanimate entities and material artifacts truly are social and evolutionary actors.

The natural history of artifacts must encompass a less restrictive definition of artifact and focus on the process of artifactual activity characterized in a more species-neutral way. This is a complex dynamic system where artifactual entities both organize and mediate the process from which they emerged. In the continuum from nonmaterial to material artifacts, cultural activity shapes and reshapes what is socially and cognitively possible through patterns of coordination, accumulation and negotiation.

The evolutionary history of our imaginings now begins baboon-like and ends with a socio-technical hybrid called modern humans. The social is complex/complicated, cognition is not just inside the head but situated in the world, and cognitive processes are distributed among actors both animate and inanimate. The evolution from baboon to modern humans is profoundly affected by the advent of material culture as society becomes a "collective" composed of diverse actors, animate and inanimate.

Material culture, as we claimed at the start, should be seen as part of a much larger, more ancient and more dynamic process. Looked at from this perspective many skills and competencies normally attributed to *Homo faber* antedate both material artifacts and the human mind. But this perspective also reveals the important limitations on what can accumulate in the socially complex world of a nonhuman primate.

Notes

[1] This form of emergence or circular causality (Kelso, 1995) is different from regular feedback (as in a thermostat, or a car) in which the "feeding back"

does not constrain the form of the elements. This is in fact the crux of the distinction from mechanistic explanations in which simple feedback and control are deemed sufficient to explain the behavior of a system.

²There has been a distinction made between tools and artifacts with a tool referring to something that extends the natural actions of the user. Other important distinctions in this context include the difference between making and using tools, techniques and technology (for an extended discussion on these terms see Ingold 1986, 1993). We respect but do not take up these very important distinctions in the discussion that follow.

³Additional interesting issues about the evolutionary continuity betwee nonhuman and human primates grow out of the framework of activity and practice theories. For example, what mediates baboon activity? what do baboons do skillfully? is there anything like apprenticeship in baboon society? is there situated learning by legitimate peripheral participation? is there distributed cognition? is there opportunistic use of structure? innovation by accident? do we see the kind of complementary participation that Reynolds described in the face-to-face task groups?

References Cited

Altmann, J.
   1980 *Baboon Mothers and Infants*. Harvard University Press, Cambridge, MA.
   1983 *The Architecture of Cognition*. Harvard University Press, Cambridge, MA.

Aureli, F.
   1992 Post-conflict behavior among wild long-tailed macaques (*Macaca fascicularis*). *Behavrioral Ecology and Sociobiology* 31:329-337.

Bates, E., and J. Elman
   1993 Connectionism and the study of change. In *Brain, Development and Cognition: A Reader*, edited by M. Johnson, pp. 623-642. Blackwell Scientific, Oxford.

Boesch, C.
   1991 Teaching among wild chimpanzees. *Animal Behaviour* 41:530-532.
   1993 Aspects of transmission of tool-use in wild chimpanzees. In *Tools, Language and Cognition in Human Evolution*, edited by K. R. Gibson and T. Ingold, pp. 171-183. Cambridge University Press, Cambridge.

Bourdieu, P.
   1977 *Outline of a Theory of Practice*. Cambridge University Press, New York.

Bradley, B., and Sampson, C. G.
   1986 Analysis by replication of two Acheulian artefact assemblages. In *Stone Age Prehistory*. edited by G. N. Bailey and P. Callow, pp. 29-45. Cambridge University Press, Cambridge.

Byrne, R., and A. Whiten (ed.).
   1988 *Machiavellean Intelligence*. Clarendon Press, Oxford.

Capra, F.
   1997 *The Web of Life*. Harper Collins, London.

Caro, T. M., and M. D. Hauser
   1992 Is there teaching in nonhuman animals? *Quarterly Review of Biology* 67:151-174.

Cheney, D., and R. Seyfarth
   1990 *How Monkeys See the World*. University of Chicago Press, Chicago.

Clancey, W.
   1993 Situated cognition: a nerupsychological interpretation on response. *Cognitive Science* 17:87-116.

Cole, M.
   1996 *Cultural Psychology: The Once and Future Discipline*. Harvard University Press, Cambridge, MA.

Cords, M.
   1988 Resolutio of aggressive conflicts by immature long-tailed macaques (*Macaca fascicularis*). *Animal Behavior* 36:1124-1135.

Datta, S. J.
   1986 The role of alliances in the acquisition of primate rank. In *Primate Ontogeny*, edited by J. Else and P. Lee, pp. 219-225. Cambridge University Press, Cambridge.

Davidson, I., and W. Noble
   1993 Tools and language in human evolution. In *Tools, Language and Cognition in Human Evolution*, edited by K. R. Gibson and T. Ingold, pp. 363-388. Cambridge University Press, Cambridge.

DeVore, I., and S.L. Washburn
   1963 Baboon ecology and human evolution. In *African Ecology and Human Evolution*, edited by F. C. Howell and F. Bourliere, pp. 335-367. Aldine, Chicago.

DeVore, I., and K. R. L. Hall
   1965 Baboon ecology. In *Primate Behavior: Field Studies of Monkey and Apes*, edited by I. DeVore, pp. 20-52. Holt, Rinehart and Winston, New York.

Dunbar, R. I. M.
- 1984a *Reproductive Decision: An Economic Analysis of Gelada Baboons Social Strategies.* Princeton University Press, Princeton.
- 1984b Use of infants by male gelada in agonistic contex: agonistic buffering, progeny production or soliciting suppor? *Primates* 25:28-35.
- 1992 Neocortex size as a constraint on group size in primates. *Journal of Human Evolution* 20:469-493.
- 1993 Coevolution of neocortical size, group size and language in humans. *Behavioral and Brain Sciences* 16:681-735.
- 1996 *Grooming, Gossip and the Evolution of Language.* Harvard University Press, Cambridge.

Edelman, G. M.
- 1987 *Neural Darwinism.* Basic Books, New York.

Edwards, S. W.
- 1978 Nonutilitarian activities in the Lower Paleolithic: a look at the two kinds of evidence. *Current Anthropology* 19:135-139.

Elman, J., E. Bates, M. Johnson, A. Karmiloff-Smith, D. Parisi and K. Plunket
- 1996 *Rethinking Innateness: A Connectionist Perspective on Development.* MIT Press, Cambridge, MA.

Fagan, B.
- 1989 *People of the Earth. An Introduction to World Prehistory* (6 ed.). Scott, Foresman and Co., Glenview, IL.

Fodor, J. A.
- 1975 *The Language of Thought.* Thomas Y. Crowell, New York.

Fodor, J. A.
- 1983 *The Modularity of Mind.* MIT/Bradford Press, Cambridge, MA:.

Forster, D., and S. C. Strum
- 1994 Sleeping near the enemy: patterns of sexual competition in baboons. In *Current Primatology, Vol.2: Social Development, Learning and Behavior*, edited by J. Roeder, B. Thierry, J. Anderson and N. Herrenschmidt, pp. 19-24. Université Louis Pasteur, Strasbourg.

Gardner, H.
- 1983 *Frames of Mind: The Theory of Multiple Intelligences.* Basic Books, New York.
- 1985/87 *The Mind's New Science: A History of the Cognitive Revolution*, paperback edition. Basic Books, New York.

Gibson, K. R.
- 1993a General introduction: Animal minds, human minds. In *Tools, Language and Cognition in Human Evolution*, edited by K. R. Gibson and T. Ingold, pp. 3-19. Cambridge University Press, Cambridge.
- 1993b Tool use, language and social behavior in relationship to information processing capacities. In *Tools, Language and Cognition in Human Evolution*, edited by K. R. Gibson and T. Ingold, pp. 251-269. Cambridge University Press, Cambridge.

Gibson, K. R., and T. Ingold (Ed.).
- 1993c *Tools, Language and Cognition in Human Evolution.* Cambridge University Press, Cambridge.

Goodall, J.
- 1990 *Through a Window: My Thirty Years with the Chimpanzees of Gombe.* Houghton Mifflin Co., Boston.

Goodwin, C.
- 1994 Professional vision. *American Anthropologist* 96:606-633.

Gouzoules, S.
- 1984 Primate mating systems, kin association and cooperative behavior: evidence for kin recognition. *Yearbook of Physical Anthropology* 27:99-134.

Gowlett, J. A. J.
- 1984 Mental abilities of early man: a look at some hard evidence. In *Hominid Evolution and Community Ecology*, edited by R. Foley, pp. 167-192. Academic Press, London.

Griffin, D.
- 1976 *The Question of Animal Awareness: Evolutionary Continuity of Mental Experience.* Rockfeller University Press, New York.
- 1984 *Animal Thinking.* Harvard University Press, Cambridge, MA.
- 1992 *Animal Minds.* The University of Chicago Press, Chicago.

Hamilton, W. C.
- 1971 Geometry for the selfish herd. *Journal of Theoretical Biology* 31:295-311.

Haraway, D.
- 1991 *Simians, Cyborgs, and Women: The Reinvention of Nature.* Routledge, New York.

Harcourt, A.
- 1988 Alliances in contests and social intelligence. In *Machiavellian Intelligence*, edited by R. Byrne and A. Whiten, pp. 132-152. Clarendon Press, Oxford.

Harcourt, A., and F. de Waal (Ed.).
  1992 *Coalitions and Alliances in Humans and Other Animals.* Oxford University Press, Oxford.
Harlow, H., and M. Harlow
  1965 The affectional systems. In *Behavior of Nonhuman Primates,* Vol. 2, edited by A. Schrier, H. Harlow and F. Stollnize, pp. 287-334. Academic Press, New York.
Hinde, R. A.
  1983 *Primate Social Relationships: An Integrated Approach.* Blackwell, Oxford.
  1985 Expression and negotiation. In *The Development of Expressive Behavior: biology-environment interactions*, edited by G. Zivin, pp. 103-116. Academic Press, New York.
  1987 *Individuals, Relationships and Culture: Links Between Ethology and the Social Sciences.* Cambridge University Press, Cambridge.
Holloway, R. L.
  1969 Culture: a human domain. *Current Anthropology* 10:395-412.
Humphrey, N.
  1978 The social function of intellect. In *Growing Points in Ethology*, edited by P. Bateson and R. Hinde, pp. 303-317. Cambridge University Press, Cambridge.
Hutchins, E.
  1995 *Cognition in the Wild.* MIT Press, Cambridge, MA.
Ingold, T.
  1986 Tools and *Homo faber*: construction and the authorship of design. In *The Appropriation of Nature: Essays on Human Ecology and Social Relations*, by T. Ingold, pp. 40-78. Manchester: Manchester University Press.
  1993 Tool-use, sociality and intelligence. In *Tools, Language and Cognition in Human Evolution*, edited by K. R. Gibson and T. Ingold, pp. 249-446. Cambridge University Press, Cambridge, UK.
Judge, P.
  1991 Dyadic and triadic reconciliation in pigtail macaques. *American Journal of Primatology* 23:225-237.
Karmiloff-Smith, A.
  1992 *Beyond Modularity: A Developmental Perspective on Cognitive Science.* MIT Press, Cambridge, MA.
Kelso, S.
  1995 *Dynamic Patterns: The Self-Organization of Brain and Behavior.* MIT Press, Cambridge, MA.
King, B. J.
  1994 *The Information Continuum: Evolution of Social Information Transfer in Monkeys, Apes, and Hominids.* SAR Press, Santa Fe.
Kummer, H.
  1967 Tripartite relations in hamadryas baboons. In *Social Communication Among Primates*, edited by S. Altmann, pp. 63-72. University of Chicago Press, Chicago.
Kummer, H.
  1973 Dominance versus possession: an experiment on hamadryas baboons. In *Precultural Primate Behavior: Symposium of the 4th International Congress of Primatology, Vol. 1*, edited by E. Menzel, pp. 226-231. S. Karger, Basel.
Kummer, H.
  1978 On the value of social relations to nonhuman primates: a heuristic scheme. *Social Science Information* 17:687-705.
Kummer, H.
  1995 *In Quest of the Sacred Baboon* (trans. by M. A. Biederman-Thorson) Princeton University Press, Princeton.
Latour, B., and S. C. Strum
  1991a A common genealogy for humans and their artefacts. *Unpublished Manuscript.*
Latour, B.
  1991b Technology is society made durable. In *A Sociology of Monsters: Essays on Power, Technology and Domination*, edited by J. Law, pp. 103-131. Routledge, London.
Latour, B.
  1994 On technical mediation: philosophy, sociology, genealogy. *Common Knowledge* 3:29-64.
Lave, J.
  1988 *Cognition in Practice.* Cambridge University Press, Cambridge.
Lave, J., and E. Wenger
  1991 *Situated Learning.* Cambridge University Press, New York.
Leonard, D.
  1995 *Wellsprings of Knowledge.* Harvard University Press, Boston.
Leont'ev, A.
  1981 *Problems in the Development of Mind.* Progress Publishers, Moscow.
Luria, A.
  1979 *The Making of Mind.* Harvard University Press, Cambridge, MA.

Mandler, J. M.
   1992 How to build a baby: on the development of an accessible representational system. *Cognitive Development* 3:113-136.

McGrew, W. C.
   1992 *Chimpanzee Material Culture: Implications for Human Evolution.* Cambridge University Press, Cambridge.
   1993 The intelligent use of tools: twenty propositions. In *Tools, Language and Cognition in Human Evolution*, edited by K. R. Gibson and T. Ingold Cambridge University Press, Cambridge.

Merriam-Webster's Collegiate Dictionary
   1993 10th ed. Merriam-Webster, Springfield, MA

Moran, G., J.C. Fentress and I. Golani
   1981 A description of relational patterns of movement during "ritualized fighting" in wolves. *Animal Behavior* 29:1146-1165.

Nardi, B. (Ed.).
   1996 *Context and Consciousness: Activity Theory and Human-Computer Interactions.* MIT Press, Cambridge, MA.

Newll, A., and H. Simon
   1972 *Human Problem Solving.* Prentice-Hall, Englewood Cliffs, NJ.

Norman, D. A.
   1991 Cognitive artifacts. In *Designing Interaction: Psychology at the Human-Computer Interface*, edited by J. M. Carroll, pp. 17-38. Cambridge University Press, Cambridge.

Parker, S. T., and K.R. Gibson
   1979 A developmental model for the evolution of language and intelligence in early hominids. *Behavioral and Brain Sciences* 2:367-408.

Parker, S. T., and C. Milbrath
   1993 Higer intelligence, propositional language, and culture as adaptations for planning. In *Tools, Language and Cognition in Human Evolution*, edited by K. R. Gibson and T. Ingold, pp. 314-333. Cambridge University Press, Cambridge.

Ransom, T. W.
   1979 *The Beach Troop of Gombe.* Bucknell Univeristy Press, Lewisburg.

Reynolds, P. C.
   1993 The complementation theory of language and tool use. In *Tools, Language and Cognition in Human Evolution*, edited by K. R. Gibson and T. Ingold, pp. 407-428. Cambridge University Press, Cambridge.

Ristau, C. A. (Ed.).
   1991 *Cognitive Ethology: The Minds of Other Animals.* Lawerence Erlbaum Associates, Hillsdale, NJ.

Rogoff, B.
   1990 *Apprenticeship in Thinking: Cognitive Development in Social Context.* Oxford University Press, New York.

Rumelhart, D. E., and J. L. McClelland (ed.).
   1986a *Parallel Distributed Processing: Explorations in the Microstructure of Cognition. Vol.1: Foundations. Vol. 2: Psychological and Biological Models.* Bradford Books/MIT Press, Cambridge, MA.

Rumelhart, D. E., and J. L. McClelland
   1986b On learning the past tenses of English verbs. In *Parallel Distributed Processing: Explorations in the microstructure of cognition. Vol.2. Psychological and biological models*, edited by D. E. Rumelhart and J. L. McClelland, pp. 216-271. MIT Press, Cambridge, MA.

Savage-Rumbaugh, E. S., and D. Rumbaugh
   1993 The emergence of language. In *Tools, Language and Cognition in Human Evolution*, edited by K. R. Gibson and T. Ingold, Cambridge University Press, Cambridge.

Schmidt, R., C. Carello and M. Turvey
   1990 Phase transitions and critical fluctuations in the visual coordination of rhythmic movements between people. *Journal of Experimental Psychology, Human Perception and Performance* 16:227-247.

Schubert, G., and R. Masters (Ed.).
   1991 *Primate Politics.* Southern Illinois Press, Carbondale.

Seyfarth, R. M.
   1978 Social relationships among adult male and female baboons II. Behaviour through the female reproductive cycle. *Behaviour* 64:227-247.

Silverberg, J., and J. P. Gray
   1992 *Aggression and Peacefulness in Humans and Other Primates.* Oxford University Press, Oxford.

Simon, H. A.
   1969 *Sciences of the Artificial.* MIT Press, Cambridge, MA.

Smuts, B.
   1985 *Sex and Friendship in Baboons.* Aldine, New York.

Stein, D.
  1984 *The Sociobiology of Infant and Adult Male Baboons*. Ablex, Norwood, NJ.
Strum, S. C.
  1975 Primate predation: interim report on the development of a tradition in a troop of olive baboons. *Science 187*:755-757.
  1981 Processes and products of change: baboons predatory behavior at Gilgil, Kenya. In *Omnivorous Primates*. edited by G. Teleki and R. Harding, pp. 255-302. Columbia University Press, New York.
  1982 Agonistic dominance in male baboons: an alternative view. *International Journal of Primatology* 3:175-202.
  1983a Use of females by male olive baboons. *American Journal of Primatology* 5:93-109.
  1983b Why males use infants. In *Primate Paternalism*, edited by D. Taub, pp. 146-185. Van Nostrand Reinhold, New York.
  1987a *Almost Human*. Random House, New York.
  1988 Social strategies and primate psychology. *Behavioral and Brain Sciences* 11:264-265.
  1994a La société sans culture materielle. In *De la Prehisorie aux Missiles Balistiques*, edited by B. Latour and P. Lemonnier, pp. 27-44. Decouverte, Paris.
  1994b Reconciling aggression and social manipulations as means of competition, part 1: lifehistory perspective. *International Journal of Primatology* 15:739-765.
Strum, S. C., and L. Fedigan
  1999 Theory, method and gender: what changed our views of primate society. In *The New Physical Anthropology: sicence, humanism and critical reflection*, edited by S. C. Strum, D. Lindburg, and D. Hamburg, pp. 67-105. Prentice Hall, Englewoods Cliffs, NJ.
Strum, S. C., D. Forster and E. Hutchins
  1997 Why Machiavellian intelligence may not be Machiavellian. In *Machiavellian Intelligence II: extensions and evaluations*, edited by A. Whiten and R. Byrne, pp. 50-85. Cambridge University Press, Cambridge.
Strum, S. C., and B. Latour
  1987b Redefining the social link: from baboons to humans. *Social Science Informtion* 26:783-802.

Suchman, L.
  1987 *Plans and Situated Actions: The Problems of Human-Machine Communication*. Cambridge University Press, New York.
Taub, D. (Ed.).
  1983 *Primate Paternalism*. Van Nostrand Reinhold, New York.
Tomasello, M., and J. Call
  1997 *Primate Cognition*. Oxford University Press, Oxford.
Toth, N., and K. Schick
  1993 Early stone industries and inferences regarding language and cognition. In *Tools, Language and Cognition in Human Evolution*, edited by K. R. Gibson and T. Ingold, Cambridge University Press, Cambridge.
Vygotsky, L.
  1978 *Mind in Society: The Development of Higher Psychological Processes*. Harvard University Press, Cambridge, MA.
de Waal, F.
  1982 *Chimpanzee Politics*. Allen and Unwin, London.
  1984 Sex differences in the formation of coalitions among chimpanzees. *Ethology and Sociobiology* 5:239-255.
  1989 *Peacemaking among Primates*. Harvard University Press, Cambridge, MA.
Western, J. D., and S. C. Strum
  1983 Sex, kinship and the evolution of social manipulation. *Ethology and Sociobiology* 4:19-28.
Whiten, A., and R. Byrne
  1988 The Machiavellian intelligence hypothesis: editorial. In *Machiavellian Intelligence*, edited by R. Byrne and A. Whiten, pp. 1-9. Clarendon Press, Oxford.
Whiten, A., and R. Ham
  1992 On the nature and evolution of imitation in the animal kingdom: Reappraisal of a century of research. In *Advances in the Study of Behavior*, edited by P. Slater and J. Rosenblatt, vol 21:239-283. Academic Press, New York.
Wynn, T.
  1993 Layers of thinking in tool behavior. In *Tools, Language and Cognition in Human Evolution*, edited by K. R. Gibson and T. Ingold, pp. 389-406. Cambridge University Press, Cambridge.

6. Archaeological Implications of Paleoneurology

Harry J. Jerison

Abstract

Intelligence evolved as the brain evolved. From fossil "brains" we know that in mammals there evolved increases in the brain's capacity to process information about the external world. The genetic blueprint for a brain to develop this intelligence-creating capacity is actually an epigenetic blueprint requiring a normal environment for the growth and development of the nervous system. In their fundamental biology, therefore, brain and intelligence result from a nature-nurture interaction. There are different intelligences (in the plural) that evolved in different species, depending on their neural specializations, and the human variety derives mainly from the evolution of language.

Prologue

Paleoneurology is the study of the fossil evidence of the brain. It increases our understanding of the evolution of the human mind as part of the evolutionary history of brain, mind, and intelligence—of the capacity to handle information. The evidence concerns the external appearance of the brain in fossils, and the analysis is based on the classic uniformitarian principle, namely, that existing relationships between the external appearance of the brain and brain structure and function in living species were also true in fossils. Uniformitarianism (Simpson 1970) is a kind of parsimony principle, in effect, a statement that the laws of nature have been the same throughout history.

I have documented elsewhere that the enlargement of the brain in mammals as a class, including the human species, evolved in relation to the brain's control of perception (Jerison 1991). More generally, the adaptation enables a species to know an external world as a real world of objects embedded in space and time. This feature of the brain's work is well enough understood to enable one to develop scenarios for the evolution of mind as the evolution of the brain's control of cognitive capacity.

Since paleoneurology provides the most direct evidence on the evolution of the brain, and since archaeology is concerned, among other things, with the artifacts created by working human minds, one must devote a few lines to the mind-brain problem. There is presently no good theory of mind-brain relations, although there is an almost universal consensus that "mind" is a creation of the working brain (Churchland 1986)

The mind-brain problem, briefly, is to how one can explain the correspondence and lack of correspondence between the real world as we experience it and the world we know from physical measurements, including measurements of the activity of a brain. The most difficult feature to compute is the invariance in experience, which has been stated by the Gestalt psychologists as the problem of "constancy": the experience of objects as the same despite the variation in sensory information about them. A coin seen on edge is known to be the same coin that one saw full-face, different only in its orientation. This pattern-recognition problem when presented to computers demands very large amounts of computational resources and remains incompletely solved (Minsky 1985). It is solved by brains in a way that makes perception seem easy, yet the amount of neural machinery that is encumbered is known to be very large (see Jerison 1991, n.d., for further discussion). The evolution of mind, as a mammalian property, is thus related to the evolution of enlarged brains, a feature of all mammalian brains.

My most reliable data on living species are on brain size and the number of processing units in mammalian brains. It is possible to measure the sizes of brain structures in living species, to count microscopic elements (nerves and synapses), and to correlate these data with overall brain size. One can also map functional systems: the body surface, which is mapped as the somatosensory and motor systems, and the visual, auditory, and olfactory systems. These are familiar maps of the functional areas in the brain, which are like those I illustrate in Figure 1, below. Applying the geologist's uniformitarian hypothesis (Simpson 1970), one recognizes that brainlike features in endocasts from living mammals can be related to brain structure and function, and one may assume that there were corresponding relationships in extinct species. We can, thus, reason from quantitative relationships and maps in living brains to the brains of fossils. With suitable speculations about brain/mind relations (Jerison 1991), the fossil evidence about the evolution of the brain can then be applied to the evolution of mind.

Brain Structure and Function: Living and Fossil Brains

The basic principle is that the brain "hangs together." Despite its complexity, the brain has many orderly features. Here are a few principles about orderliness in brains of different species that are applicable to interpretations of fossil brains in mammals: 1) In a general way, brain structure is consistent across species. For example, all vertebrate brains are divided into forebrain, midbrain, and hindbrain. These parts of the brain control

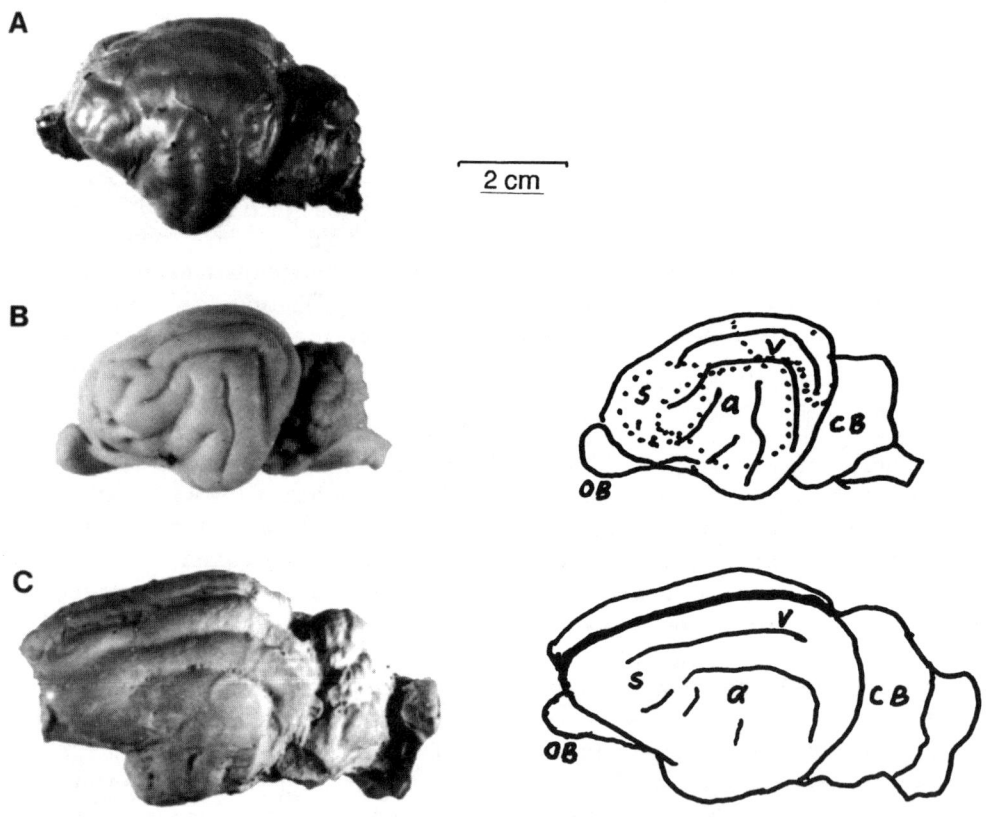

Fig. 1. Brains, endocasts, and brain maps. A. Endocast of domestic cat (vol. = 30 ml). B. (left) Brain of same cat (weight = 29.1 g). Note similarity to endocast of this cat. B. (right) Approximate locations of brain maps of projection areas for audition (a), vision (v), and somatic senses (s). Auditory and somatosensory areas overlap. Olfactory bulbs indicated as OB, cerebellum as CB. C. (left) Endocast of Oligocene sabretooth, *Hoplophoneus primaevus* of 30 million years ago (vol. = 50 ml; Specimen No. USNM 22538, at the United States National Museum, Smithsonian Institution). C. (right) Probable regions for audition (a), vision (v), and somatic senses (s), and olfactory bulbs (OB) and cerebellum (CB) by analogy with living cat brain. Cat brain is an almost perfect lateral view. Both endocasts are rotated somewhat about the anterior-posterior axis, exposing the longitudinal fissure (heavily inked in C, right). Projection areas are indicated only sketchily. See Johnson (1990) and Welker (1990) for more details. (Based on Jerison 1973, Fig. 2.2, p. 31. Reprinted by permission.)

activity in generally similar ways in all vertebrate species, and in even more similar ways in mammals. 2) The sizes of most major parts of the brain in mammals are determined almost entirely by the size of the whole brain. Because this is not usually appreciated, even by neurobiologists, I will show data to clarify this principle in the next section. 3) Differences among species in total brain size are determined mainly by body size. In statistical terms, the regression on body size accounts for about 80 percent of the between-species variance in brain size. 4) The remainder, the statistical residual variance in brain size, is called encephalization and represents the brain's enlargement that is independent of body size. It is this feature that has most often been identified with the evolution of intelligence in different species (see Weaver et al., this volume).

Not really a principle, but simply a fact of nature is that, except for the unusual encephalization, the human brain is a normal mammalian brain. It differs from other mammalian brains in its topographic organization (placement of major neocortical fissures and lobes), but in that respect it is a normal anthropoid primate brain. Other anthropoid primates (monkeys and apes) weighing 70 kg would be expected to have a brain weighing about 400 grams, about 1/3 the size of a human brain. Prosimian primates (lemurs, etc.) and familiar mammals such as wolves, sheep, horses, and so forth, were they to weigh 70 kg would be expected to have brains weighing about 200 grams. Aside from absolute brain size, however, the human brain works as expected for a large mammalian brain. Its parts are the right size for the whole. Even the prefrontal lobes, so significant as control systems in a human brain, are exactly the size expected in a primate brain of human size (Jerison 1997; Semendeferi et al. 1997). Although the evidence on prefrontal lobes in other mammals is presently limited to the laboratory rat, the human prefrontal lobe scales with gross brain size in exactly the same way as the laboratory rat's (Uylings & Van Eden 1990). These are empirical statements about comparisons among species. The comparability of brain structure across species is important in justifying the comparative method. If it were not true, we could not have learned much of what we know about human brain function, because a significant part of our knowledge of how our brains work developed from studies on other species.

To review the other principles, the second one implies that regardless of the reason why a brain reached a particular size in a mammalian species, the sizes of the parts of the brain are appropriate for overall brain size, e.g., that prefrontal cortex in humans is the right size for a 1300 g brain. One of the more unusual conclusions from the quantitative analyses that support this principle is that gross brain size in mammals can be used as a kind of statistic to estimate the total information processing capacity of the brain. I show the evidence for this conclusion in Figure 2, below. The third and fourth principles deal with how or why the whole brain reaches a particular size. Note that the quantitative analyses describe relationships among species; there are small within-species effects, appropriate for the genetic control of brain size, which may be treated as if they were "error variance" (see Jerison n.d.).

There is another functional principle, important for any neurological comparison among species though rarely applicable directly to the fossil evidence. This is the principle of "proper mass" (Jerison 1973; Butler & Hodos 1996): If species differ in their investment in behaviors controlled by known parts of the brain, there will be comparable differences in the relative sizes of the parts of the brain devoted to those behaviors.

The principle of proper mass has been applied directly to fossils in the analysis of the expansion of the inferior colliculi in the midbrain, which are acoustic analyzers in all mammals. The evidence comes from insect-eating bats (microchiropterans), species in which echolocation is an unusually important sensory adaptation (Grinnell 1995). The colliculi, in which important features of the control of echolocation are localized, are normally hidden beneath the cortex and cerebellum, but they are enlarged to so great an extent in bats that they are visible on the surface of their brains. Such exposed colliculi are present in 50 million year old fossil bats as well as in living species (Edinger 1964; Jerison 1973) and provide evidence of the early evolution of echolocation in these flying mammals.

In living brains this kind of differentiation and reorganization is not usually evident as modification of structures visible on an endocast and is demonstrable only with physiological analysis (Johnson 1990; Welker 1990). However, when unusual enlargement of the whole brain is evident, as in the human species relative to other hominoid primates, it is reasonable to attribute the enlargement to the evolution of categories of behavior known to require very large masses of brain tissue. Thus the expansion of the brain during the past 3 million years of human evolution may be attributed to the evolution of language, the control of which is known to involve enormously enlarged regions of the living human brain. The evidence for this enlargement has not been assembled in a single place, though I have reviewed the portion based on brain injury (Jerison 1976). It may be seen in the data of magnetic resonance imaging (MRI) scans made during routine analyses of verbal activity (Posner 1994) in which large fractions of a human brain distributed over much of its surface are shown as activated. The amount of brain activated during language use is much larger than the very large classic speech and language areas of Broca and Wernicke (see Lieberman, this volume), and is less restricted to the "dominant" hemisphere of the brain. The amount of fissurization in brains is related primarily to their gross size (see Fig. 2), but the identification of fissures in the equivalent of Broca's area in a specimen of *Homo habilis* (Falk 1992), suggests the likelihood of the appearance of language in some form in the human lineage, about 2 million years ago.

The Evidence

Before discussing inferences for archaeology, I must show you data to help you appreciate these principles. I illustrate the use (not the proof) of the first principle, that of uniformity of functional organization of the brain, in Figure 1 for felid brains. In this illustration I show a brain and endocast from a domestic cat and an endocast from a fossil

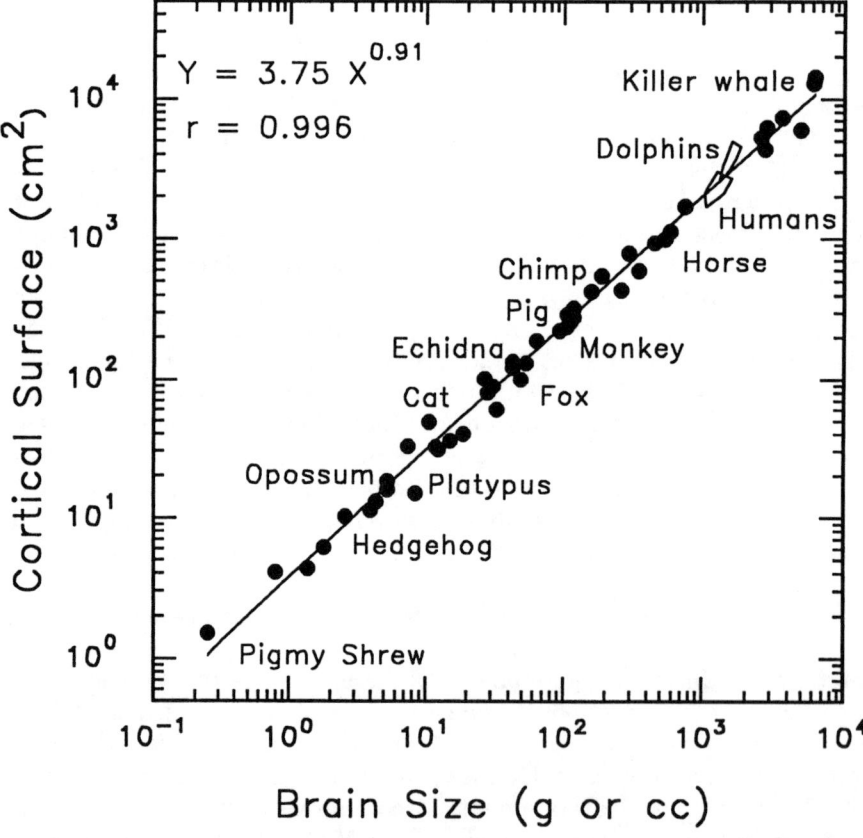

Fig. 2. The relationship between cortical surface and gross brain size "between-species" in mammals. Each point represents a species. In addition, two labeled minimum convex polygons enclose all of the presently available individual data for "Humans" (N = 23) and "Dolphins" (*Tursiops truncatus*, N = 13), and show the typically small within-species variability compared to between-species variability. They also show that highly encephalized species have approximately the amount of cortical surface area as expected for their brain sizes. The slope of the "regression line" on this double logarithmic graph is the exponent of the allometric equation shown on the graph, and the excellent fit is reflected in the very high correlation coefficient $r = 0.996$. (From Jerison 1991, Fig. 4, p. 29. Reprinted by permission.)

sabretooth. The first principle is used to draw the functional map of the sabretooth's brain as corresponding to the map of the cat's brain. The figure also illustrates how well a brain is represented by an endocast of a living species.

The brain map of the living cat in Figure 1 was determined by recording the location of electrical activity in the brain that corresponded to localized peripheral activity. For example, when a sound is presented, even to an anaesthetized animal, there is a measurable change in electrical activity in the brain, and the place where the change occurs can be marked as "auditory cortex" in Figure 1. All of the sensory and motor systems pictured for the domestic cat were mapped in this way. Such maps have been determined for other species of mammals, and although the details of the mappings differ in accordance with "proper mass," their location relative to one another is approximately the same in most species. Their features have been determined in exquisite detail (Allman 1990; Johnson 1990; Welker 1990).

The fossil endocast in Figure 1 is from a sabretooth (*Hoplophoneus primaevus*) that lived about 30 million years ago. Next to the photograph I have sketched the sabretooth's brain, with a map of its functional areas. Nobody will ever perform the necessary physiological studies to verify the sabretooth's map. I created it by following the uniformitarian hypothesis, extrapolating from the map of the cat's brain, and I believe that it is as good a guess as one can make. We do know that such an exercise would have produced a reasonably correct mapping of a dog's brain and we could verify this, because the physiological studies have been done in dogs as well as cats (Jones & Peters 1990). It is this kind of procedure that enables one to conjecture that there was a Broca's area in the brain of *Homo habilis*.

The second principle, about the orderliness of size relationships within brains, provides some of the best evidence for our understanding of the evolution of mind. The general principle is that the brain "hangs together," that is, if you know the size of the brain as a whole you can make surprisingly good estimates of the sizes of major structures within the brain (Jerison 1991; 1997).

Figure 2, which is from my monograph on brain size (Jerison 1991), is one of many sets of data suggesting this kind of relationship. It shows that if you know the gross size of a mammal's brain you can estimate the total cortical surface area of that brain with an error (logarithmic data) of about 1 percent. Based on four different research reports on 50 different species of mammals, the graph is important because the surface area of the cortex is known to be closely related to the number of neurons, the number of cortical columns, and the number of synapses in the cortex (see Jerison 1991, for details). Neurons, synapses, and cortical columns are usually identified as information processing units in the brain. The 1 % error of estimation is derived from the correlation coefficient, $r = 0.996$, and shows that brain size is an accurate estimator of the neural processing capacity of a brain. The estimate seems to work regardless of species.

Having established that brain size is a good "statistic" for facts about the brain's work in handling information, we should determine why a mammal's brain reaches a particular size. Part of the answer is in Figure 3, also from Jerison (1991), which shows brain-to-body relations in 76 species of mammals, including 27 anthropoid primates, 18 lemur-like (prosimian) primates, and 31 insectivorous species. Insectivores are often thought of as primitive because of their relatively small brains and their bodily resemblance to fossil mammals that lived during the age of dinosaurs.

You can see in Figure 3 that the main determinant of gross brain size is body size but that the three subgroups are also differentiated from one another by a vertical displacement of the lines fitted to their data. The allometric body size effect, represented by the slopes of the fitted lines, quantifies a universal biological rule that larger brains are needed to control larger bodies. (There is a technical debate about the correct slopes for the lines when a single slope is used for all [see Jerison 1991], but for our purpose it is only their displacement that is important, in order to show that the groups can be differentiated according to their encephalization.) The displacement of the lines, indicating differences in *relative* brain enlargement, or encephalization, is explained by the evolution of new adaptations for ecological niches that required more information about the external world. Insectivores as a group evidently require less brain than do lemuroid primates, which in turn require less brain than anthropoid primates. (If one performs the analysis as in Figure 3 on a larger and more diverse sample of mammals, lemuroid primates are recognized as "average" mammals, comparable to wolves, sheep, and horses in relative brain size.)

The evolution of these species for lives in their various niches involved relatively small differentiation among the groups in brain size, and we should recognize that all of the species do quite well in their niches, with brains of appropriate size for the lives they lead. An enlarged brain beyond the allometric body size requirement seems best explained by relatively small differences in adaptation. The massive enlargement of the human brain, to a size some six times larger than expected for a 70 kg mammal, is peculiar, however, and requires additional explanation.

Paleoneurology of Mind: I. Early Evolution of Encephalization

I have recently reviewed the fossil record of encephalization (Jerison 1991), and documentation for the remainder of this chapter is in that monograph, with additional material in my earlier book (Jerison 1973) and in a chapter of a text in neuropsychology (Jerison 1994). Encephalization was first evident among late Paleozoic sharks that lived about 250 million years old. These were larger-brained relative to their body sizes than any other vertebrate of their time. In their subsequent history, this group of vertebrates has continued to include many relatively large-brained species. We cannot explain it. With the possible exception of electroreception (Bullock 1986), no behavioral correlate of that evo-

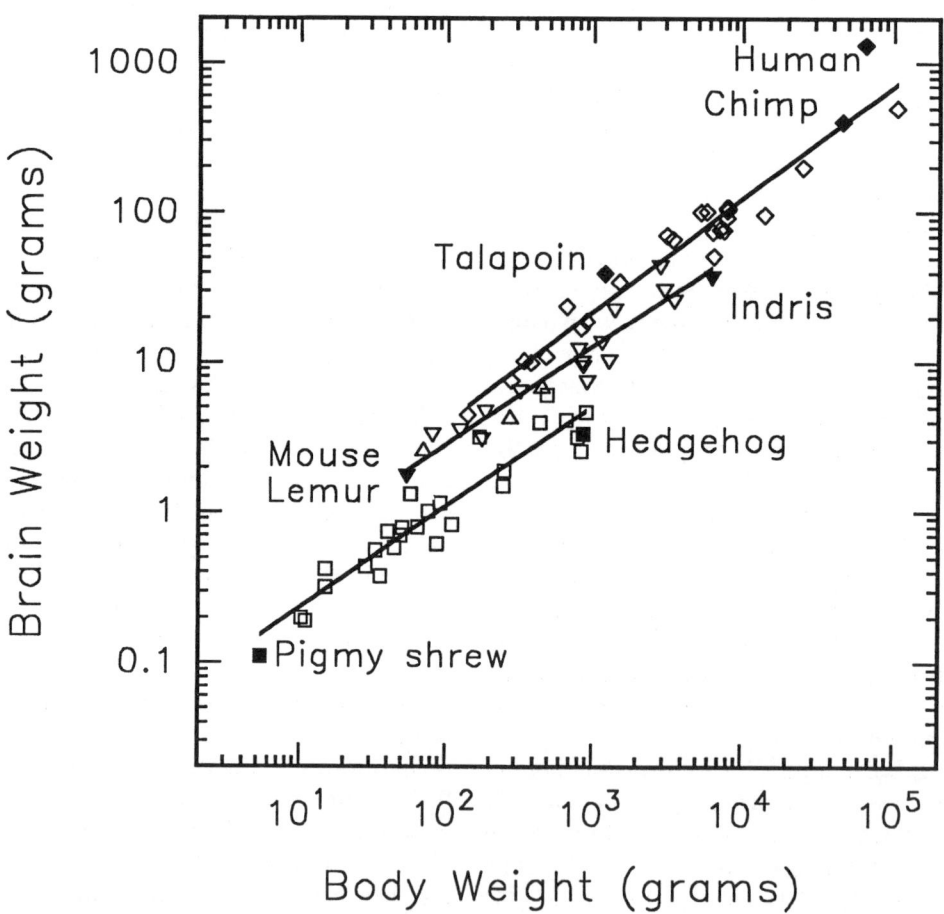

Fig. 3. Brain size as a function of body size in 76 species of mammals, with fitted regression lines to three taxa. Fitted regression lines (assuming errors only in Y) and correlation coefficients for the separate taxa: insectivores (lowest line, squares): $Y = 0.05 X^{0.67}$, $r = 0.946$; prosimians (middle line, triangles): $Y = 0.14 X^{0.66}$, $r = 0.960$; anthropoids (top line, diamonds): $Y = 0.13 X^{0.75}$, $r = 0.972$. Points for named specimens are filled. For comparisons with respect to encephalization, one must force the same slope (exponents of above equations), and although there are technical arguments with respect to the choice of 2/3 versus 3/4, either slope would indicate an approximate doubling of encephalization in each advance in encephalization among these groups. (From Jerison 1991, Fig. 13, p. 56. Reprinted by permission.)

lutionary trend has been identified.

The next evidence of encephalization is in mammals, which first appeared during the Triassic period of the Mesozoic era, about 225 million years ago. These animals were contemporaries of early dinosaurs and other large reptiles. Encephalization is one of the traits that differentiate mammals from reptiles, and the difference is by a factor of about four: the earliest known mammal brains were about four times as large as expected in reptiles. Mammalian species were small-bodied throughout the Mesozoic era; no fossil larger than a house cat has yet been found. The era ended at the well known K-T (Cretaceous-Tertiary) boundary, 65 million years ago, with the extinction of many animal species including most of the giant reptiles. Although diverse in other ways, all species of mammals appear to have been at the same grade of encephalization throughout the 160 million years of their Mesozoic evolution.

The mammalian brain first became enlarged as a result of the evolution of the cerebral cortex, a uniquely mammalian brain structure. The present consensus is that cortex probably evolved in connection with the specialization of early mammals as nocturnal land animals. The adaptational problem that had to be solved was how to live in a sunless world, an environment in which the normal Mesozoic reptilian adaptation for diurnicity was inadequate. The earliest mammals were simply deviant "reptiles" that had solved that problem. Part of their solution was miniaturization as a way to avoid predators. Another part was endothermy, the brain-controlled maintenance of stable and warm body temperature at twilight and in darkness, when the reptilian temperature control system involving movement in and out of the warmth of the sun was impossible. Endothermy also provided the necessary internal environment for supporting an enlarged brain. What about the brain?

As diurnal animals, Mesozoic reptiles could be responsive to visual aspects of their external world without significant brain enlargement because much visual analysis in all vertebrates is performed by the millions of nerve cells in the retina, which is external to the brain. To live an active nocturnal life, an animal must have information about events at a distance, which could be available from hearing, touch, and olfaction, but there are no structures peripheral to the brain to store neurons for nonvisual senses that might perform an analysis comparable to that performed by the neural retina. The necessary neurons would have to be within the brain case and be parts of the brain itself. For hearing, touch, and olfaction to serve as surrogates for normal vision a brain must be enlarged, simply to package the neural machinery that would do the work. In addition, since the information for such a nocturnal sensing system would be received through several different sensory channels, neural mechanisms to integrate multisensory and correlated motor information would also be required. When mammals evolved as the small nocturnal "reptiles" of their time, their brains became enlarged to do the additional processing, primarily by evolving cerebral cortex where the new neural systems were localized. From this perspective, mammalian encephalization may be viewed as the solution of a packaging problem of where to put the required additional neural machinery.

The beginning of the Tertiary period of the Cenozoic Era, 65 million years ago, witnessed the evolution of larger mammalian species, which probably invaded some of the diurnal niches left vacant by the great extinction. For the next 10 million years or so, except for the allometric enlargement to be expected when larger bodied species evolved, there is little evidence of encephalization in mammals beyond the initial mammalian grade. During the past 55 million years there has been further encephalization in several orders of mammals, although living hedgehogs and opossums are no more encephalized than the earliest known species. In general, when encephalization can be identified in an evolutionary lineage, it seems to have evolved rather rapidly, followed by a period of stasis. At present, average mammalian encephalization is about 3 times that of the earliest mammals. In the horses, for example, the earliest species of 50 or 55 million years ago (*Hyracotherium*) were only slightly more encephalized than the early mammals. But species of about 30 million years ago (*Mesohippus*) were at about the same level as living horses and are average living mammals in this respect. The horse brain appears to have been at equilibrium in relative size since its expansion at least 30 million years ago.

Primates are known from essentially all of the Tertiary and seem always to have been somewhat more encephalized than their contemporaries among the mammals (Simons & Rasmussen, 1996). Just as encephalization distinguishes mammals from reptiles, further encephalization appears to have been one of the distinguishing features of primates among the mammals. However, although their fossil relatives of 50-odd million years ago were the earliest mammals in which encephalization beyond the basic mammalian grade is evident, living lemur-like (prosimian) species are average in encephalization. Anthropoid primates (monkeys and apes), on the other hand, have brains 2 to 3 times as encephalized as present mammalian averages. At the present time, the most encephalized mammals are humans and dolphins (*Tursiops truncatus*) with brains about 5 or 6 times the average size as expected from body size.

The enlargement of the hominid brain during the past 3 million years, from an anthropoid grade to the present human grade, is the best-documented instance of the rapidity of encephalization. Although one assumes that hominid encephalization resulted from natural selection favoring larger brains, it is significant that the best mathematical models of evolution presently available (Lande 1976, 1979) do not rule out the possibility that this brain enlargement was an accidental result of genetic drift. This strange result follows from Lande's model which has terms for mutation rate, selection pressure, population size, and other factors known to determine the rate of evolutionary change. To *prove* that selection occurred one would have to show that entering zero-selection in the equations would result in nonsense figures for at least one of the other terms. It does not. All speculations, including mine, assume that there has been selection for larger brains, but it is worth recognizing that present quantitative theories do not make an unequivocal case for natural selection for encephalization, even in hominid history.

Encephalization and Mind

I wish now to consider the mental correlates of encephalization, which I believe to have some basic similarities in all mammals. My view (Jerison 1973, 1991) is that the integration of neural information in the mammalian cortex occurs as a transformation of the activity of nerve cells and synapses into a model of a possible world. I have not suggested how such a model might be constructed, an issue now addressed in computer science as pattern recognition by artificial "neural networks" (Rummelhart & McClelland 1986), but the model would be a translation of probabilistic, seemingly chaotic, neural activity into stable representations of objects in space and time. Stability in the model is the "constancy" of the Gestalt psychologists and may be contrasted with the continuously varying neural activity in brains generated by external stimuli. Model building of this sort, I suggest, must have characterized the work of the earliest mammalian brains and may have to occur in any brain when the amount of information that is handled becomes large enough. In the human species, the model of a possible world created by our brains is the real world as we know it, and in that sense the work of the cerebral cortex is to create our real world.

The amount of information that is involved is enormous, even in very small mammals. A mouse, for example, has about 40,000,000 cortical neurons and 80,000,000,000 cortical synapses in its 0.5 gram brain. Neurons are active at all times and average about 10 or so signals per second. That is the raw information in the brain. Model building, according to this view, is the brain's way of handling an otherwise impossible load of information and is the biological basis for mind.

The contrast is between "mindful" behavior, in which neural signals are "chunked" into the elements of a possible world, and behavior based on reflex control that uses neural signals without translating them into the chunks. (I follow Simon [1974] in the idea of chunking, to translate neural activity into *objects*, *space*, and *time* as elements of a possible world.) Such reflexes can operate on raw neural data analyzed with decision mechanisms that make connections between sensing and action. Decision making in this way is a relatively simple problem in engineering and information theory, first applied to the design of signal detecting machines (Cherry 1956). In biological terms, a group of nerve impulses received within a particular time interval at one neural center could trigger signals to another neural center that controls a program to release an appropriate movement. The mechanism for determining whether the signal count is within the right range is a kind of window governing the flow of information. Directed fly-catching by tongue-flicking in amphibians has been analyzed in these terms, with the superior colliculi as the basic control center in the brain in which the analysis is performed (Ewert 1974; Roth 1987). Such a reflex system does not appear to depend on a model of an external world, because tongue-flicking can be elicited in response to artificially generated signals to the retina, with nothing resembling a fly present in the real world.

In encephalized species, the activity that interests us and is likely to be relevant for archaeology is "mindful" behavior that results from the operation of a model of a real world constructed by the brain. This is the conclusion that I want to emphasize for you. It concerns cognition and experience rather than overt behavior. I emphasize the capacity to *know* the real world, which is the perceptual and cognitive work of enlarged brains. One categorizes the work as learning, thinking, communicating, social behavior, and so on, and these categories can be used for behavioristic as well as experiential analysis. One can observe animals communicating, for example, and measure their learning in mazes and puzzle boxes. My view is that when such an analysis is rigidly behavioristic, we can misinterpret the effect of encephalization, because behavior under all of the categories occurs in most animals, whether or not their brains are enlarged or cognitive activity is present.

In many animals, no model of reality has to be constructed by their brains. Learning can be demonstrated in invertebrate species with only a few hundred nerve cells in their "brains" (Macphail 1993). Communication and social behavior can be controlled by very little brain tissue. The familiar "language" of bees and their elaborate social systems are controlled by no more than about 400,000 nerve cells in the honeybee's brain (Seeley 1995). It is likely that even mammals depend on reflex mechanisms for important communication rather than on learned signal systems (Smith 1977). Only if a cognitive dimension is involved should there be a requirement of significant expansion of the brain to support complex behavior.

As a more elaborate example, consider throwing, which may have evolved as a result of natural selection for neural control (Calvin 1983). Such evolution as an explanation for encephalization makes sense in the present terms only to the extent that throwing requires cognitive as well as motor control. One can design accurate throwing machines by writing appropriate differential equations for feedback programs to control the machines. The fundamental theory is a half century old (Weiner 1948), though I understand that robotic devices to do this accurately are yet to be built. Nevertheless, the "behavioristic" problem of making accurate tosses could, in principle, be solved by a small "mindless" nervous system or by a relatively small computer system. Unlike amphibian flycatching by "throwing" the tongue, the human solution of the throwing problem obviously involves a translation of sensory and motor neural data into a perceived target object and suitable judgments about the forces to be applied. Our experience tells us that people are guided by some model of reality when throwing. But does it *require* a model of reality constructed by a brain? The answer is "No," because we can design machines and specify neural reflex systems that could do the job. The parsimonious solution may be the machine's or the frog's way, and that is why it makes no sense to insist on parsimony in determining what actually happens when a problem is solved by a large-brained animal. The human solution is "mindful" in spite of the possibility for reflex control. The point arises again in the analysis of the evolution of language.

Paleoneurology of mind: II. Speculations on the Beginnings of Language

Encephalization both at the beginning of mammalian evolution and early in the evolution of primates is explained as adaptation for specialized kinds of knowledge of the external world. The same kind of explanation is appropriate for the enlargement of the brain in the human lineage. One must recognize, too, that there might be different explanations for the initial enlargement in the earliest of the hominids, about 4 million years ago, and the later enlargement beginning about 2 million years ago with the appearance of the genus *Homo*. Furthermore, if encephalization was a response to a particular selection pressure, selection should have occurred in response to the environmental challenge at that time rather than in anticipation of future challenges. Finally, the enlargement in hominids should be assumed to have met a need for a specialized cognitive activity. Only secondarily would it have evolved as a specialized ability to communicate. Communication can be handled by many nervous systems and in many ways. As I argued in my analysis of the throwing hypothesis as an explanation of encephalization, communication as behavior is almost universal in animal species, and does not usually require encephalization. Cognition, on the other hand, as an adaptation of a perceptual/cognitive system in mammals, is the only kind of brain work that requires encephalization. Since much of the subsequent enlargement of the hominid brain should be related to the evolution of language functions (Jerison 1976), one is inevitably led to the view that language began as an adaptation for knowing the real world. Despite the present importance of language for human communication, it makes good sense to see language as having begun as a cognitive adaptation, which was preadaptive for the unique way to communicate that evolved in language as we now know it.

My scenario has early australopithecines successfully invading a niche comparable to that of living wolves, a niche for social predators with an extensive range. Wolves depend on olfactory information, sniffing and scent-marking to know their territories, which typically cover hundreds of square kilometers. The knowledge is a "cognitive map" (Tolman 1948; O'Keefe & Nadel 1978) that enables wolves to navigate their ranges. For a sense of the magnitude of their informational problem, recognize that our close contemporaries among large primates, gorillas and chimpanzees, typically navigate a much smaller range of no more than a few square kilometers. In their use of olfaction as a major sensory cue, wolves depend on typically mammalian olfactory bulbs as the sense organs that provide the raw information. Further analysis is performed by other brain structures that are of normal size in mammals. Some of the "primitive" brain structures that I assume are used for the analysis, such as hippocampus, are not specifically olfactory analyzers but are involved in mapping and memory functions in the brain.

The problem for hominid adaptation to such a niche results from what may have been an evolutionary accident in primate evolution. For at least the past 20 million years, anthropoid primates have had much reduced olfactory bulbs. *Aegyptopithecus* and other earlier anthropoids had relatively normal olfactory bulbs, but this structure all but disappeared in early Miocene anthropoid brains. We do not know why. Such reduction of the olfactory bulbs seems to have occurred only in anthropoids among land mammals. Olfactory information that would normally be used by predatory mammals to map an extended environmental range would, consequently, have been diminished in early hominids. Significantly, the size of "primitive" brain structures involved in the analysis of olfactory information as well as in mapping and memory, is normal in anthropoids (Jerison 1991).

As in the evolution of sensory capacities for nocturnicity, the solution for anthropoids "invading" a social predator's niche would be to evolve other sensory pathways to develop information normally available from olfaction. Knowing the significant place of vocal signals in the lives of living anthropoid primates, I have speculated that an auditory/vocal channel served in the place of olfaction in the construction of the necessary mental map of their geographical environment. This would have required some expansion of regions in the brain controlling that channel in order to handle the additional information load, and I see in such expansion the origin of language areas in the human brain. The expansion would be an addition of auditory association cortex to primary auditory cortex, and additional motor association cortex to regions of the brain involved in the control of the lips, tongue, and voice box. In living humans these areas are Wernicke's and Broca's areas of the brain.

To review the early hominid creation of the cognitive map required to navigate a range covering tens of square kilometers, consider the basic information available without the evolution of a "language sense." Present evidence is that the ancestral species had normal primate hearing and vision (presumably color vision), and could scan its range across much of its distance. The cognitive map would literally be a map of the region. The objects within its range could be remembered (the work of the memory systems of the brain), but the full range would include much that is beyond the visual field. Audition is not much needed for this part of the construction, but tactile senses would be useful. Olfaction would be almost useless in this primate, given the reduced input from the olfactory bulbs. In moving through the range, I imagine this early hominid as vocalizing when objects appeared in view and retaining a vocalized code as an aid to visual memory. The information would register in a portion of this primate brain destined eventually to evolve into a Wernicke's area, near auditory cortex, but with a small fraction of the neural machinery that would eventually be encumbered by the activity. However, the map of the range (or territory) would be enhanced by the vocal signals and their memories, even after the animal moved out of the region that it viewed. As social animals, these early ancestors could share the knowledge of the map at least to the extent that there were vocal markers added to it, even if the visible map were no longer present. They could thus communicate with conspecifics about at least some features of the maps. An adaptation of this sort would be required if the early hominid were to succeed as a predator over the larger range, and this particular adaptation would have the potential of evolving into a communication system in addition to the knowledge system required for navigation.

The adaptation in its first appearance would have required only small additional neural control tissue, and evidence from brain size indicates that the australopithecines were at most only slightly more encephalized than chimpanzees and gorillas. It would not require the lateralization of specialized features that characterizes the brain's control of language in living humans. I do not know of an evolutionary explanation for lateralization. The major feature of the adaptation to keep in mind is that it was not an adaptation for improved communication. Rather, when it first appeared, it was a cognitive adaptation to provide better information about the external world. This possibility as fundamental to the evolution of language may be the most important suggestion from paleoneurology for archaeologists.

Paleoneurology of mind: III. Language and Communication

I do not deny the major role of language in human communication. In fact, I propose that some of the peculiar features of language as communication arise from its fundamentally cognitive function and from the problems of having a cognitive system serve as a communication system as well.

If the enlargement of the hominid brain in australopithecines was primarily an adaptation of what has become the language system, and if it was specifically a cognitive adaptation related to the problem of creating a "cognitive map" of an extended territory, communication would not play a major role in the initial history of the adaptation. Nevertheless, this particular adaptation for mapping could clearly have been pre-adaptive for communication if the vocal and auditory apparatus was involved in creating the map. Since the same apparatus is involved in many communications among primates, it must have been almost immediately the case that the maps created by one individual could be shared with a second individual.

There is a fine analogy to such sharing of a cognitive map in a much smaller-brained species, the insect-eating echolocating bat. Evidently, when these animals use echolocation while foraging for insect prey, individual bats can intercept one another's signals. It would be a strange experience. Neurologically the effect would be as if we humans could connect our brains together at the level of, say, the optic nerves so that information transmitted along the optic nerve of one individual could trigger brain responses (including perception) in another individual. It is very close to a science fiction picture of telepathy in which we enter one another's minds. I like this image, because it is, in fact, very close to what we do when we use language to communicate. We construct images for one another and experience those images almost as if they were real, a virtual reality that predates the computer age by many centuries. We know that it is quite real from our experience as readers. We can live in worlds created by realistic authors and enjoy them in so special a way that we have named the experience "vicarious."

Were the creation of a cognitive map of a geographical territory from auditory/vocal neural data the only function of language, it would be insufficient to explain the extraordinary further enlargement of the hominid brain that began some 2 million years ago in the genus *Homo*. It is only to enable the communicational function to develop to its present state that additional information-processing capacity had to evolve. There must be unusual aspects of communicating with what is fundamentally a cognitive system. Unlike most communication in other species, human language works by sharing imagery rather than by releasing stereotyped behaviors. If the geographic map experienced by one individual can be shared with other individuals, other aspects of one individual's experience might be shared in the same way with another individual by using the same neural system. That is the essence of human language, as I see it, that when we communicate with language we share images and knowledge.

In most species, communication involves a set of commands that release appropriate behaviors in animals that receive the signals. The repertory of messages can be relatively small and can require relatively little neural control capacity. The messages can therefore be unequivocal and subject to little error. When a deer senses a predator and flags its flight by a raised tail, other deer can respond to the tail and join in the flight without requiring further knowledge of the predator. When hatchlings cry and open their mouths for food, the signal is unmistakable even to the human ear and unequivocal to the parent bird. A warning call by the beachmaster sea lion is equally definite and is ignored at one's peril. I argue that the brain systems controlling such behavior are designed for certainty and are effectively hard-wired in the brain. They require relatively little neural tissue for their control.

A brain system to handle non-olfactory information for creating a map of one's environment can be only partially hard-wired. The elements of the environment have to be introduced as the environment is explored and labeled. As a supplementary sensory-perceptual system, furthermore, it would have the characteristics now recognized for other senses, that is, its neural basis would be formed by the interaction between genetic and environmental events, and it would not develop to function normally without such interaction. The lengthy period of learning human language, and the diversity of languages despite some similarities in their fundamental "deep" structure argue further for its perceptual roots. However, to have such a perceptual/cognitive language system do the job it does for humans today does require an enormous investment in neural machinery. We can recognize that the expansion of the hominid brain from an australopithecine to a *Homo* grade and within the *Homo* grade might have been in connection with the development of the communicational system that enables us to share one another's experiences with the help of language.

It should not be overlooked that this evolutionary scenario is peculiar to hominids only in that they are primates lacking a normal neural system for easy access to mapping systems such as the hippocampus. The unusual adaptation in these primates, which we call language, is the unique feature of human evolution for which a morphological marker exists, namely, the unusually enlarged brain. The lateralization of language functions in the brain is not measurable morphologically, and the unusual functions of the human hand and foot could not be discovered from the fossil evidence. The enlarged brain, on the other hand, implies increased information-processing capacity, and should raise questions about the utility of that capacity. The language sense answers those questions, when it is recognized as a "knowledge" sense, that enables us to know the external world. When we communicate with this knowledge sense we literally share that knowledge, as if we could arrange for someone else to see or hear things with our eyes and ears.

Conclusion

The implication for archaeology is straightforward and simple, and is based on inferences from psychological analysis as much as from paleoneurology. Paleoneurology has provided unequivocal evidence for the brain's enlargement in the hominids. Among recent analyses of encephalization (Weaver et al., this volume; Martin 1990) the concern has been more with issues of energetics than with brain function. Brains are, after all, information-processing organs, and selection for brain size must have been selection for increased or improved information processing capacity. Energetic aspects of that evolution, such as improved blood supply for an enlarged brain, or parental investment in the energetics that enable brain growth, explain how it was possible for an enlarged brain to evolve. The reason for the enlargement, however, must be found in the requirement for more or better information. Paleoneurology tells us that the enlargement did, in fact, occur. The evolutionary scenarios to explain the occurrence depend on a psychological analysis of the information that was required and was available.

For the working archaeologist, one inference from these considerations is to seek understanding of the artifacts that have been preserved from more than the techniques required for their manufacture. One must also consider the thought that had to go into the manufacture. Technique is important, of course, but the real issue is to understand the development of the imagination of those who developed the artifacts. In short, the important conclusion is to consider the minds of those who created the artifacts as well as the techniques they used.

References Cited

Allman, J.
    1990   Evolution of neocortex. In *Cerebral Cortex, Vol. 8A: Comparative Structure and Evolution of Cerebral Cortex, Part I*, edited by E. G. Jones and A. Peters, pp. 269-283. Plenum Press, New York.

Butler, A. B. and W. Hodos
    1996   *Comparative Vertebrate Neuroanatomy*. Wiley-Liss, New York.

Calvin, W. H.
    1983   *The Throwing Madonna*. McGraw-Hill, New York.

Cheney, D. L. and R. M. Seyfarth.
    1990   *How Monkeys See the World: Inside the Mind of Another Species*. University of Chicago Press, Chicago.

Cherry, C. (Ed.)
    1956   *Information Theory*. Butterworths, London.

Chomsky, N.
    1972   *Language and Mind*. Harcourt Brace Jovanovich, New York.

Churchland, P. S.
    1986   *Neurophilosophy: Toward a Unified Science of the Mind-Brain*. MIT Press, Cambridge, MA.

Dennett, D. C.
    1991   *Consciousness Explained*. Little Brown, Boston.
    1995   *Darwin's Dangerous Idea*. Simon & Schuster, New York.

Donald, M.
    1991   *Origins of the Modern Mind*. Basic Books, New York.

Edelman, G. M.
  1992 *Bright Air, Brilliant Fire: On the matter of the mind.* Basic Books, New York.

Edinger, T.
  1964 Midbrain exposure and overlap in mammals. *American Zoologist* 4:5-19.

Ewert, J. P.
  1974 The neural basis of visually guided behavior. *Scientific American* 230:34-42.

Falk, D.
  1992 *Braindance.* Henry Holt, New York.

Fodor, J. A.
  1983 *The Modularity of Mind.* MIT Press, Cambridge, MA.

Greenfield, P.
  1991 Language, tools and brain: the ontogeny and phylogeny of hierarchically organized sequential behavior. *Behavioral and Brain Sciences* 14:531-595.

Grinnell, A. D.
  1995 Hearing in bats: an overview. In *Hearing by Bats*, edited by R. R. Fay and A. M. Popper, pp. 1-36. Springer Verlag, Heidelberg.

Hailman, J. P.
  1967 How an instinct is learned. *Scientific American* 221 (6):98-106.

Jerison, H. J.
  1967 Activation and long term performance. In *Attention and Performance I (A Special Edition of Acta Psychologica, Vol. 27)*, edited by A. F. Sanders, pp. 373-389. North-Holland, Amsterdam.
  1968 Attention. In *International Encyclopedia of the Social Sciences*, edited by D. L. Sills, 1:444-449. Macmillan, New York.
  1973 *Evolution of the Brain and Intelligence.* Academic Press, New York.
  1976 Paleoneurology and the evolution of mind. *Scientific American* 234(1):90-101.
  1977 Vigilance: Biology, psychology, theory, and practice: keynote address. In *Vigilance: Theory, Operational Performance, and Physiological Correlates*, edited by R. Mackie, pp. 27-40. Plenum, New York.
  1985 Issues in brain evolution. *Oxford Surveys in Evolutionary Biology* 2:102-134.
  1990 Fossil evidence on the evolution of the neocortex. In *Cerebral Cortex*, Vol. 8A, edited by E. G. Jones and A. Peters, pp. 285-309. Plenum, New York.
  1991 *Brain Size and the Evolution of Mind: 59th James Arthur Lecture on the Evolution of the Human Brain.* American Museum of Natural History, New York.
  1994 Evolution of the brain. In *Neuropsychology (Handbook of Perception and Cognition)*, 2nd ed., edited by D. Zaidel, pp. 53-82. Academic Press, New York.
  1997 Evolution of prefrontal cortex. In *Development of the Prefrontal Cortex: Evolution, Neurobiology, and Behavior*, edited by N. A. Krasnegor, G. R. Lyon and P. S. Goldman-Rakic, pp. 9-26. Paul H. Brookes Publishing Co., Baltimore.
  n.d. The evolution of intelligence. In *Handbook of Human Intelligence* (2nd edition), edited by R. J. Sternberg. Cambridge University Press, New York & London.

Johnson, J. I.
  1990 Comparative development of somatic sensory cortex. In *Cerebral Cortex*, Vol. 8B, edited by E. G. Jones and A. Peters, pp. 335-449. Plenum, New York.

Jones, E. G. and Peters, A. (eds)
  1990 *Cerebral Cortex*, Vols. 8A & B. Plenum, New York.

Koffka, K.
  1935 *Principles of Gestalt Psychology.* Harcourt, New York.

Köhler, W.
  1935 Physical Gestalten. In *A Source Book In Gestalt Psychology*, edited by W. D. Ellis. Humanities Press, New York. [Translated excerpts from *Die physischen Gestalten in Ruhe und im statonären Zustand, Eine naturphilosphische Untersuchung.* Erlangen, 1920].

Lande, R.
  1976 Natural selection and random genetic drift in phenotypic evolution. *Evolution* 30:314-334.
  1979 Quantitative genetic analysis of multivariate evolution, applied to brain:body size allometry. *Evolution* 33:402-416.

Leakey, R. E. and Lewin, R.
  1992 *Origins Reconsidered: In Search of What Makes us Human.* Doubleday, New York.

Lieberman, P.
  1984 *The Biology and Evolution of Language.* Harvard University Press, Cambridge, MA.

Lumsden, C. J. and Wilson, E. O.
  1981 *Genes, Mind, and Culture: The Coevolutionary Process.* Harvard University Press, Cambridge, MA.

Macphail, E. M.
   1982 *Brain and Intelligence in Vertebrates.* Clarendon, Oxford.

Macphail, E. M.
   1993 *The Neuroscience of Animal Intelligence: From the Seahare to the Seahorse.* Columbia University Press, New York.

McCulloch, W. S.
   1965 *Embodiments of Mind.* MIT Press, Cambridge, MA.

Minsky, M.
   1985 *The Society of Mind.* Simon and Schuster, New York.

O'Keefe, J. and Nadel, L.
   1978 *The Hippocampus as a Cognitive Map.* Oxford University Press, Oxford.

Plotkin, H. C.
   1994 *The Nature of Knowledge: Concerning Adaptations, Instinct, and the Evolution of Intelligence.* Penguin Press, London.

Posner, M. I.
   1994 *Images of Mind.* Scientific American Library, New York.

Pribram, K. H.
   1991 *Brain and Perception: Holonomy and Structure in Figural Processing.* Erlbaum, Hillsdale, NJ.

Rescorla, R. A.
   1988 Pavlovian conditioning: it's not what you think it is. *American Psychologist* 43:151-160.

Roth, G.
   1987 *Visual Behavior in Salamanders.* Springer-Verlag, Berlin & New York.

Rumelhart, D. E. and McClelland J. L.
   1986 *Parallel Distributed Processing: Explorations in the Microstructure of Cognition.* MIT Press, Cambridge, MA.

Russell, B.
   1912 *The Problems of Philosophy.* Home University Library, Oxford. [Reissued, Oxford University Press, 1959]

Seeley, T. D.
   1995 *The Wisdom of the Hive: The Social Physiology of Honey Bee Colonies.* Harvard University Press, Cambridge, MA.

Simon, H. A.
   1974 How big is a chunk? *Science* 183:482-488.

Simons, E. L. and Rasmussen, D. T.
   1996 Skull of *Catopithecus brownni,* an early Tertiary catarrhine. *American Journal of Physical Anthropology* 100:261-292.

Simpson, G. G.
   1970 Uniformitarianism: an inquiry into principle, theory, and method in geohistory and biohistory. In *Essays in Evolution and Genetics in Honor of Theodosius Dobzhansky,* edited by M. K Hecht and W. C. Steere, pp. 43-96. North-Holland, Amsterdam.

Smith, W. J.
   1977 *The Behavior of Communicating: An Ethological Approach.* Harvard University Press, Cambridge, MA.

Von Neumann, J.
   1958 *The Computer and the Brain.* Yale University Press, New Haven.

Weiner, N.
   1948 *Cybernetics.* MIT Press, Cambridge, MA.

Welker, W. I.
   1990 Why does cerebral cortex fissure and fold? A review of determinants of gyri and sulci. In *Cerebral Cortex*, Vol. 8B, edited by E. G. Jones and A. Peters, pp. 1-132. Plenum Press, New York.

Wilson, E. O.
   1978 *On Human Nature.* Harvard University Press, Cambridge, MA.

Yellin, A. M. & Jerison, H. J.
   1980 Photically evoked potentials and afterpotentials recorded from the visual cortex of unanesthetized hedgehog. *Brain Research* 182:79-94.

Zeki, S.
   1993 *A Vision of the Brain.* Blackwell, London.

7. Intellectual Surplusage: The Role of Bipedalism and Neonatal Head Trauma

Sean C. Hogan and Gordon G. Gallup, Jr.

Abstract

Since early *Homo*, human evolution has been associated with a continued increase in brain size. This feature of human evolution has influenced several morphological features of human infants that are relatively unique when compared to other great apes and monkeys. Although the human infant's poorly ossified skull permits the delivery of a large brained neonate and accommodates a rapidly expanding brain during postnatal development, it has the disadvantage of increasing the risk of birth related and postnatal induced brain damage. We suggest that a larger adult brain may have been selected for in *Homo* to compensate, in part, for early incidents of brain trauma. The hypothesis is supported by research in a number of areas including child neurology, primatology, and obstetrics.

Intellectual Surplusage: The Role of Bipedalism and Neonatal Head Trauma

Our ability to think in abstraction is unparalleled in other species. We use the same brain to perform complicated mental feats such as mathematical reasoning and building spacecraft, as our hunter gatherer ancestors used for millennia to fabricate primitive tools, hunt and procure food. The basic question that arises from such comparisons is how it is that humans have come to possess a "surplus" (Boice 1977) of mental capacities that appear to be unrelated to the selective pressures encountered during our evolutionary history. Stated another way, what mechanism could favor the evolution of abilities not directly necessary for our day to day adaptation and survival?

The evolution of the human brain has been characterized by an overall increase in size that has occurred throughout the last several million years (Leigh 1992). With a brain of approximately 1400 cm^3, modern humans possess a brain that is more than twice the size of the mean cranial capacity of *Homo habilis*, which is reported to have been approximately 650 cm^3 (Falk 1992). Some of the factors suggested as prime movers for the development of a larger, more complex brain include the development of tool use (Washburn 1960), climactic versatility (Calvin 1991), and selection for greater intelligence (i.e. language acquisition) (Falk 1992).

Evidence from the fossil record, including that of early *Homo* shows that there has been an increase in bipedal efficiency coupled with an overall increase in brain size (Washburn 1960). These two factors may well have had, and continue to have, important implications for both human parturition and the skull morphology of the human infant. Together, they have resulted in a situation in which, as compared to monkeys and other great apes, infantile head/brain injury was very likely both during and after parturition.

Bipedalism as an exclusive means of locomotion is a defining characteristic of humans. Although some primates, such as *Pan paniscus*, can walk bipedally for short spans of time (Abitbol 1993), they have not undergone the morphological changes that would allow them to use it as a primary means of locomotion. The restructuring of the pelvis has had important implications not only on the posture and locomotive abilities of the mother, but also on the infant, since it must pass through this altered pelvis during birth.

Skull development in the human infant has been modified considerably to accommodate the pelvic changes that have occurred to support the development of bipedalism. The increasingly restrictive size of the obstetrical canal has had two important effects. It has placed an upper limit on the size of the head at birth (Trevathan 1987) and secondly, the close fit between the skull and the pelvis has led to the evolution of a biomechanical design which can accommodate varying degrees of compression (Rosenberg 1992). Essentially, the benefit of greater efficiency in bipedalism, and the resulting vulnerability of the infant skull, has led to a situation in which there is an increased vulnerability to brain damage during parturition, and especially during the first two years after birth.

We hypothesize that the ever-present risk of infant head/brain trauma in evolutionary history may have favored the production of humans who would eventually attain a greater adult brain volume than was actually necessary for normal functioning. Given the relatively high risk of early head trauma, infants would have been more likely to survive into adulthood if they had enough residual, compensatory, undamaged brain tissue to deal with the demands of adaptation and survival. Therefore, we suggest that infants with a larger potential for compensatory growth in overall brain size following birth were selected for. In support of this position, humans experience nearly a fourfold increase in postnatal brain size, from approximately 400 cm^3 in infancy to 1400 cm^3 in adulthood (Leutenegger 1972). Other primates do not share this same dramatic increase. An infant chimpanzee has a brain volume of approximately 155 cm^3 (Leutenegger 1972) and an adult has a volume of nearly 394 cm^3 (Tobias 1971). In other words, individuals with sufficient compensatory neural tissue would be less likely to suffer cognitive deficits due to the prevalence of early head injuries.

In our view, the development of a larger brain provides a biological mechanism which can better accommodate the injuries incurred during birth and infancy. Throughout hominid evolution, the constant threat of brain injury due to both the birth process and the early postnatal environment, would have created selective pressure for generally larger adult brains. Perhaps, in part because of this surplus of neural tissue and a reduction in early head trauma with the emergence of civilization, *Homo sapiens* and earlier hominids became the dominant species of primate on the planet. A number of recent studies indicate that the size of the adult brain bears a significant and very substantial correlation to standardized measures of intelligence (Willerman et al. 1991; Andreasen et al. 1993).

Structures

Although humans, chimpanzees, and other great apes share a common ancestor, humans differ in several significant ways due to their evolved reliance on bipedal locomotion (Spuhler 1988). These differences include elongated hindlimbs, an altered pelvis for better upper body support and leverage, altered feet for more stable support, and shortened toes (Rosenberg 1992). These changes make it possible for humans to remain in an upright posture for long periods of time.

The pelves of all great apes (see Figure 1) are in many respects quite similar. In both humans and other apes, the pelvis is composed of two hip bones (each comprised of the ischium, the ilium, and the pubis), the sacrum, and the coccyx. Together these bones create a fairly rigid basin for abdominal support, muscle attachment, as well as an obstetrical outlet (Tague and Lovejoy 1986).

However, the size of the pelvic outlet in the chimpanzee and other great apes (see Figure 2) is not as constrained as it is for humans since they do not depend on the pelvis for support of the upper body (Leuteneggar 1982). For example, the human ilium is much shorter, broader, and expanded front to back, and the ischium is also shorter and broader. These differences contribute to our ability to remain in an upright posture. The enlargement in the transverse dimension allows for a wider stance in erect posture and places the ischia and related musculature in a better position for functioning bipedally (Rosenberg 1992). The entire pelvic brim is inclined toward the sacrum, resulting in improved transmission of weight from the spine to the legs. Additionally, the sacrum is directly opposite the pubic symphysis, contributing further to a smaller pelvic inlet (Trevathan 1987).

The Passenger

As detailed by Oxhorn (1985), the base of the human infant's skull is composed of ossified, firmly united, and non-compressible bone that protects the vulnerable areas of the hindbrain responsible for basic homeostatic functions. However, the cranium, or cranial vault, is composed of several poorly ossified bones connected by elastic membrane. As shown in Figure 3, these include the two parietal bones, and the two temporal and the two frontal bones (located anteriorly). At birth, these bones slide loosely over one another when exposed to pressure and provide the required structural flexibility so that the head of the neonate can be molded as it passes through the bony constraints of the pelvic canal.

Cranial molding in infants is possible due to the presence of two types of soft tissue areas, referred to as suture zones and fontanelles. Suture

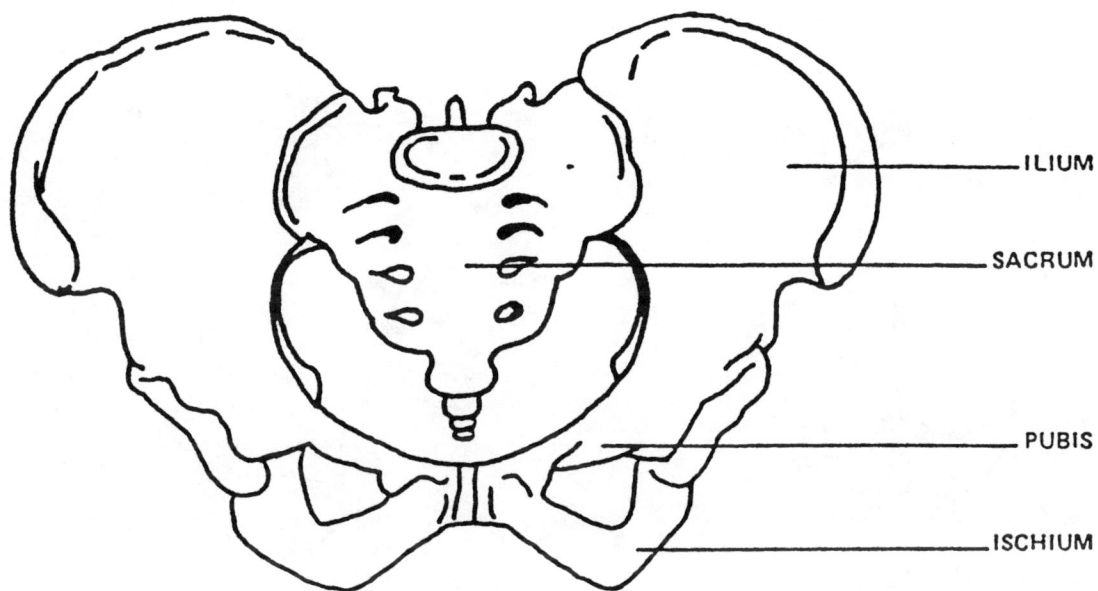

Fig. 1. The human pelvis (Trevathan 1987:18).

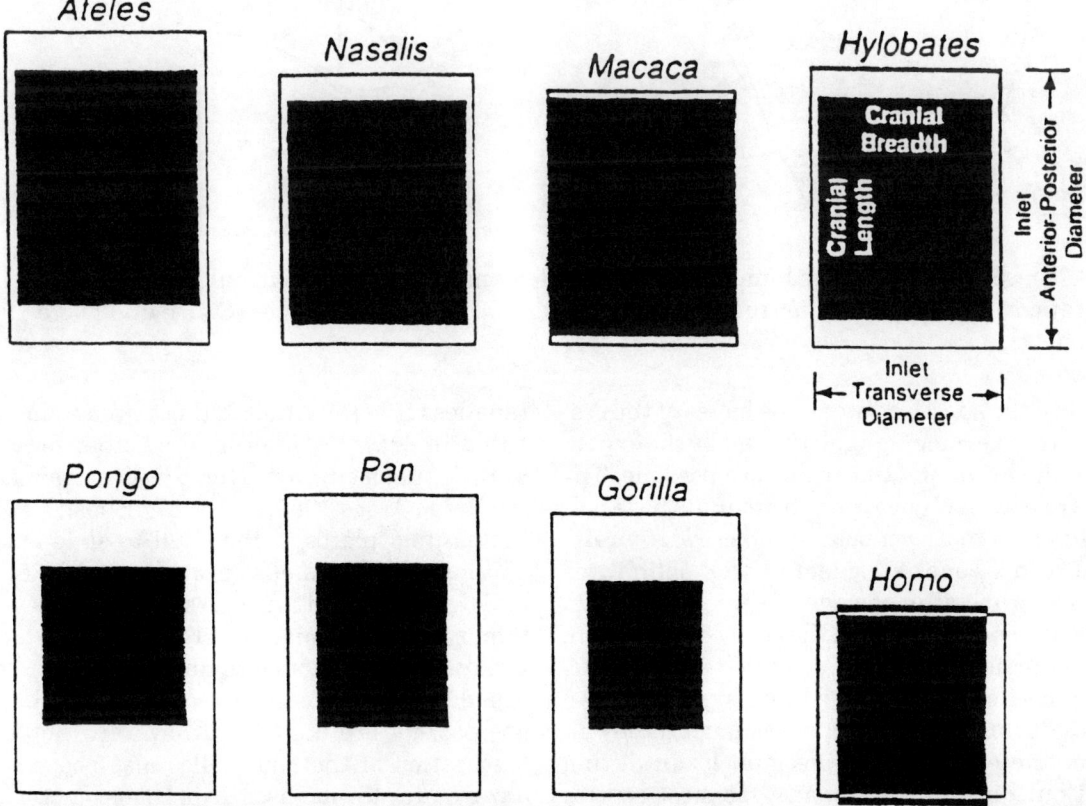

Fig. 2. The above figures depict the relationship between the size of the female pelvis and the head of the average infant born to several representative primate species. As shown, the dimensions of the human infant cranium can actually exceed those of the maternal pelvis along some dimensions (Rosenberg 1992:96).

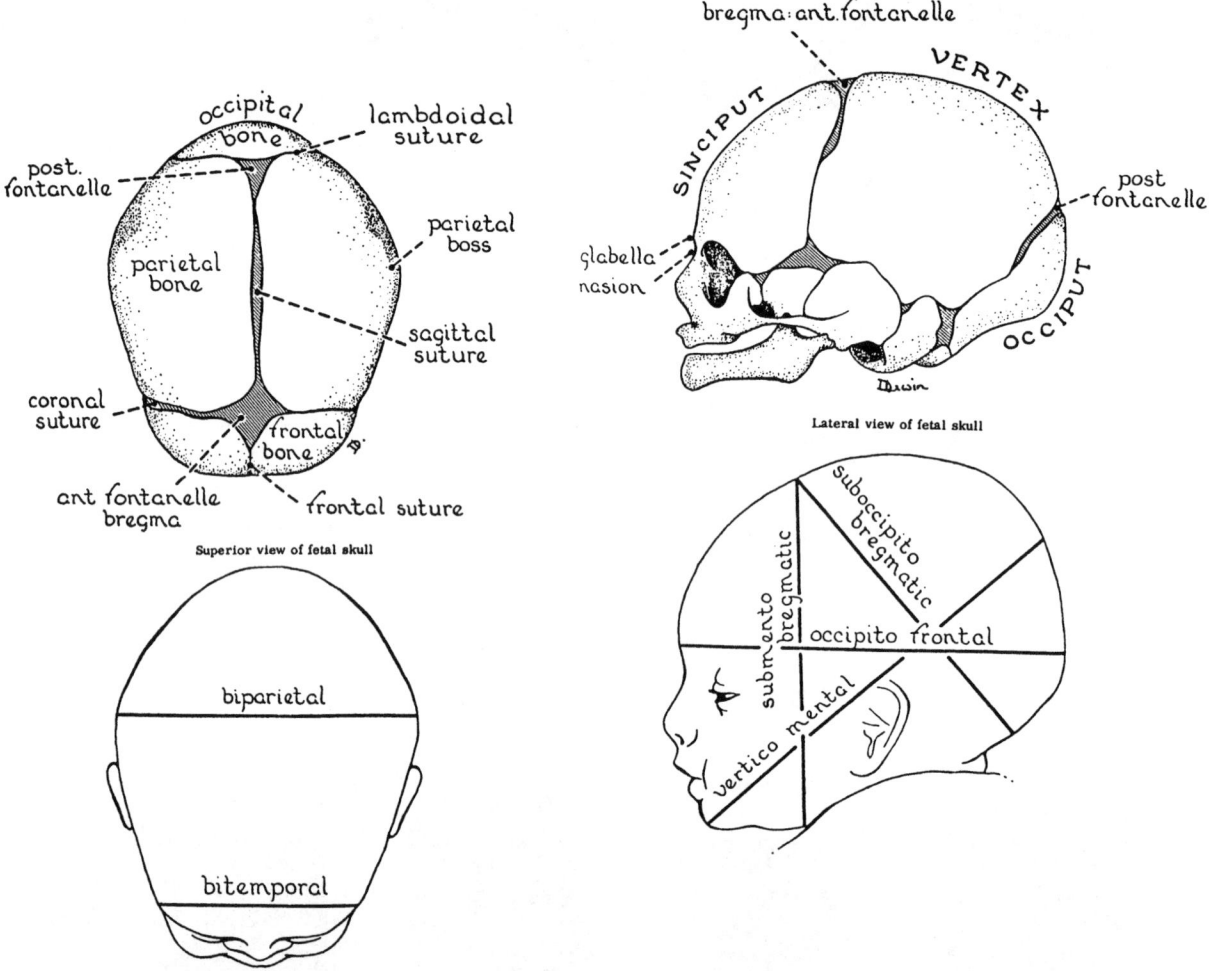

Fig. 3. These diagrams of the human infant skull demonstrate the relative vulnerability of the infant brain. Notice, in particular, the large fontanelles and open suture zones (Oxorn and Foote 1968:39 and 41).

zones are the spaces between the bones of the cranium on the developing skull that make the expansion of the brain and the brain case possible. The chief areas of new bone growth occur at these borders in order to accommodate the growing brain. These suture zones are generally not obliterated until sometime after the age of three (Gill 1987). The relatively large gaps on the surface of the skull, referred to as fontanelles, are filled with fibrous connective tissue. Possessed by many mammalian species during normal fetal development, they allow for the expansion of the fetal brain during gestation. But, unlike humans, for most species these structures are completely closed prior to birth.

For humans the fontanelles are very prominent at birth. In fact there are a number of these open spaces located across the cranium, the largest being the anterior (2-4 cm) and posterior fontanelles (1-3 cm). At birth these areas can be palpated in order to determine whether or not the brain is subject to any adverse internal pressure or swelling (Oxorn 1985; Gill 1987). Since they allow the ossified plates of the skull to slide over one another, the fontanelles provide much of the cranial flexibility that is necessary for the head of the infant to pass through the restricted birth canal. Because they do not close until nearly the second year of life, their presence also makes possible rapid postpartum brain growth. However, the delayed maturation of the fontanelles also poses certain hazards due to increases in infant locomotor development. Infants between 8 and 14 months of age spend a great deal of time falling and bumping into objects during repeated attempts to become proficient at walking and moving about their environment (Craft 1975). This overlap in development makes the infant particularly vulnerable to head injury during this period.

The neonatal skulls of the other great apes possess tiny fontanelles relative to those found in humans (Rosenberg 1992) since the neonates of chimpanzees, gorillas, orangutans, and bonobos are not exposed to the extreme molding pressures that result from a bipedal pelvis. The fact that all other primates are neither bipedal nor possess large, late closing fontanelles, suggests that much of human infant cranial morphology, and the overall size of the adult brain may have been influenced by the refinement of bipedalism. Thus human evolution, with it's premium on brain size, has exploited a strategy of having infants born with immature brains and relatively imperfectly formed skulls, which not only help to accommodate a difficult birth process, but provide expansion to allow for considerable postnatal brain growth and development.

The Big Squeeze

Gestation times are remarkably similar across great apes. For Pongidae (including chimpanzees, orangutans, and gorillas) the average is approximately 250 days, while for humans it is 267 days (Harvey and Clutton-Brock 1985). However, humans differ dramatically from other great apes in terms of their level of development at birth. Comparative data show that we take nearly twice as long to reach the same levels of neural and motor development as chimpanzees (Gould 1977). This extended period of relative immaturity is sometimes referred to as exterogestation, meaning that in some ways humans infants have a period of gestation that continues outside the womb (Montagu 1961). The immaturity of the human infant, in addition to its lack of skull development, makes it especially vulnerable to postpartum brain trauma.

A constraint for all great apes, including humans, is that the neonate skull must be small enough to make its way through the obstetrical canal. In cases where the skull is larger, the death of the infant and/or the mother may result. For the most part, this type of danger is rare among chimpanzees, gorillas, and orangutans because (as shown in Figure 2) the cephalo-pelvic ratios in these animals are fairly large (Leuteneggar 1982). As a result there are very few accounts reporting difficulties in parturition for any of these species. Field studies indicate that these animals give birth relatively quickly in comparison to humans (Beck 1984;

Galdikas 1982). In most cases, apes wander off on their own to give birth and, unlike humans, receive no assistance from conspecifics.

Human females experience an intense, prolonged birth process which can endanger the lives of both mother and child. Labor usually lasts many hours, and can lead to complications rarely seen in other large bodied primates.[1] For these reasons most human births involve the presence of birth attendants (Trevathan 1988).

The human female pelvis is different from other great apes in that the pelvis is widest in the transverse dimension at the pelvic inlet, and widest in the sagittal dimension at the pelvic mid-plane. The skull of the human infant is most narrow in the sagittal dimension. Therefore, in order to make its way through the canal a human infant must go through a series of rotations until it finally emerges from the mother in the occiput-anterior position (Zlatnik 1990).

As shown in Figure 2, the human pelvic canal is also quite different from other great apes in that its size is nearly equal to that of the neonate skull (Oxorn 1985). Along some dimensions the neonate skull actually exceeds the diameter of the maternal pelvis. In fact, they are so similar that humans possess one of the lowest cephalo-pelvic ratios found in any primate (Trevathan 1987). This has led to the necessity of structural modifications to the infant skull such as relatively large late closing fontanelles.

Problems and Injuries Associated with Birth

Two of the more serious outcomes associated with difficult human births are asphyxia and birth trauma, which have been reported to be the leading causes of mental handicaps in developing countries. It is estimated that 3.7 million newborns (or 2.6% of all global births as of 1989) suffer from the effects of moderate or severe asphyxia which often lead to brain damage. In fact, 1.2 million (or 0.8% of all global births as of 1989) infants die each year due to these types of injuries (Shah 1991).

There are many factors that contribute to the difficulty and danger that characterize the process of human birth. Some of the more common problems include various types of breech presentation, cephalo-pelvic disproportion, general abnormalities of the fetus, as well as excessive molding of the skull (Steer and Danielian 1994). All of these increase the risk of head trauma or anoxia, or both, in the neonate. While many of the difficulties can be avoided, or minimized through medical intervention today, they most certainly had important roles in natural selection during the past several million years of human evolution.

Breech presentations differ from normal births because the fetal pelvis, and not the head, enters the canal first. The specific orientation of the pelvis varies across births. Factors which can lead to a breech birth include prematurity, excess amniotic fluid, multiple pregnancy, a contracted pelvis, and an overly large fetus. There are a number of breech presentations possible. Many of these presentations can lead to problems such as fetal asphyxia, aspiration of the amniotic fluid, and prolapse of the umbilical cord (Oxorn, 1985).

Feto-pelvic disproportion is another fairly serious problem associated with human birth (Steer and Danielian 1994). There are generally two types of disproportion. The first is absolute, and refers to the inability of the fetus to pass safely through the birth canal. In the second there is relative disproportion, which involves other factors that contribute to the disproportion. These include pelvic size, pelvic shape, and abnormalities of the soft tissue areas. Pelvic contraction is the most common and means that the area is smaller than normal. The constriction can occur at the inlet, the mid-pelvic area, or at the outlet. The particular ramifications for the fetus resulting from a contracted pelvis depend on which area of the canal is affected.

Another factor is abnormality of the soft tissue areas (Oxorn 1985). In some females, normal dilation of the membranes is not possible and makes the passage of the fetus difficult. For the most part, problems with the soft tissue areas only become critical when they are observed in conjunction with other problems such as android pelves or breech presentations. Characteristics of the infant that influence disproportion include the size of the fetus, its attitude and position, and the moldability of the skull. An excessively large human fetus is defined as weighing over 4,500 grams (9 pounds, 15 ounces). For infants in this category there is an increased chance of mortality, a higher incidence of severe injury, skull fracture, and brain damage. (Pritchard et al. 1985)

Finally there is the issue of skull moldability. Although fetal cranial molding is an important adaptive response to a tight fit in the birth canal, excessive molding is linked to a number of brain disorders in infants, including cerebral trauma, mental retardation, cerebral palsy and other forms of brain damage (McPherson and Kriewall 1980a, 1980b). In a normal birth there is a flexion of the fontanelles in response to the maternal pelvic canal. However, a reduction of more than 0.5 cm in the biparietal diameter is considered dangerous and can lead to laceration of the underlying brain tissues (Craft 1975).

Over the course of the last several decades many of the problems associated with birth have been avoided through cesarean section, which involves the delivery of a fetus weighing 500 grams or more by abdominal surgery that requires an incision through the wall of the uterus. Therefore, both mother and infant are able to avoid many difficulties associated with the impending vaginal birth. Rates for cesarean section increased dramatically during the 1970's and 1980's (Flamm and Quilligan 1995). For example, in 1970, the incidence of cesarean section in the US was approximately 5%. However, by the late 1980's the rate had risen to between 20-25%, so that nearly one in four infants was born through surgical intervention (Bureau of Biostatistics 1987; Dunn 1990). There are basically three reasons for this remarkable change: 1) an increase in the number of fetal problems believed to be better managed by abdominal vs. vaginal delivery, 2) a high incidence of abdominal delivery in women who have previously had cesarean-sections, and 3) a rise in ratio of primigravid to multiparous parturients. Primary cesarean section is approximately nine times more common in primigravids than in multiparous females.

Head Injury and Recovery in the Infant

Although human birth involves a molding process that can lead to injury of the underlying brain tissues, the risk of head injury is even higher after birth. Head trauma is a common cause of morbidity and mortality among children in the United States (Rosenthal and Bergman 1989). Although most of the nearly 5 million pediatric head injury cases are minor, approximately 200,000 children are hospitalized each year with head trauma, and 15,000 require prolonged hospitalization (Di Scala et al. 1991).

Being less than 2 years of age has been identified as a risk factor, in and of itself, for significant head injury. This increased risk is related to the developmental immaturity of the skull. The infant skull can be characterized as a very pliant collection of loosely connected semi-ossified bones held together by connective tissue. The underlying brain tissue is characterized by smooth fossae, unmyelinated fibers, and shallow subarachnoid spaces.

Such structural immaturity makes the infant brain especially susceptible to some injuries (Ward 1995). Because of this vulnerability pediatric patients are often evaluated with specialized care. For example, while the presence of a cephalohematoma (bump on the head) is not a serious concern in older children and adults, for an infant it often predicts underlying tissue injury (Pietrzak et al. 1991).

Considered as a whole, the incompletely fused bones of the infant skull confer both advantages and disadvantages as a form of protection against injury compared to the rigid skull of the adult. For instance, in the case of a blow to the head, injury due to the acceleration of the brain within the skull is reduced by the elasticity of the bones which serve to absorb part of the force generated on impact. However, this same flexibility of the skull with underdeveloped bony buttresses at the base, combined with shallower convolutions on the surface, leads to a much greater deformation of the brain on impact. Further, shearing forces are set up within the brain, which may lead to secondary degeneration of nerve fibers which have been stretched or torn (Craft 1975; Ward 1995).

The primary injuries sustained by the infant brain are contusions and lacerations. Contusions occur when the brain comes into contact with the walls or bony protuberances of the skull. After healing, gliosis produces scar tissue in the injured areas. Small lacerations or cuts are more common in the traumatized infant. They occur in the white matter and superficial cortical layers (Ward 1995).

The eventual outcome of head injury in children depends not only on the severity of the trauma, but also on the age at which the incident occurs, and the time of the recovery. Recent studies have indicated that certain injuries are more damaging to overall cognitive development in very young children or infants, as compared with older children and adults (Kolb 1989). In addition, the infant brain, at this point in development, is extremely vulnerable to other types of injuries such as traumatic, infective, or hypoxic (Craft 1975).

Neurological plasticity refers to the ability of the brain to recover from traumatic injury and other parts of the brain to take over functions of damaged areas, such that normal motor and cognitive functioning are not impaired. Recovery of function was first demonstrated earlier this century with lower primates (Kennard 1936, 1938) and again in more recent years with humans through the use of clinical studies (Klonoff et al.1971). However, recent evidence has shown that the specific course of recovery can be affected by any number of factors.

These include the nature of the injury, the age of the brain at injury, handedness, the size of the brain, as well as other modulating factors in the environment. We contend that the limitations of the infant brain to recover from some types of injuries, in combination with the increased vulnerability of the relatively undeveloped skull, may have provided the selective pressure for an overall increase in **post-natal** brain growth to compensate for these eventualities.

Surplusage

Human infants are born with unusually flexible skulls that permit the passage of a large brained neonate through a narrow obstetrical canal. As with any evolved system, there are advantages and disadvantages associated with this adaptation. We contend that the refinement of bipedalism, which has continued throughout the evolution of the *Homo* lineage, has resulted in an increased risk of early childhood brain damage.

Over the last 2 million years, beginning with early *Homo*, the ever-present risk of postnatal, as well as birth induced head trauma and brain damage may have led to selective pressure for the postnatal elaboration of extra brain tissue to compensate for neonatal loss. This surplus tissue would have allowed for the recovery of functions that might otherwise have been lost, and is supported by the fact that the human brain is somewhat limited in its ability to recover from early traumatic brain injury. Consistent with our hypothesis, plasticity appears to be more limited in infants and young children, than it is in older children and adults. Thus the development of surplus cortical tissue extending beyond the critical first two years of postnatal development, may have evolved in humans as a unique neural recovery mechanism for the relatively common occurrence of infantile head trauma. Under modern conditions, in which early brain damage is less likely, a side-effect of this adaptation would be a surplus of neural tissue and a corresponding intellectual capacity that has surpassed what would have been typically attained under previous conditions.

As further support for the impact that our peculiar vulnerability to early brain damage may have had on the evolution of various design features of the human brain, consider the cerebral cortex. In addition to differences in overall brain size, one of the most striking changes in the appearance of the mammalian brain as one moves from rodents to monkeys and to humans is that the cortex becomes more convoluted. In the case of humans, the brain is so folded that much of the surface of the cortex is buried in between these folds or what are called sulci. Markowitsch and Tulving (1995) have concluded, on the basis of a review of the evidence concerning cortical folding in humans, that areas at the bottom of these sulci, referred to as fundi, have special functional significance. In light of diverse evidence including results derived from positron emission tomography, they theorize that sulcal and especially fundal regions of the cortex are more likely to be involved in higher mental processes and cognitive functions.

An obvious extension of the Markowitsch and Tulving hypothesis, based on our account of surplusage, would be to suggest that this feature of human cortical architecture may be a byproduct of natural selection wrought by early head injury. Simply stated, those areas of the cortex buried deep within these sulci would be less vulnerable to head trauma. Many of the most common head injuries incurred by infants and young children, for example, are surface haematomas. Thus, to the extent that human evolution put a premium on cognitive and intellectual functioning, the ever-present threat of head injury early in life could have favored selection for organizational changes that further embedded or recessed these areas deep within sulcal and fundal zones thereby rendering them less susceptible to damage.

It seems reasonable to suppose that the primitive, and at the same time highly mobile lifestyle of our ancestors, may have exposed infants to a substantial risk of environmentally induced head trauma as compared to the infants of modern day. Nowadays, people have gone to great lengths to provide infants with protective measures to keep them from being injured (i.e., cribs, pillows, baby seats, and carriages). But even with all of these safety devices in place, children are occasionally dropped or take injurious blows to the head. For example, blunt head injury continues to be a leading cause of morbidity and mortality among infants that have sustained trauma in the United States today (Rosenthal and Bergman 1989).

If our account of this feature of human brain evolution is correct then several testable predictions should be upheld. One of the most powerful means of assessing our position would involve a post-mortem comparison of the brains of individuals born and raised in primitive, pre-agricultural societies with the brains of infants born and raised in industrialized countries. We would predict significantly more evidence of early brain damage in pre-agricultural populations due to their lifestyle and lack of devices designed to protect vulnerable infants. This type of evidence would help demonstrate that post-natal head trauma, although a fairly common occurrence even for modern human societies, may serve as the general rule in those that are less technologically advanced. Additionally, this type of examination would provide important information concerning the relative location and consequences of various injuries incurred by all infants regardless of cultural differences.

Another test of this hypothesis would involve a long-term follow-up comparison of various methods of birth. The remarkably high levels of cesarean section in the United States provide a population of infants with which vaginally delivered infants could be easily compared. Such a comparison could provide useful insights into the effects that birth experience, and more specifically the obstetrical canal, has on the neonatal brain. Longitudinal studies of cesarean vs. vaginal delivered children could provide information on the cognitive correlates of delivery method. Although it would be expected that neural recovery would occur in both cesarean as well as vaginal births, it is anticipated that vaginal births may incur higher risks due to cranial compression by the birth canal. In the case of planned cesarean sections (as opposed to those that are impromptu, in which the incidence of other complications would be much higher) our model would predict a higher level of overall intellectual functioning in later life, due to a lower incidence of birth related brain trauma, as compared to children born vaginally.

Finally, it is important to acknowledge an alternative and more parsimonious account of intellectual surplusage. Whereas we have postulated an evolved increase in the potential for postnatal brain growth in humans to compensate for the high probability of incurring brain damage both during partuition and in the first postnatal year of life, there is another possibility. If, as we have argued, one simply assumes that early infant neurological trauma has been dramatically reduced in recent times as a consequence of technological developments, then it follows that people today probably sustain less brain damage than used to be the case. Thus, the contemporary ability to excel at intellectual activities, which at best seem remotely related to the adaptive features of life during most of human evolutionary history, might simply be a byproduct of the fact that we now have less damaged and therefore more functional brain tissue.

In other words, rather than being a byproduct a compensatory increase in brain mass coupled with a reduction in early head trauma, surplusage may simply be a consequence (or, if you like, artifact) of a relatively recent species-specific reduction in brain damage in early life.

Note

[1] A notable exception to this rule of relatively easy non-human primate parturition is the squirrel monkey. Squirrel monkeys experience prolonged labor similar to that found in human females. They also experience a fairly high rates of infant mortality due to difficulties in birth (Trevathan 1987). This is primarily due to the close fit between the pelvic outlet of the female squirrel monkey and her infant's head. It is for this reason that these monkeys may serve as the best non-human primate models for studies of human birth difficulties.

References Cited

Abitbol, M. M.
   1993 Quadrapedalism and acquisition of bipedalism in human children. *Gait and Posture* 1:189-195.

Andreasen, N. C., M. Flaum, V. Swayze II, D. S. O'Leary, R. Alliger, G. Cohen, J. Ehrhardt, and W. T. C. Yuh
   1993 Intelligence and brain structure in normal individuals. *American Journal of Psychiatry* 150:130-134.

Beck, B. B.
   1984 The birth of a lowland Gorilla in captivity. *Primates* 25:378-383.

Boice, R.
   1977 Surplusage. *Bulletin of the Psychonomic Society* 9:452-454.

Bureau of Biostatistics
   1987 *Cesarean Childbirth in New York State: A Statistical Profile*. New York State Department of Health, New York.

Calvin, W. H.
   1991 *The Ascent of Mind: Ice Age Climate and the Evolution of Intelligence*. Bantam Books, New York.

Craft, A. W.
   1975 Head injury in children. In *Handbook of Clinical Neurology*, vol. 23, edited by P. J. Vinken and G. W. Bruyn, pp. 445-458. American Elsevier, New York.

Di Scala, C., J. S. Osberg, B. M. Gans, L. J. Chin and C. C. Grant
   1991 Children with traumatic head injury: morbidity and postacute treatment. *Archives of Physical Medicine and Rehabilitation* 72:662-666.

Dunn, L. J.
   1990 Cesarean section and other obstetric operations. In *Danforth's Operations in Obstetrics & Gynecology*, 6th ed., edited by J. R. Scott, P. J. DiSaia, C. B. Hammond and W. N. Spellacy, pp. 639-658. J. B. Lippincott, Philadelphia.

Falk, D.
   1992 *Braindance*. Henry Holt and Co., New York.

Flamm, B. L. and E. J. Quilligan
   1995 *Cesarean Section: Guidelines for Appropriate Utilization*. Springer-Verlag, New York.

Galdikas, B. M. F.
   1982 Wild orangutan birth at Tanjung Puting Reserve. *Primates* 23:500-510.

Gill, W. L.
   1987 Essentials of normal newborn assessment and care. In *Current Obstetric and Gynecologic Diagnosis and Treatment*, edited by M. L. Pernell and R. C. Benson, pp. 204-215. Appleton & Lange, East Norwalk.

Gould, S. J.
   1977 *Ontogeny and Phylogeny*. Harvard University Press, Cambridge, MA.

Harvey, P. H., and T. H. Clutton-Brock
   1985 Life history variation in primates. *Evolution* 39:559-581.

Kennard, M. A.
   1936 Age and other factors in motor recovery from precentral lesions in monkeys. *American Journal of Physiology* 115:138-146.

   1938 Reorganization of motor functions in the cerebral cortex of monkeys deprived of motor and premotor areas in infancy. *Journal of Neurophysiology* 1:477-496.

Klonoff, H., M. D. Low, M. D. and C. Clark.
   1971 Head injuries in children: a prospective five year follow-up. *Journal of Neurology and Neurosurgical Psychiatry* 40:1211-1219.

Kolb, B.
   1989 Brain development, plasticity, and behavior. *American Psychologist* 44:1203-1212.

Leigh, S. R.
  1992 Cranial capacity evolution in *Homo erectus* and early *Homo sapiens*. *American Journal of Physical Anthropology* 87:1-13.

Leuteneggar, W.
  1972 Newborn size and pelvic dimensions of *Australopithecus*. *Nature* 240:568-569.
  1982 Encephalization and obstetrics in primates with particular reference to human evolution. In *Primate Brain Evolution*, edited by E. Armstrong and D. Falk, pp. 85-89. Plenum, New York.

Markowitsh, H. J. and E. Tulving
  1995 Cognitive processes and cerebral cortical fundi. *Neuroreport* 6:413-418.

McPherson, G. K., and T. J. Kriewall
  1980a Fetal head molding: An investigation utilizing a finite element model of the fetal parietal bone. *Journal of Biomechanics* 13:9-16.
  1980b The elastic modulus of fetal cranial bone. A first step towards an understanding of the biomechanics of fetal head molding. *Journal of Biomechanics* 13:9-16.

Montagu, A.
  1961 Neonatal and infant immaturity in man. *Journal of the American Medical Association* 178:56-57.

Oxorn, H. and W. R. Foote
  1968 *Human Labor and Birth*, 2nd ed. Appleton-Century-Croft, New York.

Oxorn, H.
  1985 *Oxorn-Foote Human Labor and Birth*, 5th ed. Appleton and Lange, New York.

Pietrzak, M., A. Jagoda, and L. Brown
  1991 Evaluation of minor head injury in children younger than two years. *American Journal of Emergency Medicine* 9:153-156.

Pritchard, J. A., P. C MacDonald and N. F. Grant
  1985 *William's Obstetrics*, 17th ed. Appleton-Century-Croft, Norwalk, CT.

Rosenberg, K. R.
  1992 The evolution of modern human childbirth. *Yearbook of Physical Anthropology* 35:89-124.

Rosenthal, B. W., and I. Bergman
  1989 Intracranial injury after moderate head trauma in children. *Journal of. Pediatrics* 115:346-350.

Shah, P. M.
  1991 Prevention of mental handicaps in children in primary health care. *Bulletin of the World Health Organization* 69:779-789.

Spuhler, J.
  1988 Evolution of mitochondrial DNA in monkeys, apes, and humans. *Yearbook of. Physical Anthropology* 31:15-48.

Steer, P. J. and P. J. Danielian
  1994 Fetal Distress in Labor. In *High Risk Pregnancy - Management Options*, edited by D. K. James, P. J. Steer, C. P. Weiner and B. Ganik, pp. 1077-1100. W. B. Saunders Co. Ltd., London.

Tague, R. G., and C. O. Lovejoy
  1986 The obstetrical pelvis of A. L. 288-1 (Lucy). *Journal of Human Evolution* 15:237-255.

Tobias, P.
  1971 *The Brain in Hominid Evolution*. Columbia University Press, New York.

Trevathan, W. R.
  1987 *Human Birth: An Evolutionary Perspective*. Aldine De Gruyter, Hawthorne.
  1988 Fetal emergence patterns in evolutionary perspective. *American Anthropologist* 90:674-681.

Ward, J.
  1995 Craniofacial injuries. In *Management of Pediatric Trauma*, edited by W. L. Buntain, pp. 177-188. Saunders, Philadelphia.

Washburn, S. L.
  1960 Tools and human evolution. *Scientific American* 203(3):3-15.

Willerman, L., R. Schultz, J. N. Rutledge, and E. D. Bigler
  1991 In vivo brain size and intelligence. *Intelligence* 15:223-228.

Zlatnik, F. J.
  1990 Normal labor and delivery and its conduct. In *Danforth's Operations in Obstetrics & Gynecology*, 6th ed., edited by J. R. Scott, P. J. DiSaia, C. B. Hammond and W. N. Spellacy, pp. 161-188. J. B. Lippincott, Philadelphia.

# 8. Before or After the Split? Hominid Neural Specializations

Katerina Semendeferi

Abstract

Studies of material culture can provide powerful insights into human cognition and its evolution. Studies of the brain reveal ways in which the organ of the mind regulates cognitive behaviors in humans and other primates. What is the underlying neural circuitry responsible for the incredible complexity of human behavior? Which anatomical features of the human brain are shared with the rest of the hominoids and which ones constitute a novelty? Positive identification of uniquely human neural substrates of behavior can only be accomplished by analyzing extant hominoid comparative material. Studies of the evolution of human cognition based on the archeological and paleoneurological records of the Plio/Pleistocene can build on prior knowledge of species-specific cognitive behaviors and the underlying neural circuits. Tracing the evolution of cognition in the hominid line after the split from the African apes can thus be largely facilitated by contributions from comparative neuroanatomy.

Developments in the application of noninvasive techniques to the study of the human brain have contributed to a revived interest in the evolutionary study of the brain based on comparisons of extant species. An increase in the size of the hominid brain has been accompanied by a reorganization of larger and smaller neural sectors. Differentiation of cortical regions within each sector of the brain, and a richer interconnectivity among those areas might be responsible for the remarkable cognitive differences between humans and other hominoids. Analysis of large sectors of the hominoid brain such as the frontal lobes, temporal lobes, and the cerebellum point to differential sizes for some, but not all, of the sectors of the human brain when compared to the rest of the hominoids. Some of the neural circuits involved in cognitive and emotional processes were present in the African hominoid stock prior to the split of the hominid line, while others constitute real novelties whose expression can be searched for in the archeological record.

Introduction

The reconstruction of the evolution of the human mind depends largely on information on the neural substrates that govern human and nonhuman primate cognition. The evolutionary study of the hominid brain, very much like most other fields of evolutionary anthropology, draws evidence from the fossil record and the comparative anatomy of extant humans, apes and other anthropoids.

The fossil record provides valuable information on the presence or absence of certain features of the brain, as well as on the timing of their onset or offset as detected in endocasts. One such important feature is size. Based on the fossil record we know that australopithecines had a small brain size, similar to that of the great apes, both in absolute and relative (to body size) terms and that a substantial increase in size took place with the early appearance of the genus *Homo*. We also know that later species of *Homo*, like archaics and Neanderthals, had brain sizes close to or within the range of modern humans (Tobias 1995).

Information coming from paleoneurology also concerns features other than size that are detectable on the surface of the endocast of a particular species. These include the position and orientation of major sulci (e.g. lunate sulcus), the presence or absence of gyral and sulcal complexes (e.g. inferior frontal gyrus), the identification of asymmetries (e.g. petalias) and the size of various lobes or large sectors of the brain (e.g. frontal lobe, cerebellum). All of these features are used in attempts to reconstruct patterns of organization that would place a hominid species closer to either humans or great apes in terms of brain structure and subsequently cognitive abilities (Falk 1991; Holloway 1995).

The analysis of a small sample, as provided by the fragmentary fossil record, is in itself a challenging task, but the efforts of paleoneurologists to interpret the hominid endocasts are further constrained by a series of problems related to the features themselves whose significance they try to interpret. Each feature needs to be studied in its proper context in terms of structural and functional inter- and intraspecific variation as evidenced in extant species. This type of information is drawn from the comparative analysis of the extant hominoid brains and recent advances in this field will be presented here.

What do we know about the anatomy of the human and ape brains, the functional significance of the brain features and the inter- and intraspecific variation? It may come as a surprise to the non-specialist that we know very little. The last decades of the twentieth century saw major developments in the field of experimental neuroscience. New techniques assisted in unraveling many of the principles that govern the cortical organization and neural connectivity of the mammalian brain. These mostly invasive techniques were applied on rats, cats, and rhesus monkeys that served as models of the "typical" mammalian brain. Availability of experimental subjects, regardless of species, and the excitement of the potential of the new techniques led to an emphasis on the investigation of the basic common features of the mammalian brain. Possible differences in the neural structure between different species were largely ignored, rather the emphasis was placed on the structures investigated (Kandel and Schwartz 1985).

The direction taken by experimental neuroscience has been useful in the sense that similarities of the primate brain had to be understood and described in order to gain an appreciation of the principles governing the structure of the human brain. Work on the above taxa did indeed provide an enormous amount of information on basic brain structure and function, but how did this emphasis on the homogeneity of the mammalian brain affect the field of evolutionary neuroanatomy?

Anatomists have long been fascinated by questions pertaining to the evolution of the human brain and a large amount of experimentation and analysis took place in the first part of the twentieth century based on a variety of animal species including the great apes. Nevertheless, only few studies in the last decades addressed comparative issues of the organization of the primate brain (Allman 1977; Kaas et al. 1979; Stephan, Frahm and Baron 1981) and discussions on the evolution of the human brain focused for the most part on issues of size (Jerison 1973; Aiello and Dean 1990; Changeux and Chavaillon 1996). The data available included body weight, brain weight and the size of the endocranial space of adult animals. Humans were shown to have the largest absolute brain size and a larger brain than expected for a primate or a hominoid of the human body size.

An important question is whether or not brain size alone can account for all the cognitive differences and the complexities that characterize the human mind. The human brain is three to four times larger than the brain of any of the great apes and more than ten times the size of the brain of the rhesus monkey. Can species-specific cognitive, social and behavioral specializations be attributed exclusively to differences in overall absolute and relative size of the brain? How about the organization of the brain and the concept of mosaic evolution?

An inspection of the gross anatomy of primate brains reveals the presence of a series of common patterns in gyrification, sulcal configurations and lobular subdivisions in each hemisphere. The similarity in primate brains is particularly notable when apes are compared to humans (Figure 1). The brain of the macaque, although much smaller and less gyrencephalic than that of the hominoids, includes all of the main neural structures. Nevertheless despite the common basic organization of the primate brain, a closer examination of even the external gross anatomy of the hominoid brains reveals a mosaic of differences. The brain is a composite of many larger and smaller sectors at macroscopic and microscopic levels. What are the differences in the organization of the brain of humans and apes and when did these differences appear in the hominid line? The question of whether size or reorganization or both are responsible for the cognitive adaptations exhibited by some hominids and by modern humans cannot be adequately addressed unless the above questions are answered.

Non-invasive techniques commonly used in the analysis of human post-mortem brain tissue are now being applied to the brains of ape subjects as well. Lately there is a renewed interest among evolutionary anthropologists trained in the neurosciences to investigate issues of the evolution of the brain through the comparative neuroanatomy of hominoid brains and the results of these recent efforts will be discussed here.

Shared Neural Characters

Macroscopic features

Next to overall brain size, the frontal lobe (Figure 2) of the human brain served as the principal

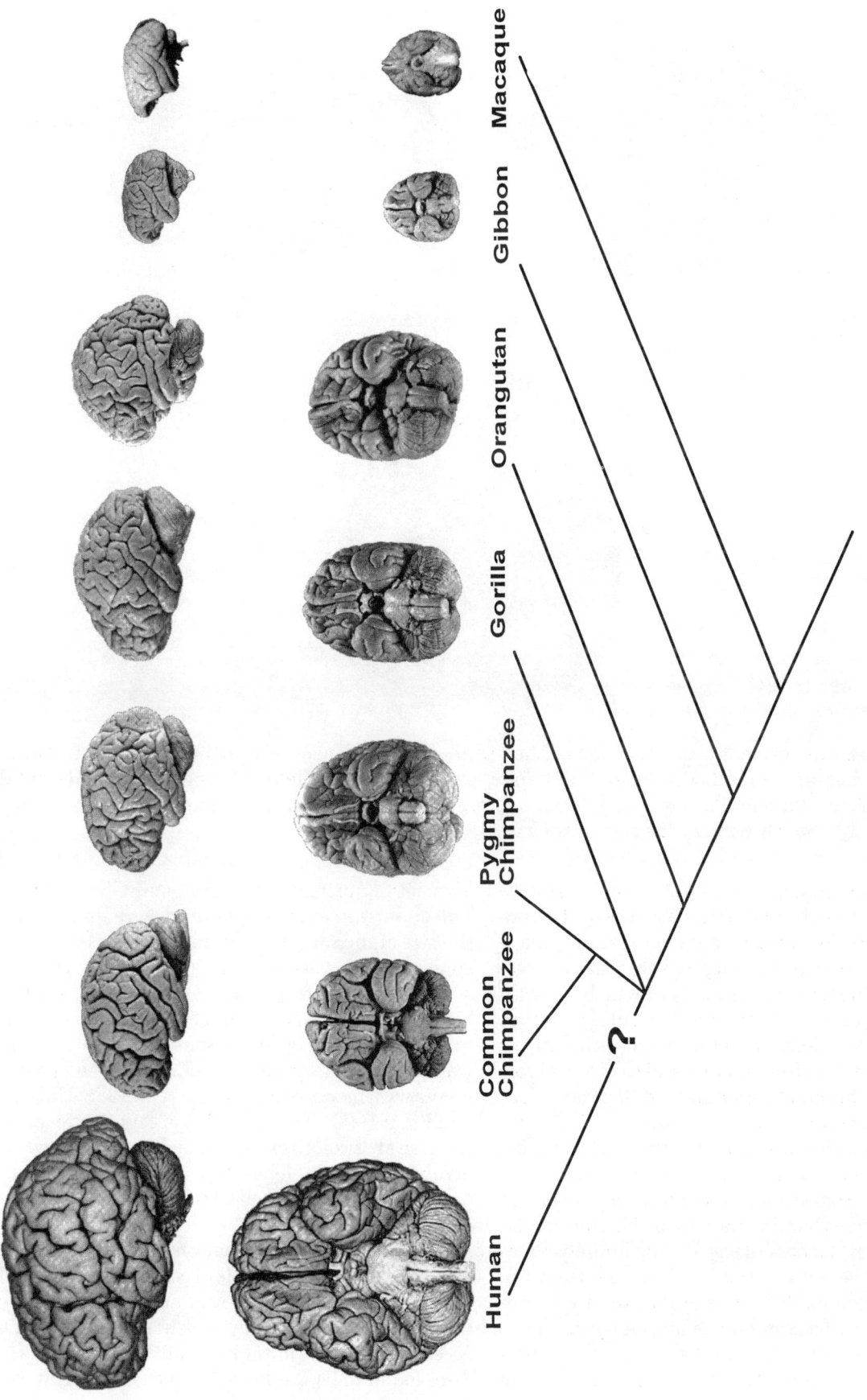

Fig. 1. Photographs of the lateral view of the left hemisphere (top) and the orbital view (bottom) of the brain of postmortem specimens of all extant hominoids and the macaque. Note that the brains, obtained after the natural death of the subjects, are in approximate scale only.

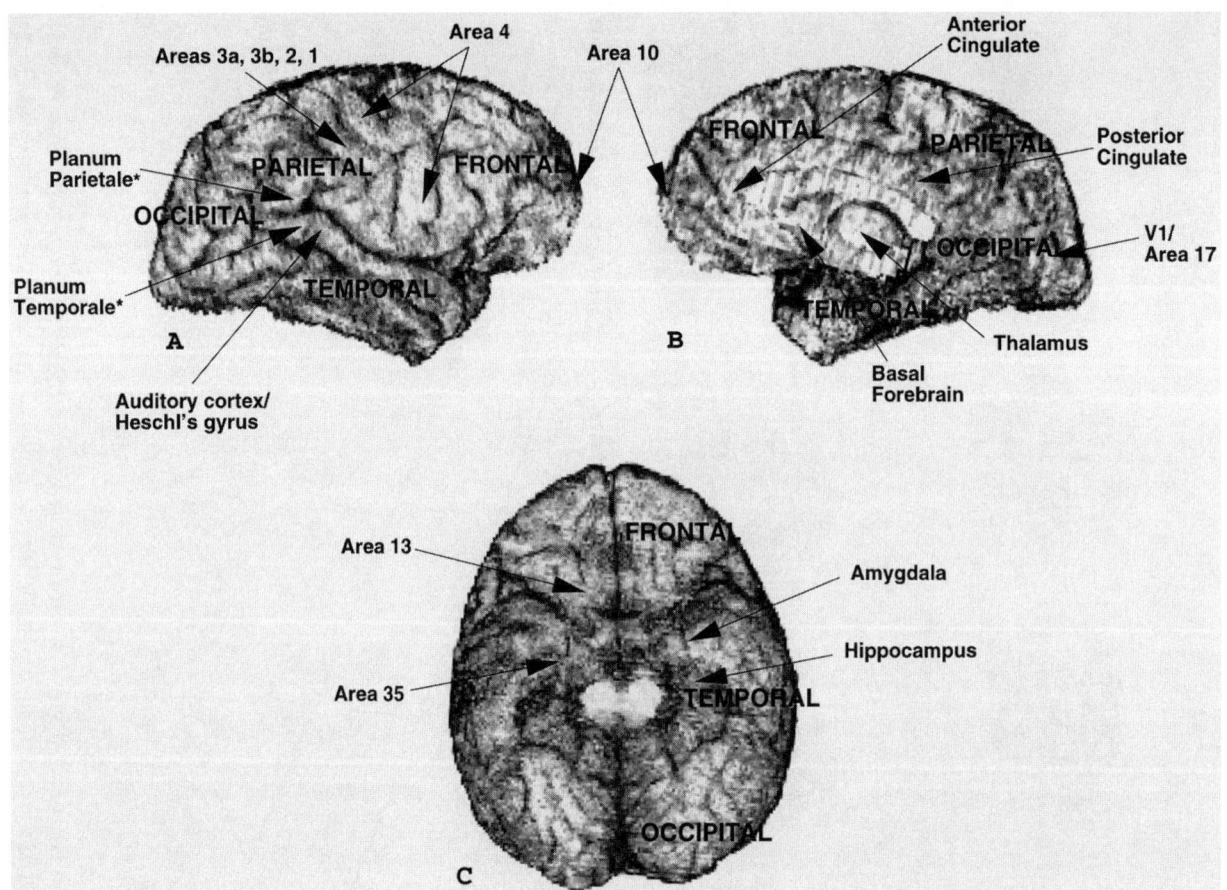

Fig. 2. Three-dimensional reconstruction of the brain of a chimpanzee scanned with Magnetic Resonance imaging. Areas discussed in this chapter are marked on the later (A), mesial (B), and orbital (C) surface. *Planum Parietale and Planum Temporale are not seen on the surface, but are buried within the convolutions in the region shown.

candidate for interpreting most of our cognitive differences with the rest of the natural world. Frontal lobes are involved in decision-making, planning of future actions, language, artistic expression and other functions, many of which are considered to be uniquely human. A popular idea that dominated the literature of the twentieth century has been that the frontal lobes and the frontal cortices of the human brain enlarged disproportionately during human evolution (Deacon 1988; Finger 1994), but few comparative studies have tried to test this notion on original, primary data.

Are the frontal lobes disproportionately larger in our species than in the rest of the hominoids? Did the frontal lobes enlarge during human evolution after the split of the hominid line from the other African apes? Earlier in the twentieth century comparative anatomists attempted to investigate the size of individual lobes and the organization of the gray matter (mostly cortical gray) on a variety of primates (Blinkov and Glezer 1968). The quantitative tools they used were limited and the size of the sample very small (one or two hemispheres from selected primate species). Despite such limitations the results of these early studies became the primary source of information regarding the evolution of the primate brain and rarely was there an attempt to replicate the measurements with modern tools (Zilles et al. 1988; Uhlings and Eden 1990). The evolution of cognitive capacities like abstract thinking and language, known to involve the frontal lobes, have been mostly attributed to an ill-documented disproportionate enlargement (Finger 1994).

The size of the hominoid frontal lobe was investigated in a small sample of post-mortem ape and human brains that were scanned with Magnetic Resonance (MR) imaging (Semendeferi et al. 1997). As expected, the frontal lobes of the human brain were found to be the largest in absolute terms followed by those of the great apes, then the gibbon and the macaque. Contrary to expectation, in relative terms human values do not stand out among the hominoids. As a percentage of the total hemisphere the chimpanzee frontal lobe value falls within the range of the human values and the rest of the apes follow closely. The frontal lobe of humans is not larger than expected of an ape brain of human size.

The cortex of the frontal lobes was further subdivided into three sectors known to have distinct functional attributes: dorsal (manipulation of space, numbers and language), mesial (attention and autonomic mechanisms), and orbital (social behavior). The three sectors of the frontal cortex (dorsal, mesial, orbital) also form three plateaus with the same general distribution of their absolute values as seen in the case of the hemispheres and the frontal lobes. The macaque and gibbon have the smallest values. They are followed by the great apes. Human brains, have the largest values. The only noticeable exception is the small size of the orangutan orbital sector, whose value lies between that of the gibbon and of the other great apes. In all species the orbital sector is the smallest followed by the mesial and the dorsal sectors. The relative values of the three sectors (calculated as a ratio of the volume of the cortex of the frontal lobe and immediately underlying white matter) are quite similar across all hominoids and the only value that stands out is that of the orangutan orbital sector.

The human frontal lobes and their main subdivisions are not larger than expected for an ape brain of human size. Humans and great apes (but not other primates) have overlapping relative values (33-39% of the volume of the hemispheres).

In a more recent study (Semendeferi and Damasio 2000) a larger sample of living human and ape subjects were used for better resolution of the brain scans and to test the preliminary findings of the previous study. Again, humans were shown to have the largest frontal lobes *only* in absolute terms, followed by the great apes and then the gibbons. Any relative measure of the size of the human frontal lobe underscores the fact that human frontal lobes are *not* larger than expected of an ape brain of human size. The size of the frontal lobes as a percentage of the total size of the two hemispheres does not stand out either, and the range of these values in individual human and great ape subjects overlaps. The mean relative values for these taxa lie between 35-37%.

The possibility that it may not be a disproportionate increase in the size of the frontal lobes that distinguishes the human brain from the great and lesser apes, but rather a disproportionate increase in the *cortex* of the frontal lobes is now being explored. Preliminary results show that contrary to traditional wisdom, human frontal cortices are not larger than predicted for an ape brain of human size. The values for the relative size of the frontal cortex as a percentage of total cortex are similar in humans and great apes. The range of relative values of individual subjects overlaps in humans and great apes, but is smaller in lesser apes (Semendeferi et al. n.d.a).

Disproportionately large frontal lobes and frontal cortices are not a hallmark of hominid brain evolution, but are shared features that may have appeared during the Miocene. Smaller sectors of the frontal lobes became specialized during ape and human evolution. One of those sectors is the orbitofrontal sector that appears to have specialized in the line leading to modern-day orangutans. Furthermore, individual cortical areas of the prefrontal cortex seem to have reorganized with respect to size and structural features as will be discussed later.

The parietal and occipital cortices integrate sensory and visual information, and are involved in functions such as visuospatial coordination and attention mechanisms. This sector as a whole, just like the frontal lobe, does not stand out in the human brain, but is as large as expected for an ape brain of human size (Semendeferi and Damasio 2000).

The planum temporale, an area involved in language functions in humans, is located at the posterior part of the temporal lobe buried within the Sylvian fissure. The planum temporale has long been known to present a left-right asymmetry in the human brain favoring the left hemisphere (Geschwind and Levitsky 1968). This asymmetry has been associated with the fact that in humans the left hemisphere is heavily involved with language functions and thus it was assumed to be for the most part a uniquely human feature. This assumption has been partially challenged in the past by studies investigating the length of the Sylvian fissure in ape brains (LeMay and Geschwind 1975; Yeni-Komsian and Benson 1976). The Sylvian fissure was found to end higher on the right hemisphere than on the left in both apes and humans and this asymmetry was correlated with the asymmetry present in the planum temporale.

Recent studies of the planum temporale in human and ape brains have demonstrated that the asymmetry found in human brains is also present in chimpanzees (Gannon et al. 1998), as well as in other great apes (Hopkins et al. 1998). The planum temporale could not be identified in the brains of lesser apes or other anthropoids (Hopkins et al. 1998). The identification of this structure as a common great ape-human feature challenges the idea of a simple relationship between the asymmetry in this part of the brain and language function in humans.

The planum parietale is also involved in language tasks and is located in the ascending limb of the posterior Sylvian fissure and the associated inferior parietal intrasylvian cortex. It was recently shown that this structure presents a similar asymmetry at the gross level in both humans and chimpanzees (Kheck et al. 1998). The planum parietale is larger on the right hemisphere than on the left, an asymmetry that is independent of the left-right asymmetry in the planum temporale. The presence of independent asymmetries in the two plana in both humans and chimpanzees suggests that the asymmetries may have existed at least in the last common ancestor and that other, or more, features have to be responsible for the hominid neural specialization involved in human language.

Right-handed humans exhibit a right frontal/left occipital petalia, which means that the posterior portion of the left hemisphere is wider and protrudes further than the right, while the anterior portion of the right hemisphere is wider and protrudes further than the left. Petalias are also known to be present in the brain of the great apes, but occur to a lesser extent (LeMay 1976 LeMay and Geschwind 1975; Holloway and De La Coste-Lareymondie 1982). Recent studies by Zilles et al. (1996) and Gilissen et al. (1996) did not reveal any statistically significant asymmetry in the right frontal/left occipital petalia in chimpanzees in contrast to the condition they identified in the human sample. The authors note that unpublished observations on gorilla and orangutan brains argue for the presence of petalias in these species and that the chimpanzee may exhibit a species-specific pattern.

In contrast, both human and chimpanzee cortical shape was found to be regionally inhomogeneous and the patterns are similar in both species (Gilissen et al. 1996). The highest variability in both hemispheres exists in regions that process higher sensory and associative functions and motor programming including frontal, temporal and parietal lobe areas.

As shown above, humans and some or all of the great apes share a variety of gross anatomical features, some of which were previously thought to be uniquely human. These include the size of the frontal lobe, the frontal cortices, and the parieto-occipital lobe, the planum temporale and planum parietale, the frontal/occipital petalias and the degree and region of variability in cortical shape.

How can these findings be related to the fossil record and the evolution of hominid cognitive specialization? Regarding the frontal lobe, possible relative increases ought to be searched for in the endocasts of species existing prior to the Plio-Pleistocene. On the basis of the comparative material of extant hominoid brains we hypothesized (Semendeferi et al. 1997; Semendeferi and Damasio 2000) that australopithecines and representatives of the genus *Homo* are unlikely to have had a disproportionate enlargement of the frontal lobe. Our hypothesis has been confirmed at least regarding hominids of the Mid-Pleistocene to the present. A recent paleoneurological study that investigated the frontal cranial profiles of Homo *heidelbergensis*, Neanderthals and modern humans (Bookstein et al. 1999), concludes that anterior brain morphology has been stable over the last 300,000 years, a period of time during which modern human cognitive capacities appeared.

Microscopic features

The posterior cingulate cortex, located on the mesial surface of the hemisphere, is involved in visuospatial processing and is composed of isocortical and allocortical areas. Anthropoids differ from prosimians in isocortical regions (areas 30, 23, 31) in terms of cellular organization. Catarrhines (humans, apes, and Old World monkeys) have similar cytoarchitectonic organization in allocortical regions (area 29) that is different from New World monkeys (Armstrong et al. 1986). Humans and other hominoids do not stand out in any of the areas investigated in the posterior cingulate cortex.

A uniquely hominoid feature is to be found in the anterior cingulate cortex, a cortical area of the mesial part of the hemisphere that belongs to the limbic lobe and limbic system. A particular class of neurons, spindle cells, have been known to be present in in layer Vb of the anterior cingulate cortex in humans, but their existence was not certain in other primates. In layer Vb spindle cells have now been demonstrated to be abundant in the human, bonobo and chimpanzee brains, present but not plentiful in the brain of the gorilla and sparse in the orangutan. No spindle cells could be identified in either lesser apes or other anthropoids (Nimchinsky et al. 1999). The same study reports that the pyramidal cells in layer Va of the anterior cingulate cortex exhibit high levels of a calcium-binding protein, calretinin, in humans and chimpanzees followed also by gorillas and orangutans. The anterior cingulate cortex is involved in autonomic functions, emotion and vocalization and the unique presence of this class of neurons in great apes and humans is thought to express some uniqueness related to linguistic/vocalization functions and emotions.

The organization of the auditory cortex presents similarities in humans, chimpanzees and macaques and it is suggested that early stages of auditory cortical processing are comparable in these primates (Hackett et al. 1998). Also, an examination of several visual cortical areas in humans, chimpanzees and macaques revealed a comparable organization in the three species. The location of these areas in relation to sulcal patterns was found to be more similar between humans and chimpanzees. Visual areas V1, V2, and MT are similar in all three species examined, while areas V3, V4, LIP, and VIP are more similar in chimpanzees and macaques (Gattas et al. 1996).

The somatosensory areas (3a, 3b, 1) are located in the most rostral parts of the parietal lobe in the primate brain (Figure 2). Use of conventional and immunocytochemical techniques revealed that, although the laminar patterns and densities of staining vary across taxa, they are more similar in humans and chimpanzees than between either one of the two hominoids and the macaque (Qi et al. 1998).

The primary motor cortex (area 4) has recently been reported to include three subdivisions in humans, chimpanzees, and macaques (areas 4c, 4i and 4r). The topographical relationship between these subdivisions differ between humans and chimpanzees. The chimpanzee motor cortex extends around the precentral gyrus, while the human motor cortex occupies only the posterior bank of this gyrus and is confined within the central fissure (Preuss et al. 1997). Nevertheless, features other than this topographical difference in the extent of the primary motor cortex, like common patterns of laminar distribution of immunoreactivity, are reported to be similar in chimpanzees and humans in contrast to the macaques.

The neuropeptide galanin (GAL) is involved in modulating cholinergic basal forebrain (CBF) neurons that provide major cholinergic innervation of the cortex and the hippocampus associated with cognitive functions. In the macaque all CBF neurons (in nucleus basalis, septal diagonal band, etc.) contain the peptide and gene for GAL. Apes (both lesser and great) and humans have a small population of noncholinergic, galaninergic interneurons in the basal forebrain and a dense galaninergic fiber plexus that innervates CBF neurons. Such a difference possibly indicates a greater control of CBF neurons in the hominoid brain (Benzing et al. 1993).

As is the case with the gross neuroanatomical features common in all hominoids, the microscopic features described above cannot be exclusively associated with hominid neural specializations. Their first appearance can be hypothesized to have been present since the common hominoid, catarhine, anthropoid or even primate ancestor. Shared characters among catarrhines are present in the posterior cingulate allocortical regions. Shared characters among anthropoids are present in the isocortical areas of the posterior cingulate, and in the auditory and visual cortices. Whether or not the rest of the features are shared anthropoid characters or shared-derived characters present exclusively in the hominoid brains needs to be investigated in several extant primate species.

Derived Neural Characters

Macroscopic features

The temporal lobe is heavily involved in recognition and memory and it was recently shown that it is well developed in the human brain (Semendeferi and Damasio 2000). The insula, an area involved in autonomic functions, processing of internal stimuli and taste, appears to be large in the human brain as well. Nevertheless, in neither neural sectors do differences between humans and apes reach statistical significance; a larger sample is necessary to document a possible human specialization in the gross size of the temporal lobe and the insula.

The cerebellum is involved in fine motor tuning, balance and also aspects of cognition. The human cerebellum proved to be smaller than expected of an ape brain of human size (Semendeferi and Damasio 2000) and the difference reaches statistical significance. Furthermore, the cerebellum of New and Old world monkeys is smaller than that of the apes in relative terms. Apes have a larger cerebellum than expected of a non-hominoid primate of their brain size (Rilling and Insel 1998). The relationship between the cerebellum and the cerebrum changed in favor of a larger cerebrum and a smaller cerebellum during hominid evolution.

Recent evidence suggests that this structure is not only involved in motor-related functions as previously thought, but also in cognitive aspects of behavior (Fiez 1996; Muller et al. 1998). It is most likely that different components of this structure evolved at different rates during primate evolution. The studies of Matano et al. (1985) and Matano and Hirasaki (1997), who studied the cerebellar system and its components on histological sections in several mammals supported the idea of differential enlargement of certain parts of the system. The lateral cerebellar system, including the ventral pons, is larger in primates like gibbons and chimpanzees and the dentate nucleus, which is connected with the lateral cortex of the cerebellum, is larger in humans than in apes.

The cortex forms a large segment of the primate brain, but the increase in its size during primate evolution does not keep up with the increase in the size of the white matter (Hofman 1989; Ringo 1991). Intraconnectivity within each cerebral hemisphere, as expressed by the amount of white matter, is larger in larger brains. In contrast, interhemispheric connectivity as expressed by surface area of the corpus callosum is smaller in larger brains, like humans, than in smaller primate brains and does not keep pace with increasing brain size (Rilling and Insel 1999).

Few macroscopic features have been identified so far as potential candidates for hominid neural specializations. It is clear from the brief discussion above that marginal quantitative differences, like those in the temporal lobe and the insula, can only be documented on larger samples that are not available yet.

Microscopic features

The thalamic nuclei have been studied comparatively in humans and apes (Armstrong 1980 1991). One of them, the anterior principal (AP) nucleus of the thalamus, appears to be specialized in some respects in the human brain. This nucleus has more neurons than expected of an ape or anthropoid brain of human size, while the numbers of neurons in this structure in apes are predicted by scaling a monkey's AP to ape brain size. AP is also composed of an anteromedial and an anteroventral component and the size of the anteromedial component is smaller in humans and apes and larger in gibbons and monkeys. AP is connected with the cingulate, prefrontal, parietal, and inferior parietal cortices, the hippocampal region, the subiculum, and the mammilary bodies.

It has been suggested that the human AP may be important for particular aspects of human behavior, like encoding more pieces of information and sustaining attention to sensory stimuli (Armstrong 1991).

Another nucleus of the thalamus, the mediodorsal nucleus, is bigger and has more neurons in humans than is predicted for an ape brain of human size. This nucleus is heavily interconnected with the prefrontal cortices and thus its relatively large size in humans has been associated with the idea that the prefrontal cortex or some of its areas are particularly enlarged in humans.

The prefrontal cortex is a large cortical territory in hominoids that includes higher order association cortices. Two of its cortical areas, area 10 and area 13 have been analyzed in humans, chimpanzees, bonobos, gorillas, orangutans, gibbons and macaques from a comparative perspective (Semendeferi 1994). Very much like most histological studies of the brain of the great apes, the results of this study are based on a small sample and should thus be considered preliminary.

Area 10 is located in the frontal pole and is heavily involved with the planning of future actions and the undertaking of initiatives. This area appears to have similar cytoarchitectonic features in all hominoid brains. The size of the cells, their cell-packing density and staining intensity do not differ considerably across the species. Nevertheless, other aspects of its organization, such as the space available for connections and the relative width of its cortical layers, vary across species.

One structural parameter that seems to have become specialized in the hominid line concerns the space available for connections in area 10. Apes have less space for connections in their cortical layers than humans, following a well known trend of smaller primate brains having less space available for connections than larger ones (see for example Armstrong et al. 1986). Nevertheless, in area 10 humans have, in addition to the expected difference, even more space available for connections in the supragranular layers. These layers are involved in short and long association connections and it is thus suggested that at least this part of the prefrontal cortex in humans is more heavily interconnected with other higher order association areas than is the case with the apes. Nevertheless not all aspects of the organization of area 10 distinguish the human brain from that of the apes. Humans and the two chimpanzees share a similar ratio in the relative width of the cortical layers. They have larger supragranular layers at the expense of the infragranular layers that are involved in connections with subcortical structures.

Human area 10 appears to be highly specialized in another aspect as well. Its size in the human brain is larger not only in absolute terms, but also in relative terms. Area 10 is almost twice as big as area 10 in the two chimpanzees or the other hominoids and a larger sample may establish that it is larger than expected for an ape brain of human size.

This area may have undergone a shift in its location, extent and orientation during hominoid evolution. In the human, chimpanzee, bonobo and orangutan brains area 10 occupies the entire frontal pole (dorsal and orbital components). In the gibbon brain it occupies only the orbital half of the frontal pole (on the orbital side of the principal sulcus). The extent and orientation of area 10 in the gibbon is similar to that of the macaque brain where the cortex homologous to the hominoid area 10 appears to occupy only a small segment of the orbital sector of the frontal pole. The cortex forming the frontal pole of the gorilla appears highly specialized (Semendeferi et al. n.d.b).

Area 10 in the human brain is larger than it is in the brain of the apes, which suggests that functions associated with this part of the cortex have increased considerably during hominid evolution. Planning of future actions and the undertaking of initiatives are indeed hallmarks of human behavior and although present to some extent in other hominoids and possibly even other primates they must have become fully expressed at some point after the split from the common chimpanzee/human ancestor. It was hypothesized (Semendeferi 1994) that the enlargement of area 10, that forms the frontal pole, may have begun early on in hominid evolution, possibly with the appearance of the first representatives of the genus *Homo* and the associated first evidence of material culture. A paleoneurological study of endocasts of *Australopithecus*, *Paranthropus* and extant hominoids (Falk et al. 2000) has just revealed an increase in size of the frontal pole in *Australopithecus*, but not in *Paranthropus*, which they do not consider as part of the hominid line leading to *Homo sapiens*. The authors suggest that the similarity between *Australopithecus* and modern human brain morphology in the frontal pole only underscores the ancestral–descendant relationship of these hominids to the exclusion of *Paranthropus*. Future paleoneurological studies will document possible additional changes in size of the frontal pole in representatives of the genus *Homo*.

Area 13 is another cortical area of the frontal lobe that belongs to the orbitofrontal cortex. The orbitofrontal cortex of primates is not a classical association cortex, but has been linked more closely to what could be termed "limbic" or "emotional" mechanisms (Damasio et al. 1990; Kling and Steklis 1976). Area 13 has similar qualitative features in all hominoids examined (cell types, types of borders between cortical layers) (Semendeferi et al. 1998). All hominoids have a similar distribution of space available for connections across the cortical layers and humans as expected have, overall, decreased neuronal density and more space for connections. Humans and African apes are particularly close in the relative size of the cortical layers.

Nevertheless, humans and bonobos have a highly diversified orbitofrontal cortex that includes numerous subdivisions of cortical areas and a small area 13. In contrast the orangutans have a short orbitofrontal cortex in length (anterior-posterior), a small total volume of orbitofrontal cortex, and a large area 13. Area 13, along with the surrounding posterior orbitofrontal cortex, constitutes a significant part of the neural substrate responsible for behaviors related to responses to social stimuli and social cognition (Damasio and Van Hoesen 1983) and it is suggested that this region became highly specialized during hominid evolution. The issue of whether or not bonobos evolved independently a species-specific organization in these cortices, separate from the common chimpanzees, should be investigated using a larger sample.

Located in the temporal lobe, Wernicke's area or Tpt cortex is known to be present in human, ape and monkey brains in the posterior superior temporal gyrus and is involved in language comprehension in humans. Tpt cortex was investigated in human (fetal and adult), chimpanzee and macaque brains in terms of cellular architecture and vertical organization of its cellular composition. It is part of the extended primate neocortex that is organized in six horizontal layers and also in vertical orientation in cell columns. Buxhoeveden et al, (1996) used as a template the extremely linear and vertical arrangement of cells in the pre-laminated fetal cortical plate in order to address measures of linearity, cell distribution and column size in a comparative perspective in the three species investigated.

This study found that cell density decreased in species with large brains mostly due to increased spacing between columns rather than among cells within columns. Columns in bigger brains exhibit less vertical orientation, and absolute measures of linearity in chimpanzees and rhesus monkeys resemble each other more closely than humans and chimpanzees. In layer III orientation of columns is more vertical in non-humans than humans and greater differences were observed between human versus non-human brains in layer II. Nevertheless, in layer V, human and chimpanzee are similar. The authors suggest that changes in human evolution may have affected more the integrative, supragranular layers (II and III) of this part of the cortex than the specialized receptive IV or corticofugal layers V and VI.

In area V1 or area 17 (primary visual cortex) certain differences and similarities were identified across primates using both conventional histochemical techniques, like Nissl and cytochrome oxidase (CO), as well as immunocytochemistry staining for calbindin (a calcium binding protein) and nonphosphorylated neurofilament protein (NPNF) (Preuss et al. 1998). Layer IVa stains densely for CO in Old and New World monkeys, but lightly in humans and apes. The same layer has decreased/light staining in calbindin in Old World monkeys, but increased/dark staining in chimpanzees and humans. In humans only is this layer increased in NPNF. Layer IVb is increased in NPNF in all taxa and layer IVc stains densely for CO in all taxa as well. It is suggested that the above patterns of staining reflect differential patterns of connectivity and specifically that the parvocellular inputs involved with fine spatial resolution and color are decreased in hominoids. In contrast, the magnocellular inputs that are involved with motion detection are increased in humans.

Also in area 17 a particular class of neurons was studied in humans, chimpanzees and macaques (Mori 1996). Neuropeptide Y-like immunoreactive neurons exhibit similar morphological features in the three taxa for the most part. Nevertheless, this class of neurons is different between chimpanzees and macaques in regard to their size, while they differ between chimpanzees and humans in regard to their distribution.

Similar to the unique topographic relationship that appears to exist between the central sulcus and the extent of the primary motor cortex in humans versus chimpanzees is the situation in a part of the mesial temporal cortex (Figure 2). In this part of the brain, a region belonging to the limbic system involved in memory function, the position and relation between area 35 and the rhinal sulcus is similar in monkeys and apes, but different in humans (Augustinack et al. 1997).

The basal forebrain includes many magnocellular nuclei that are an extension of the reticular formation. The largest entity is the nucleus basalis of Meynert (NB) that sits below the lentiform nuclei and anterior commissure. NB is prominent in primates (this may be related to the expansion of the cortex) and is composed of many divisions; one cell group is the nucleus subputaminalis (NSP) that is a component of the basal forebrain cholinergic system. NSP found in humans and apes may be related to the innervation of frontal lobes. Although monkeys have a prominent NB, they lack a well-developed NSP. Furthermore, the most rostrolateral part of NSP might be present only in humans (Simic et al. 1999). NSP provides cholinergic innervation to the inferior frontal gyrus, where Broca's area is located. The cholinergic input to the cerebral cortex is known to have modulatory roles, including memory-learning and attention-arousal. It is thus possible that a more unique human organization of NSP is significant for the speech area.

A variety of microscopic features have so far been identified as being distinct in humans. Some of them belong to the limbic system and involve neural circuits that regulate emotions and complex reactions to social environments. Others involve higher order association areas or subcortical structures that are highly interconnected with such areas. These systems underlie complex cognitive functions, like planning and language. In order to relate these findings to gross anatomical landmarks that are identifiable in the fossil record, more research needs to be done on extant species that will address issues of inter- and intraspecific variability of structures of interest.

Concluding Remarks

A remarkable homogeneity is known to exist in the organization of the primate brain (Jerison 1973; Stephan, Frahm and Baron 1981; Finlay and Darlington 1995). Nevertheless, examination of more than one subject in closely related taxa reveals the presence of intraspecific and interspecific variation in certain sectors of the brain. Such information is essential in any attempt to reconstruct the evolution of the human mind and its underlying neural circuitry.

Among the hominoids differences are identifiable even at the gross level. The orangutan has a remarkably smaller orbitofrontal sector than the rest of the hominoids and the brain of the gorilla seems to be specialized among the great apes, with a larger cerebellum, a smaller temporal lobe, and a larger parieto-occipital sector (Semendeferi and Damasio 2000). Specializations of the human brain have been shown so far to be present in parts of the prefrontal, temporal and occipital cortices, the thalamic nuclei, and the cerebellum.

What can be said so far about the evolution of the neural circuitry that underlies cognitive and emotional functions in the hominid line? Higher cognitive functions involve heavily the frontal, temporal and parietal cortices. After the split of the hominid line from the rest of the hominoids the overall relative size of the frontal lobe and the parieto-occipital region does not seem to have increased in hominids (Semendeferi et al. 1997; Bookstein et al. 1999; Semendeferi and Damasio 2000). In contrast, the temporal lobe seems to have increased somewhat, while the cerebellum, which is also involved in some cognitive functions, has not kept up (as a whole) with the increases in size of the cerebrum. These differential enlargements took place at some point in the Plio-Pleistocene along with other reorganizational events that were discussed here. Future research in paleoneurology may reveal whether these events took place after the appearance of the genus *Homo* or as early as the first australopithecines.

Specific neural areas within the above large sectors have become specialized in one or more parameters. Area 10 of the frontal lobe, involved in abstract thinking and planning of future actions, became much larger after the split from the African apes. The mediodorsal nucleus of the thalamus that is heavily interconnected with prefrontal cortices has also enlarged in the Plio-Pleistocene hominids and so did another thalamic nucleus, the anterior principal, that is important for encoding pieces of information and sustaining attention to sensory stimuli. The orbitofrontal cortices involved in emotional processing are highly diversified in humans and bonobos and more research will reveal whether this is a condition unique to the two species or whether it represents a shared derived condition present in all African apes. Some, but not all, of the areas involved in language processing appear to be specialized in humans. Aspects of the organization of those temporal cortices involved in language perception have changed in the course of hominid evolution. On the other hand, the cortical areas of the frontal lobe involved in language expression are affected by changes that seem to have happened in subcortical nuclei that are connected with them.

It appears that one of the biggest challenges in reconstructing the evolution of the human brain is to shift our focus from theorizing about it through the reanalysis of old and nonreplicable information to collecting new data from the few species that are relevant to our questions. The brain of the African great apes, the orangutan and the lesser apes, if compared to the human brain, can provide the appropriate source of information regarding what is or is not uniquely human. Although information from a variety of other primates is valuable in appreciating species-specific patterns of organization and filling in the larger picture of primate brain evolution, it is the comparative study of the hominoids that holds the biggest promise. As soon as such information becomes available, the fossil record and the reconstruction of hominid endocasts can assist in placing neural events in the proper temporal and taxonomic context.

Human brains are the largest primate brains, but our cognitive abilities and behavioral specializations cannot be attributed exclusively to overall brain size. There is accumulating evidence that several re-organizational events took place in the hominid brain during the Plio-Pleistocene. On the other hand, simplistic interpretations like the traditional view that relatively large frontal lobes are accountable for most of our cognitive differences do not seem to hold true either. Frontal lobes have a similar relative size in hominoids and are as large as expected of an ape brain of human size. So, it is not the relative enlargement of the frontal lobes during human evolution that made us *Homo sapiens*. A detailed analysis of many parts of the human and ape brain at many different levels using the newly available non-invasive techniques holds the promise of revealing the secrets of our uniqueness.

References Cited

Aiello, L., and Dean, C.
   1990 *An Introduction to Human Evolutionary Anatomy*. Academic Press, New York.

Allman, J.
   1977 Evolution of the visual system in the early primates. In *Progress in Psychobiology and Physiological Psychology*, edited by J.Sprague and A.Epstein, pp. 1-53. Academic Press, New York.

Armstrong, E.
   1980 A quantitative comparison of the hominoid thalamus: II. Limbic nuclei anterior principalis. *American Journal of Physical Anthropology* 52:43-54.

1991 The limbic system and culture: An allometric analysis of the neocortex and limbic nuclei. *Human Nature* 2:117-136.

Armstrong, E., K. Zilles, S. Gottfried and A. Schleicher
1986 Comparative aspects of the primate posterior cingulate cortex. *The Journal of Comparative Neurology* 253:539-548.

Augustinack, J. C., J. M. Dierking, S. J. Redman, A. Solodkin and G. W. Van Hoesen
1997 The rhinal sulcus and perirhinal cortex in humans, anthropoid apes and monkeys. *Society for Neuroscience Abstracts* 23(1-2):498.

Benzing, W.C., J. H. Kordower and E. J. Mufson
1993 Galanin immunoreactivity within the primate basal forebrain: evolutionary change between monkeys and apes. *Journal of Comparative Neurology* 336:31-39.

Blinkov, S. M. and I. I. Glezer
1968 *The Human Brain in Figures and Tables: a Quantitative Handbook.* Basic Books, Inc., Publishers, Plenum Press, New York.

Bookstein, F., K. Schaefer, H. Prossinger, H. Seidler, M. Fieder, C. Stringer, G. W. Weber, J. L. Arsuaga, D. E. Slice, F. J. Rohlf, W. Recheis, A. J. Mariam and L. F. Marcus
1999 Comparing frontal cranial profiles in archaic and modern *Homo* by morphometric analysis. *The Anatomical Record (New Anat.)* 257:217-224.

Buxhoeveden, D., W. Lefkowitz, P. Loats, E. Armstrong
1996 The linear organization of cell columns in human and nonhuman anthropoid Tpt cortex. *Anatomy and Embryology* 194(1): 23-36.

Changeux, J. P. and J. Chavaillon (eds.)
1996 *Origins of the Human Brain.* Clarendon Press, Oxford.

Damasio, A. R. and G. W. Van Hoesen
1983 Emotional disturbances associated with focal lesions of the frontal lesions of the frontal lobe. In *Neuropsychology of Human Emotion: Recent Advances,* edited by K. Heilman and P. Satz, pp. 85-110. Guilford Press, New York.

Damasio, A. R., D. Tranel and H. Damasio
1990 Individuals with sociopathic behavior caused by frontal damage fail to respond autonomically to social stimuli. *Behavioural Brain Research* 41:81-94.

Deacon, T. W.
1988 Human brain evolution: II Embryology and Brain Allometry. In *Intelligence and Evolutionary Biology,* edited by H. J. Jerison and I. Jerison pp. 383-415. Springer-Verlag, Berlin and Heidelberg.

Falk, D.
1991 3.5 Million years of hominid brain evolution. *The Neurosciences* 3:409-416.

Falk, D., J. C. Redmond, J. Guyer, G. C. Conroy, W. Recheis, G. W. Weber and H. Seidler
2000 Early hominid brain evolution: a new look at old endocasts. *Journal of Human Evolution* 38:695-717.

Fiez J. A.
1996 Cerebellar contributions to cognition. *Neuron* 16:13-15.

Finger, S.
1994 *Origins of Neuroscience: A History of Explorations into Brain Function.* Oxford University Press, Oxford.

Finlay, B. L. and R. B. Darlington
1995 Linked regularities in the development and evolution of mammalian brains. *Science* 268:1578-1583.

Gannon, P. J., R. L. Holloway, D. C. Broadfield and A. R. Braun
1998 Asymmetry of Chimpanzee Planum Temporale: Humanlike Pattern of Wernicke's Brain Language Area Homolog. *Science* 279:220-222.

Gattass, R., M. M. Adams, P. R. Hof, L. G. Ungerleider
1996 Parcellation of visual cortical areas in the chimpanzee using neurofilament and calcium-bnding proteins. *Society for Neuroscience* 22:421.4

Geschwind, N. and W. Levitsky
1968 Human brain: left-right asymmetries in temporal speech region. *Science* 161:186-187.

Gilissen, E., A. Dabringhaus, G. Schlaug, T. Schormann, H. Steinmetz, and K. Zilles
1996 Structural asymmetries in the cortical shape of humans and common chimpanzees: a comparative study with magnetic resonance tomography. *Proceedings of the 3rd Joint Symposium On Neural Computation* 6:89-102.

Hackett, T. A., T. M. Preuss and J. H. Kaas
1998 Architectonic identification of core and belt auditory cortex in macaque monkeys, chimpanzees and humans. *Society for Neuroscience Abstracts* 24(1-2):1880.

Hofman, M. A.
  1989 On the evolution and geometry of the brain in mammals. *Progress in Neurobiology* 32:137-158.
Holloway, R. L.
  1995 Toward a synthetic theory of human brain evolution. In *Origins of The Human Brain*, edited by J. P. Changeux and J. Chavaillon, pp. 42-60.
Holloway, R. L. and M. C. De La Costelareymondie
  1982 Brain endocast asymmetry in pongids and hominids: some preliminary findings on the paleontology of cerebral dominance. *American Journal of Physical Anthropology* 58:101-110.
Hopkins, W. D., L. Marino, J. K. Rilling and L. A. MacGregor
  1998 Planum temporale asymmetries in great apes as revealed by magnetic resonance imaging MRI. *Neuroreport* 9:2913-2918.
Jerison, H. J.
  1973 *Evolution of the Brain and Intelligence*. Academic Press, New York.
Kaas, J., R. Nelson, M. Sur, C. Lin and M. Merzenich
  1979 Multiple representations of the body within the primary somatosensory cortex of primates. *Science* 204:521-523.
Kandel, E. R. and J. H. Schwartz
  1985 *Principles of Neural Science*. Elsevier, New York.
Kheck, N. M., P. J. Gannon, P. R. Hof, A. R. Braun, J. M. Erwin, D. C. Broadfield, M. Yuan and R. L. Holloway
  1998 Brain language area evolution II: human like pattern of hemispheric asymmetry in planum parietale of chimpanzees. *Society for Neuroscience Abstracts* 24(1-2):160.
Kling, A. and H. D. Steklis
  1976 A neural substrate for affiliative behavior in nonhuman primates. *Brain, Behavior and Evolution* 13:216-238.
LeMay, M.
  1976 Morphological cerebral asymmetries of modern man, fossil man, and nonhuman primate. *Annals of the New York Academy of Sciences* 280:349-367.
LeMay, M. and N. Geschwind
  1975 Hemispheric differences in the brains of great apes. *Brain, Behavior and Evolution* 11:48-52.
Matano, S., H. Stephan, and G. Baron
  1985 Volume comparisons in the cerebellar complex of primates. I. Ventral pons. *Folia Primatologica* 44:171-181.
Matano, S. and E. Hirasaki
  1997 Volumetric comparisons in the cerebellar complex of anthropoids, with special reference to locomotor types. *American Journal of Physical Anthropology* 103:173-183.
Mori, S.
  1996 Neuropeptide Y-like immunoreactivity in area 17 of chimpanzee. *Okajimas Folia Anatomica Japonica* 73:219-227.
Muller R A., E. Courchesne and G. Allen
  1998 The cerebellum: so much more. *Science* 282:879-80.
Nimchinsky, E. A., E. Gilissen, J. M. Allman, D. P. Perl, J. M. Erwin and P. R. Hof
  1999 A neuronal morphologic type unique to humans and great apes. *National Academy of Sciences of the United States of America* 96:5268-5273.
Preuss, T. M., H. X. Qi, P. Gaspar and J. H. Kaas
  1997 Histological Evidence for Multiple Divisions of Primary Motor Cortex in Chimpanzees. *Society for Neuroscience Abstracts* 23(1-2):1273.
Preuss, T. M., H. X. Qi and J. H. Kaas
  1998 Chimpanzees and humans share specializations of primary visual cortex. *Society for Neuroscience Abstracts* 24(1-2):645.
Qi, H. X., N. Jain, T. M. Preuss and J. H. Kaas
  1998 Comparative architecture of areas 3a, 3b and 1 of somatosensory cortex in chimpanzees, humans and macaques. *Society for Neuroscience Abstracts* 24(1-2):1125.
Rilling J. K and T. R. Insel
  1998 Evolution of the Primate Cerebellum: differences in relative volume among monkeys, apes and humans. *Brain, Behavior and Evolution* 52:308-314.
Rilling, J. K. and T. R. Insel
  1999 Differential expansion of neural projection systems in primate brain evolution. *NeuroReport* 10:1453-1459.
Ringo, J. L.
  1991 Neuronal interconnection as a function of brain size. *Brain, Behavior and Evolution* 38:1-6.
Semendeferi, K.
  1994 Evolution of the hominoid prefrontal cortex: a quantitative and image analysis of area 13 and 10. Ph.D. Dissertation, University of Iowa.

Semendeferi, K. and H. Damasio
   2000 The brain and its main anatomical subdivisions in living hominoids using magnetic resonance imaging. *Journal of Human Evolution* 38:317-332.

Semendeferi, K., H. Damasio, R. Frank and G. W. Van Hoesen
   1997 The evolution of the frontal lobes: a volumetric analysis based on three-dimensional reconstruction's of magnetic resonance scans of human and ape brains. *Journal of Human Evolution* 32:375-388.

Semendeferi, K., A. Lu, A. M. Desgouttes and H. Damasio
   n.d.a The evolution of the frontal cortices in the Plio-Pleistocene: new evidence. *American Journal of Physical Anthropology* in press.

Semendeferi, K., A. Schleicher, K. Zilles, E. Armstrong and G. W. Van Hoesen
   n.d.b Evolution of the hominoid prefrontal cortex: Imaging and quantitative analysis of area 10. in preparation.

Semendeferi, K., E. Armstrong, A. Schleicher, K. Zilles and G. W. Van Hoesen
   1998 Limbic frontal cortex in hominoids: A comparative study of area 13. *American Journal of Physical Anthropology* 106: 129-155.

Simic, G., L. Mrzljak, A. Fucic, B. Winblad, H. Lovric and I. Kostovic
   1999 Nucleus subputaminalis (ayala): The still disregarded magnocellular component of the basal forebrain may be human specific and connected with the cortical speech area. *Neuroscience* 89:73-89.

Stephan, H., H. Frahm and G. Baron
   1981 New and revised data on volumes of brain structures in insectivores and primates. *Folia Primatologica* 35:1-29.

Tobias, P. V.
   1995 The brain of the first hominids. In *Origins of The Human Brain*, edited by J. P. Changeux and J. Chavaillon, pp. 61-83.

Uylings, H., and C. G. van Eden
   1990 Qualitative and quantitative comparison of the prefrontal cortex in rat and in primates, including humans. *Progress in Brain Research* 85:31-62.

Yeni-Komshian, G. and D. Benson
   1976 Anatomical study of cerebral asymmetry in the temporal lobe of humans, chimpanzees, and rhesus monkeys. *Science* 192:387-389.

Zilles, K., E. Armstrong, A. Schleicher and H. J. Kretschmann
   1988 The human pattern of gyrification in the cerebral cortex. *Anatomy and Embryology* 179:173-179.

Zilles, K., A. Dabringhaus, S. Geyer, K. Amunts, M. Qu, A. Schleicher, E. Gilissen, G. Schlaug and H. Steinmetz
   1996 Structural asymmetries in the human forebrain and the forebrain of non-human primates and rats. *Neuroscience and Biobehavioral Reviews* 20:593-605.

9. Multilevel Information Processing, Archaeology, and Evolution

Philip G. Chase

Abstract

The behavior that produced the Paleolithic archaeological record was controlled by information processing at a number of levels: genetically determined, learned, socially learned, social, and symbolic. The story of human evolution is in large measure the story of the appearance and increasing importance of the higher levels of information processing. This depended fundamentally on genetic evolution, but as the higher levels came to be more important, the ability to change behavior independently of genes increased. This means that archaeologists cannot assume *a priori* that any given change in behavior reflected in the Paleolithic record, even a major one, necessarily indicates a genetic change in hominid "intelligence."

Introduction

Archaeologists deal not with intelligence itself, but with the material correlates of behavior, and most of our time is spent trying to find theoretical and empirical grounds for inferring specific behaviors from specific patterns in the archaeological record, taking into account questions of taphonomy, site formation process, equifinality, etc. Nevertheless, as Paleolithic archaeologists we cannot ignore the implications of our findings for the evolution of human intelligence. We must address the problem of determining whether or not those changes in behavior we do infer from the archaeological record are indicative of evolutionary changes in intelligence. Obviously, during the Pliocene and Pleistocene, evolutionary changes in the hominid gene pool resulted in changes in behavior. Yet we all also know that there can be major developments in human behavior that are entirely unassociated with any concomitant genetic evolution. Which behavioral changes, then, indicate changes in intelligence?

Actually, given the way in which modern humans process information, the question is bigger than this. As I will explain below, there are several levels of information processing that control behavior, and each level has a different relationship to the underlying genetic code. Therefore, we must first ask what level of information processing is involved in any given change in behavior. Only then can we turn to other disciplines such as neuroanatomy, comparative psychology, or primate ethology to help us determine whether genetic changes were involved.

It is my purpose here (1) to explain what I mean when I say that human information processing takes place at several levels, (2) to explain briefly why I think that each of these levels has a different relationship to the underlying gene pool of a population, and (3) illustrate very briefly what the relevance for this view is to archaeological problems. In the process, I will discuss some of the structural changes in information processing that have to have taken place in the course of hominid evolution.

In order to illustrate these points, especially the third, I will use two examples from the Paleolithic. The first comes from the very earliest part of the archaeological record. The *terminus post quem* for our discipline is when hominids first made flaked stone tools often enough so that an archaeological record exists.[1] Because chimpanzees make and use simple tools, but do not make flaked stone tools, it is natural to ask whether or not the initiation of this practice indicates a new level of intelligence.

The second example is from late in the Paleolithic record, some two and a half million years later. The prehistoric natural geographic distributions of some materials, such as certain rocks used as raw material for tool making or some species of mollusks, can be determined geologically or paleontologically. For these materials, the distance between a site in which they are found as artifacts and their nearest natural source can be measured.

The archaeological record shows a general increase through time in the maximum distances over which such materials were transported (Roebroeks et al. 1988). However, I wish to concentrate here on only one very specific part of this record of change, an apparently rather dramatic increase between the Middle and Upper Paleolithic in the distances over which materials were transported within Europe, Because of all the other changes that took place in Europe about this time — the appearance of bone industry, blade tool technology, art, and anatomically modern humans — it is tempting to attribute this change to some genetic enhancement of intelligence. That is, the data may represent greater distances of transport by a single group of people due to better economic foresight or geographic abilities due to a genetically programmed change in neuroanatomy.

I will use these two phenomena as examples of the kinds of archaeological problem that I have in mind when I argue for the multilevel nature of human information processing. I must emphasize that my main purpose is *not* to make a thorough investigation of or to provide a definitive answer to either problem. I will use them only rather cursorily for illustrative purposes. I must also emphasize that I am *not* advancing any theory or hypothesis of hominid evolution. I am not proposing an explanation of why any given human trait evolved, or how it evolved, or even when it evolved.

What I am doing is presenting a view of the nature of human information processing and its relevance to how we interpret the archaeological record. I will use the term "information processing systems" instead of "intelligence" for three reasons. First, even among specialists on the subject, there is considerable disagreement about what "intelligence" means (e.g. Kail and Pellegrino 1985, Howe 1988, Parker and Baars 1990, Sternberg and Salter 1982). This is probably because there is little consensus about how the mind or brain works, or even the extent to which these two can be considered one and the same thing (see Bechtel 1988, Churchland 1986, 1988). Fortunately, there is no need to enter into this debate, which is basically irrelevant to the points I wish to make.

Second, the word "intelligence" generally implies, at least in common parlance, some level of genetic determination. However, it is my purpose to investigate this very question of genetic determination or non-determination.

Third, the term intelligence is usually used to apply to the mental capacities of individuals. Yet as will be seen shortly, one of my main arguments is that among humans information processing takes place at the social as well as the individual level, so a term that applies only to the individual would be out of place. For these three reasons, then, I will use the term "information processing system" instead of intelligence.

By information processing I mean the ways in which an animal or group of animals perceives its environment, interprets that environment, and analyzes it in a way that produces behavior. There is considerable difference among schools of psychology as to how this takes place at the level of the individual. However, choosing one of these perspectives is not essential to the point I wish to make. All schools would agree that animals have mechanisms for gathering information from their environment and then using this information in such a way as to produce behavior. This is what I mean by information processing, but with the added stipulation that among primates it is not only individuals but also groups of interacting individuals that process information.

A central feature of the evolution of our species seems to be that our behavior has freed itself from the shackles of genetic determination to a far greater extent than is the case for any other species. This is a bit ironic, given that it is changes in the genetic code that have made this freedom possible. Nonetheless, the fact that human cultural or social systems may change, even in very important ways, in the absence of any concomitant genetic evolution is a fundamental tenet of most anthropological thought. For example, major changes in the way humans make a living from their environment occurred with the appearance of agriculture, of urban craft specialization, and of industrialization. None of these changes, however, was the result of a change in the genetically determined competence of the peoples responsible.

It is not unreasonable, therefore, to argue that in the past changes in the nature, scope and complexity of information processing systems may have occurred independently of genetic changes. However, it is also true that in the course of the Paleolithic there *were* important genetic changes involved in the evolution of hominid information processing. These two apparently conflicting ideas are nevertheless compatible because information processing takes place at several levels. Although genes structure information processing in all species, at the learned and social levels information processing can change independently of genetics, albeit within limits established genetically. Moreover, at the social level new information processing systems emerge, which in some respects differ in nature from systems confined to a single brain or mind.

Levels of Human Information Processing

That the information processing that governs human behavior takes place at several levels has been explicitly described by Plotkin (1988). He points out that different levels of information processing respond to input at different rates, and therefore each permits a species to adapt to environmental changes of different velocities. For example, very long term environmental changes are best met by changing the genetic code, while rapid changes can only be met by changing learned behavior. Plotkin points out, however, that all levels of information processing are adaptive in nature, and that all levels can be studied as evolutionary phenomena.

This multilayered nature of information processing is of enormous importance to our understanding of the relationship between genetic evolution and the evolution of information processing systems, since each level of information processing has a different relationship to the underlying genetic code. For the purposes of the current discussion, I shall distinguish five such levels: (1) genetically determined information processing, (2) individually learned information processing, (3) socially learned information processing carried out at the individual level, (4) information processing carried out at the social level, and (5) the extension of symbolism beyond reference to provide a new "medium" for information processing.

(1) The lowest level includes all those information processing systems residing in the individual nervous system whose structure and operation are determined genetically, so that the organism is obliged by its genotype to process certain kinds of information in specific ways. These include systems that are generally lumped under the term "involuntary." Examples would be the neural mechanisms that provoke perspiration or shivering to control body temperature, as well as the neural mechanisms for perceiving color or for sensing the position in space of one's extremities.

(2) The next level involves information processing that is learned[2] by an individual. It is extremely important to emphasize that this involves not just the learning of information (e.g., the locations of resources in an organism's territory) but also the learning of *ways of processing* information to produce behavior. For example, running cockroaches normally move three legs simultaneously: the front and back legs on one side of the body and the middle leg on the other side. This permits them to keep a stable triangular base on the other three legs. However, in one experiment, after the middle legs on both sides were amputated, cockroaches adopted a gate that is typical among quadrupeds because it affords the best balance possible with four feet. The left front and right rear legs moved together, and *vice versa* — in spite of the fact that these legs do not normally move simultaneously in a cockroach (Gallistel 1980 in Roitblat 1985:139-40). That is to say, the cockroaches learned a new algorithm for controlling behavior, not just a new set of data concerning their environment. It is the ability to learn ways of processing information, rather than merely processing new data in genetically determined ways, that constitutes this level of information processing.[3]

(3) The next category of information involves learning from conspecifics and passing on to conspecifics both information and ways of processing information. However, the information processing in this category still takes place at the individual level. There are many ways in which such sharing can take place, and to a great extent the evolution of hominid information processing is the story of the appearance of new ways of sharing information. I will go into this in some detail below.

(4) Information processing can also take place at the social level. The design of a modern passenger aircraft is a complex example of such information processing. A large team of specialists work, at an individual level, on different parts of the design. However, such a project also involves the coordination of all these individual efforts, with decisions made jointly, and with a large number of specialists whose job is simply the management and coordination of the work of individual engineers. That is, not only does information processing involve many individuals, but much of the information processing *consists of* the interactions among individuals. Thus, a whole new system of information processing is involved that exists at a different level.

(5) At some point in the past, symbolism extended beyond a purely referential role to create a new kind of information and a new environment and medium for both action and information processing. What I mean by this will be clearer, however, if I first discuss the nature and evolution of the first four levels of information processing listed above.

The relevant point about this stratification of information processing is that as soon as information processing systems can be learned, they can also change without changing the underlying genetic structure. Thus, for any given species, the greater its reliance on learned ways of processing information, the greater its freedom to change ways of thinking without changing the genetic code. When information processing begins to take place at the social level, this freedom from the genetic code is potentially even greater, because a change in social relationships will almost inevitably entail a change in the way information is processed.

When I say that, at a given level, information processing and the behavior produced could change independently of genetics, I do not mean that information processing is free of genetic control. At the simplest level, the genotype places constraints on what can be learned. Humans cannot learn to use ultraviolet light to recognize different species of flower with the naked eye, as some insects do, for the very simple reason that we do not have the mechanisms necessary to perceive ultraviolet light. Likewise, presumably on account of the structure of our brains, it is impossible for us to *envision* space in more than three (or perhaps four) dimensions, even though the concept of multidimensional space is a simple one mathematically. Even the ability to share information is structured genetically – the use of full-scale modern human language depends on having a human brain.[4]

Nor does the statement that information processing systems can evolve independently of genetic evolution mean that this is always the case. The fact that the underlying framework for information processing is genetically determined has two consequences. First, changes in genotype that affect this framework will inevitably change the nature of information processing systems. Second, certain kinds of information processing cannot have evolved without genetic evolution. The most obvious example is language. Yet at the same time, the invention of writing, which added a new tool to human information processing, required no change in the human gene pool.

Thus information processing systems are always intimately linked to genes and the two kinds of evolution are therefore associated. Yet at the same time information processing at the learned and social levels can change independently of the genes. As I see it, one of the major features of human evolution is that as genetic evolution provided frameworks for more elaborate systems of information processing at these levels, the more freedom such systems had to evolve on their own. Moreover, as genetic evolution permitted higher-level information processing systems to become more elaborate and more powerful, the more important was their role in the sum total of hominid adaptation, so that an increasingly large *proportion* of hominid information processing became free to change independently of genetics. Thus it is perfectly understandable that anthropologists working with extant cultures see even very major differences among societies as being independent of genetic variation, while at the same time it is also perfectly understandable that paleoanthropologists would interpret major changes in the behavioral repertoires of early hominids as evidence of changes at the genetic level. It makes sense, as a hypothesis, that the frameworks for information processing possessed by the earliest hominids, although they provided for considerable freedom in learned behavior, still did not provide enough freedom so that major changes in the general nature of information processing systems could take place independently of genetics. Yet at the same time, we can look at developments during the Holocene and find abundant evidence that new information processing systems could arise independently of genetic change.

One such example clearly is the emergence of complex systems of economic specialization. This is important from an information processing point of view because it permits a society to accumulate a much larger body of information than would otherwise be the case. A society where many individuals learn complementary bodies of knowledge can control a greater sum of knowledge than can a society where all its members must master approximately the same things. Moreover, if a complex social system is to work, a whole new kind of structure must be developed to coordinate the interdependent specialists. This combination of specialized knowledge and the complex interaction of various specialists constitutes a new kind of information processing, one responsible for major expansion of technical knowledge. A second example is the development of writing, which permitted the storing and transmission of vast amounts of information through both time and space. This development has been extended by the inventions of moveable type, of telegraph and telephone, of radio and television, and most recently of computer networks. Finally, one cannot ignore the invention of tools for manipulating (as opposed to transmitting or storing) data. These include at a minimum the invention of symbols for mathematical analysis, inventions such as double-entry bookkeeping and flow charts, and tools such as abacuses, calculators, and computers.

All of these represent new ways of processing information, and none of them owes anything to genetic changes. All of them transformed the ways the peoples involved interacted with and adapted to their environment. They thus attest to the great freedom that recent information processing systems have to evolve independently of genetics. All of them, however, are essentially Holocene developments, even if precursors can be found in the late Pleistocene. The problem then is to analyze how such a freedom to evolve could itself have evolved. To do this, we need to look briefly at some fundamental changes that have taken place during the course of hominid evolution.

Socially Learned Information Processing

We can take it as a given that all mammals depend heavily on learning to control behavior, and this tendency is especially marked among primates. Much primate learning takes place individually, in the sense that each individual learns directly from observation of or interaction with the environment, and without any transmission of knowledge from one individual to another. At this level, modern human learning is similar in overall nature to that of other primates. Where the first major differences appear are at the level of social learning, where individuals learn from conspecifics. Among other primates, and to a considerable degree among humans, much social learning takes place either by observation or by observation combined with trial and error on the part of the learner. For example, when chimpanzees learn to fish for insects by using twigs, pieces of grass, etc., they do so by observing their mothers and then trying to do the same thing (van Lawick-Goodall 1970:229, 243). What is missing in such a case is any evidence that animals are intentionally sharing information with others. The learners may be observing conspecifics rather than other aspects of their environment, but otherwise the process does not differ greatly from individual learning. Social learning among humans goes beyond this, in two respects. First, humans teach others, sometimes formally, more often very informally, but they make a deliberate effort to help others learn. Second, humans try to change what other individuals think, not just what they do, a fact that opens up both the need and the possibility for new kinds of communication.

Recognition of Mental States in Others

Somewhere in the past, specific evolutionary developments took place that greatly facilitated the sharing of information and made teaching and deliberate sharing of information possible. The first of these developments was the ability to attribute mental states to other individuals, something Premack and Woodruff (1978) call theory of mind and Dennett (1987) calls second-order intention. That is, one must be aware that others have minds and that their behavior is controlled by their minds. Communication can take place without this awareness. It is sufficient that a message be sent and that there be some change in the mental state of the receiver as a result of the message. It is not necessary that the sender realize that such a mental change will take place. For example, a dog's growling at a postman can be explained in one of three ways. (1) *The dog growls simply because it is angry, without any expectation that its growl will have any effect.* It should be noted, however, that there is good evidence that non-human primates do send out signals such as alarm calls because they expect a result (Cheney and Seyfarth 1990:144-149). (2) *The dog thinks that if it growls the postman will go away.* The dog has no thoughts concerning motivation or any other mental state on the postman's part. Its attitude toward cause and effect is identical whether a postman or an inanimate object is involved – as in: *The dog thinks that if it bites a marrow bone it will break.* (3) *The dog thinks that if it growls, the postman will fear the dog and will go away.* In this case, the dog has an awareness that the postman has a mind and that it can change his behavior by changing his mental state.

The difference between (2) and (3) is a critical distinction, because such an awareness of mental states in others is necessary to two crucial aspects of human information sharing, teaching and language. Clearly, if one is unaware that another individual has mental states and that it is these mental states that determine behavior, then one will certainly not try to teach him or her anything, since teaching consists of trying to change another's mental states (knowledge, beliefs, etc.). For example, we can imagine a chimpanzee mother watching her child attempting unsuccessfully to use a crooked twig to fish for termites. If all she thinks is, "It won't work because the twig is crooked," she will not be motivated to teach the proper way of choosing a twig. For that, she would need to think, "She doesn't understand that the twig has to be straight." The first thought attributes no mental state to the young chimpanzee; the latter does. That does not necessarily mean that having recognized her child's ignorance, the mother will automatically instruct her on how to choose a twig, but it is a necessary precondition for teaching.

By the same token, while communication is possible without attribution of mental states, full-scale modern human language cannot exist without it. Linguistic utterances can take one of several forms: commands, statements of fact, promises, attempts to persuade, etc. That is to say, for each utterance the speaker has a different purpose. The purpose of a command is to make another individual do something. As we have seen, a speaker can in theory accomplish this purpose without any awareness that the hearer's behavior is controlled by a mind. However, this is the only purpose for which this is true. The purpose in stating a fact, for example, is to give the hearer information, and such a purpose cannot exist without an awareness that the hearer has a mind. The purpose of a promise is to make the hearer believe that the speaker will do something, and again such a purpose cannot exist without an awareness that the hearer has a mind. The same is true for attempts to persuade, to arouse an emotion (such as sympathy), and so forth. Thus the evolution of language beyond a very rudimentary imperative form depends on such an awareness.

There is some evidence, although perhaps not conclusive evidence, that chimpanzees can attribute mental states to others, although monkeys appear to be unable to do so (see discussion in Cheney and Seyfarth 1990, chapt. 8). However, chimpanzees make virtually no use of it to engage in deliberate teaching. There are only two observed examples of possible pedagogy (Boesch 1993). This is so rare as to imply that pedagogy plays no significant role in chimpanzee adaptation. Thus, while the ability to attribute mental states to others may have evolved during the course of hominoid evolution, its application to information sharing appears to have developed primarily in the course of hominid evolution.

Language

Another major development during the course of hominid evolution was the appearance of human language. "Language," like "intelligence" or "culture," is a word open to many interpretations or definitions. Certainly primate communication has much in common with human language. Nevertheless, language as it exists among humans today is characterized by a number of features that make it a very efficient tool for sharing information. The first of these is symbolic reference. By this I mean that language is based on signs whose relationship to their referents is determined by arbitrary convention.[5] A sign is something (in the case of language, a sound or series of sounds) that points to or refers to something else (an object, concept, etc.). Sometimes, the reference is based on association, as smoke indicates fire or fever indicates illness. Sometimes the reference is based on resemblance, as with onomatopoetic words, representational art, and many gestures. Sometimes, especially in non-linguistic communication, reference seems to be at least in part genetically determined: human screams of pain or fear, for example. In language, however, the association between sign and referent is based entirely on arbitrary convention. Thus the sound combination /tu/ can in English refer to a number (two) or to a direction of movement (to) and in French to the concept of all (tout). At the same time, two different languages can use completely different signs (words) to refer to the same thing.

Symbolic reference as so defined is of major importance to information sharing and to information processing at the shared level, because as soon as signs need no longer be associated with their referents by genetics, association, or resemblance, it is much easier to invent new signs to refer to new concepts. The vocabulary is free to expand tremendously, and communication of ideas based on that vocabulary becomes both much more efficient and much more useful, in that it can cover a much broader range of topics. In addition, reference, and therefore communication, can be displaced; reference can easily be made to objects, situations, etc. that are not in the immediate environment of either speaker or hearer.

The advantages conferred by symbolic reference are extended by another feature of modern language, its duality of patterning (Hockett and Ascher 1964, Martinet 1960). In any modern language a limited number of sounds (phonemes) are used as the building blocks of speech, and these sounds are recombined in novel combinations to form new signs. Thus a relatively small number of phonemes can produce an almost infinite number of signs (words, morphemes, etc.). The burden placed on memory by a large vocabulary is reduced because all signs have at least one thing in common, they are constructed from the same set of phonemes.

The last major characteristic of modern human language is syntax, a set of arbitrary conventional rules that relate symbols to one another in a way that makes it possible to express the logical relationships among their referents. For example, the utterances, "The dog bites the postman," and, "The postman bites the dog," have the same basic symbols and the same referents but the syntax (in this case, word order) has been changed to indicate a different relationship between the dog, the postman, and the act of biting. Clearly, syntax is of critical importance to human information sharing and information processing at the shared level, because of both the complexity of ideas that can be expressed and the precision with which they can be expressed. However, syntax is not a unitary phenomenon, and there is nothing that demands *a priori* that modern syntax developed full blown. A number of scholars have argued for an evolution of language through a series of stages, each of them involving more complex structures (e.g., Bickerton 1990, Gibson 1990, 1994, Lieberman 1984, 1994, Whallon 1989).

There is evidence from ape language experiments that at least the ability to understand and use symbols is to be found among chimpanzees, as well as an ability to learn and use rudimentary syntax, although both claims have been challenged (Bronowski and Bellugi 1970, Fouts and Rigby 1977, Gardner and Gardner 1969, 1978, Miles and Harper 1994, Ristau and Robbins 1982, Sanders 85, Savage-Rumbaugh, Rumbaugh and Boysen 1978, 1980, Sebeok 1980, Sevcik and Savage-Rumbaugh 1994, Terrace 1979, Terrace, Petitto and Bever 1979). Whatever their abilities, there is at present little evidence that apes make use of symbolic reference in communications in the wild. It appears, therefore, to play no significant part in their adaptation. Like deliberate teaching, therefore, the actual use of symbolic language appears to have been a hominid development, whatever the underlying genetic abilities.[6]

Information Processing at the Social Level

Humans and other primates differ markedly in the ways that they process information at the social level. This is not to say that no information processing takes place on the social level among other species, but human systems differ in an important respect. This can best be illustrated by considering the difference between the aircraft design team cited above and cooperative hunting groups in other species.

Lions, wolves, and chimpanzees have all been observed to coordinate their movements in the course of hunting in order to minimize the chances that their prey will escape. This may take the form of fanning out over a broad front (in lions), surrounding a tree or other area so as to cut off escape routes (among chimpanzees), or even waiting while one member of the group moves to the other side of the prey so that the prey flees in the direction of a waiting hunter or hunters (observed among lions and wolves) (Goodall 1986:286-7, Griffin 1992:64-5, Kelsall 1968:252 in Mech 1970:230, Schaller 1972:249-51, Teleki 1973:131-2). What these strategies require on the part of a hunter is the ability to monitor and predict the locations and movements of the prey and of fellow hunters. It does not require (and apparently generally does not involve [Schaller 1972:251]) any communication on the part of the participants. Yet the result is that the movements of the hunting group is more than the sum of its parts.

This is reminiscent of a computer program written by Craig Reynolds (Waldrop 1992:241-243, see also Roetzheim 1994:66-81 and enclosed programs) that defines a series of virtual birds (called "boids"), each of which moves about the screen according to a three simple rules that attract it to other boids, make it try to move at the same speed as nearby boids, and make it try to maintain a minimum safe distance from other boids and from fixed obstacles in the environment. The result is a flock of boids that behaves very much like a flock of real birds, coming together, moving together, changing direction or splitting up to avoid an obstacle, and then reforming into a single flock.

This flock of boids is more than just the sum of its parts; it is also the patterning of their movements, patterning that can be explained by knowing the rules governing the movement of each boid. The case of hunting chimpanzees is similar. Both the flock of boids and the hunting groups are made up of individuals (boids and chimpanzees respectively), and these individuals all adjust their behaviors in response to the actions and locations of the individuals around them. This is still very different from modern human social information processing systems such as the team of aircraft designers cited above. Just where the difference lies between a chimpanzee hunting group and a human engineering team can be clarified by analogy with the difference between a flock of boids and a brain.

Both a flock of boids and a brain are information processing systems made up of individual units: boids in the first case and neurons in the second case. The individuals of both react to information from other individuals. Boids change their direction and speed; neurons transmit information to other neurons and also change the nature of their synaptic connections, thereby changing the movement of information through the neural network. This last point is important, because it means that neurons do something boids do not do: they change the information-bearing relationships among themselves in order to bring about a valid (adaptive) outcome for the system as a whole. That is, by processes not yet fully understood, neural networks change their configuration in response to feedback from the environment in such a way that they can "learn" to produce an appropriate output given a certain set of inputs (Churchland 1988). Boids, by contrast, cannot do this. They can only produce appropriate behavior at the individual level: proximity to other boids combined with an absence of collisions.

Hunting chimpanzees fall in between these two configurations. Unlike boids, they think and are therefore able to adjust their spatial relationships in order to achieve a purpose for the system as a whole.[7] What they do not do, however, is to adjust the *information-bearing* relationships among themselves in order to improve or redirect the *information processing* functions of the group (as opposed to its prey-capturing functions). Thus information processing systems among chimpanzees are limited to configurations of the social structure produced by factors other than the need or desire to process information. Modern humans, however, frequently configure social relationships in order to process information effectively. That is to say, they assign complementary information gathering, information transmitting, or information analyzing roles to different individuals. This is often a very simple and temporary matter. Lookouts may be posted with the agreed-upon responsibility for spotting either danger or game and then informing the rest of the group of what they have seen. Individuals or parties may be sent out in different directions to find and evaluate resources, and the group as a whole may then meet to decide on what actions to take on the basis of these reports. A certain individual may be responsible for coordinating a cooperative hunting drive while other individuals are assigned to locate the prey herd, to inform those members of the party out of sight of the herd of what is happening, etc. At the other end of the scale of complexity, this organization of social roles may take the form of establishing and organizing engineering teams, universities, or government bureaucracies.

The attribution of mental states to others is crucial to the reorganization of social structures in order to enhance information processing, because this cannot be done (other than by accident) without the recognition that information is passing from one individual to another. This can be clearly seen by considering the group that agrees to send out information-gathering parties in different directions in order to pool their knowledge later. This requires that each individual recognizes that members of other parties will gain information that will be useful but will differ from his or her own. More to the point, it requires that members of the returning parties recognize that they have information that members of the other parties do not have but would like to have. Finally, it requires that each individual understands that all the other individuals understand their responsibilities. All of these requirements involve a recognition that other individuals have minds and that what they know or think may differ from what one knows or thinks oneself.

By the same token, it is clear that beyond a very rudimentary level, the development of information processing at the social level will require at least some of the attributes of modern human language. Thus these two traits, attribution of mental states and at least a partially developed human language, must necessarily have evolved either before or along with information processing at the social level. Both traits would have been useful, albeit not necessarily essential, at the next lower level of information processing (information sharing), and so may well have evolved earlier. It is not clear to me that social information processing would have required, *a priori*, the evolution of any other new mental abilities. I will, however, leave this question to those in disciplines better equipped to address it.

Information Processing and Extended Symbol Systems

A fifth development took place at some point in the course of hominid evolution: symbolism was extended to include much more than reference. Up to this point, I have used the word symbolism in a limited sense, to mean reference by convention. Language has been presented as referential in nature. This is obvious in my description of the way in which arbitrary symbols are used semantically to refer to things or concepts. It is also true, however, of the way I have presented syntax. It is true that syntax involves the manipulation of relationships among symbols, but this manipulation is itself essentially referential in nature in that the relationships among the symbols are used to reflect or refer to the relationships in the mind of the speaker among the concepts to which they refer (as in the two sentences "The dog bit the postman," and "The postman bit the dog."). Presented in this manner, language and symbolism are merely tools by which hominids were able to express ideas that in their basic nature were similar to those of other primates.

However, as Byers (1994, see also Byers, this volume) has pointed out, in modern human life symbolism goes well beyond just reference. Symbols are used to define entities that owe their existence to symbolism and have no reality except in a symbolic context. The game of chess is an excellent example of such an entity. It is something that does not exist in the natural world, but instead consists of rules, objects, actions, etc. that are defined and agreed upon entirely by arbitrary convention. The purpose of the game is not to refer anything else; it may occasionally be used in this way (although it is hard to think of an example), but the game was not invented for such a purpose. Yet chess could not have been created without symbolism, it could not exist without symbolism, and it would have no meaning without symbolism.

Symbolically defined entities such as this take on myriad forms: concepts (paradise, justice), institutions (clans, secret societies), acts (initiation, coronation), rules for behavior, values, social roles, and so forth. Moreover, the lives of modern peoples are enmeshed in such symbols, and their behavior is to a very large extent governed by them. Symbolic entities are organized into all-encompassing symbol systems, systems that may be quite coherent and stable, or may be filled with contradictions and internal stresses and very unstable. However, almost everything that humans see, hear, say, or do involves them. Every act comes to have symbolic significance and is judged in terms of its symbolic meanings and the values assigned to it within the symbol system. The chimpanzee's life differs from ours in that while she may worry about whether her twig is straight enough to fish for termites, she need not worry about whether its shape will get her branded as a hopelessly out of step with fashionable society.

For the sake of convenience, I will use the term "symbolic culture" to refer to the extension of symbolism beyond reference to the creation of symbolic entities organized into all embracing symbolic systems. In doing so I make no claim that this is the true meaning of "culture" or that I reject any other concept of culture. However, I need a term with which to refer to this phenomenon, and it seems to me that this will fit as well as any other.

Symbolic Culture and Behavior

Saying that behavior is judged in terms of a symbol system is another way of saying that the information processing that controls behavior now takes place largely in terms dictated by symbolic culture. Perhaps it is not quite accurate to say that in this context behavior is controlled by a new level of information processing; information processing is still taking place at the social level. However, information processing is now taking place in a new medium, that of symbolically defined categories, values, and rules. This means that a major change in behavior can only take place when the symbolic culture or ideology is altered enough to permit it. It also means that culture may dictate behavior that would be rejected if only its concrete consequences were considered. Among extant human groups, the information processing that controls behavior must take place within the realm of symbolic culture. Information processing now not only exists at the first four levels described above, but also takes place in a new "medium" as well.

Symbolic Culture and Genetic Evolution

To the best of my knowledge, the extension of symbolism beyond reference to the creation of symbolic culture would demand nothing of the human brain that would not be demanded by the use of referential symbolism in language. This is not to say that no such trait or capacity exists, but as archaeologists we are not justified in assuming that the appearance of archaeological evidence for symbolic culture implies, *a priori*, genetic evolution. It is quite possible, even probable, that the origins of symbolic culture, like the origins agriculture millennia later, owed nothing to genetic change but were something that could be "invented," independently, in different places at different times in response to local conditions or to local historical facts.

Once information processing takes place in the context of symbolic culture, of course, it has extraordinary independence from genetic control. Symbolic entities are free to be invented, their meaning is free to change, and the relationships among them are also free to change. In fact one of the most remarkable things about culture, as opposed to the adaptations of most species, is its great variability in both time and space. Thus at this level, the freedom of information processing from genetic control is quite marked in two ways: changes in behavior can be caused by changes in a symbolic culture that have nothing to do with genetics, and changes in behavior can only take place when the culture, a social construct, changes as well.

Recapitulation

It would be useful here to recapitulate the main points made so far. I have argued that the information processing that governs behavior takes place at several levels. While the ability to process information at each of these levels owes its existence to the species' genetic code, the ability to construct more elaborate information processing systems at the higher levels also liberates from genetics the development of new information processing systems and, therefore, of new behaviors. The ability to attribute mental states to others, along with various components of language, underlie the development of modern information-processing systems. Finally, the extension of symbolism beyond reference to include almost all things and all actions in comprehensive symbol systems provides the medium in which much, probably most, human information processing occurs today.

It should be noted that I have said almost nothing about how or why such changes came about. I have not, for example, presented hypotheses to explain the emergence of language or any part of it. I have not discussed the mechanisms by which non-symbolic primate communication could have evolved into symbolic language. These are topics for linguists, psychologists, neuroscientists and the like. Nor have I discussed the timing of any of these changes in information processing. A few things are arguable *a priori*. Syntax could not have appeared before referential symbols for the simple reason that syntax *is* the relating of symbols one to another. Symbolic culture could not have existed without both the ability to attribute mental states to others and referential symbolism. Beyond that, however, little can be stated *a priori*. Did the ability to attribute mental states to others, referential symbolism, and deliberate teaching all develop simultaneously or separated from one another by hundreds of millennia? If simultaneous, was the evolution sudden, gradual, or punctuated? Did symbolic culture appear suddenly or bit by bit in conjunction with different aspects of language? These are questions that archaeologists must try to address, since after all it is our discipline that most directly studies what happened in the past. It may be, in fact, that some of the developments discussed predate our separation from other primate lineages, but they are all essential steps to the modern human condition.

Implications for Archaeology

This brings us back to the problem of how we are to interpret the archaeological record in terms of changes in information processing systems. It is tempting to attribute to genetic evolution any behavioral innovation that seems to involve a new kind of information processing. However, because information processing systems can change independently of genes, we need to look carefully at the nature of the changes in the underlying information processing systems. The two examples from the Paleolithic archaeological record cited at the beginning of this paper serve to illustrate how interpretation of the archaeological record relates to the framework described here.

The first example concerns the origins of the habitual flaking of stone tools. Enough experimental work has been done so that we can specify what a flintknapper needs to understand in order to remove flakes from a core on a regular basis (Dibble and Whittaker 1981, Dibble and Pelcin 1995). There must be a striking platform available with an angle between it and the exterior surface of the flake to be removed. This exterior surface should be at least somewhat convex in two directions (parallel and at right angles to the platform) and the platform must be adjacent to this convexity. Moreover, the flintknapper must be able to determine how far from the edge to strike the platform, taking into account the angle with the exterior platform, and must strike at an appropriate angle and with at least a minimum amount of force (although the latitude available for these last two is much greater than is generally believed). Thus, we know that once hominids were making stone tools regularly, they must have mastered these mental skills. We do not know whether or not these skills already existed, as preadaptations, or if they evolved in response to selective pressures generated by the need for reliable manufacture of stone tools.

However, there has been a very interesting experiment along these lines with a bonobo or pygmy chimpanzee (*Pan paniscus*) (Toth et al. 1993). The bonobo was motivated to produce flakes by being trained to use them to cut a rope or membrane to get access to food inside a box. He was also shown by example how to remove flakes from a core by striking a nodule with a hammerstone. While he did learn to do this, it is fairly clear from the experimental results that he did not master all of the skills listed above. For example, "Once he successfully detached a flake from an edge of the core, he tended to concentrate his blows in that same general area in attempts to produce more flakes. He sometimes was very persistent in this: dozens or even hundreds of subsequent blows might be directed towards one general area of a core" (Toth et al. 1993:85). In fact, what he seemed to have understood were the necessity for percussion, the amount of force necessary to detach a flake of useful size, and that where one hits the core may be important. However, he does not appear to have understood what criteria to use in judging this last point. In fact, he invented a method of flaking by throwing cores against a hard surface. This may indicate a certain ingenuity, but it also seems to indicate an inability to learn a more reliable technique. If this is correct, it would indicate that our closest living relatives lack some or all of the mental abilities that archaeological experiments have shown are necessary to produce flakes on a regular basis. Thus the appearance of a stone tool industry in the archaeological record does appear to reflect a genetic change that made possible new mental skills. Presumably, flintknapping was a learned skill for Oldowan hominids. That is, flintknapping undoubtedly involved both individual trial and error and observation of others. However, there was probably a change at the very lowest, genetically-controlled level of information processing as well. The data suggest that the neural anatomy of *Homo habilis* differed from that of other hominoids in ways that made it possible for them to recognize angles between surfaces, judge the relationship between platform edge and flake thickness, etc. This is just what we might expect of an early development in hominid information processing.

The second example, the increase between the Middle and Upper Paleolithic in the distances over which materials were transported within Europe, is somewhat less straightforward. As I mentioned above, a number of factors make it tempting to interpret this change as well to a change in the gene pool of European hominids about this time. However, in light of the above analysis, an alternative explanation presents itself. The movement of goods over greater distances may reflect the establishment of trading networks, which is likely to have been a purely a social development.

During the Lower and Middle Paleolithic, what was transported were raw materials for stone tools and the tools themselves. In the Upper Paleolithic, these materials were transported over distances equivalent to those of the later Middle Paleolithic (up to about 300 km. [Roebroeks et al. 1988, Table 1]). What were transported over much longer distances were not ordinary materials but mollusk shells. These are used in the European Upper Paleolithic for decorative and presumably symbolic purposes. This implies that both the choice of the materials transported and the distances over which they were transported reflect the organization of large-scale social networks and a demand generated by the appearance of symbolic culture. This social and cultural versus genetic interpretation is reinforced by the fact that the distances over which mollusk shells were transported continued to increase to the Neolithic (ibid., Fig. 2).

This is completely in line with the other changes that mark the beginning of the Upper Paleolithic in Europe, the frequent appearance in the archaeological record of art, decorative objects, and even standardized stone tools.[8] Phenomena such as the transport of items like shells over long distances is exactly what one would expect in the archaeological record as a reflection of the appearance of symbolic culture, that is, of the appearance of the final level of information processing. As I stated above, I know of no solid grounds for attributing a genetic cause to such an extension of symbolism beyond pure reference.

Conclusion

The story of human evolution appears to have been in large part the story of the progressive liberation from strict genetic control of both behavior and the information processing behind that behavior. This has been possible because information processing structures exist at a number of different levels. At each higher level, behavior gained more independence from the underlying genetic code. Of course, the ability to learn or to construct new information processing structures depended on the evolution of other abilities, such as the ability to attribute mental states to others or to make use of symbolic reference.

At the same time, however, this increasing freedom from genetic control means that archaeologists must be wary of attributing new behaviors manifested in the archaeological record to changes in the underlying gene pool. The more the burden of information processing is transferred up through the various levels from learned to social to symbolic, the more likely it is that major changes can occur without any concomitant genetic evolution. Thus we must turn to other disciplines to help us identify the social, psychological or genetic changes that could account for such changes.

We Paleolithic archaeologists are relative newcomers to the systematic study of the evolution of human information processing. There are good reasons for this. We have been busy constructing chronologies and establishing grounds in middle range theory, taphonomy, and in lithic and faunal analysis for inferring specific behaviors from patterns in the archaeological record. I believe that we are now in a position to turn at least some of our attention to the information processing systems behind those behaviors. However, we must do so carefully and systematically, working along with, rather than in isolation from, other disciplines, such as psychology and neuroscience, that are already fully engaged in the problem.

Acknowledgments

I would like to thank April Nowell for bringing about both the publication of this volume and the symposium on which it is based. I have received very helpful comments and suggestions from Igor Kopytoff, Harold Dibble, Shannon McPherron and Dorothy Cheney, who read earlier versions of the manuscript.

Notes

[1] Raymond Dart (1949, 1957, 1958, 1960) argued that before the appearance of habitual stone-tool making, hominids made and used an "osteodontokeratic" tool kit, that is, one making use of animal bones, teeth, and horns. This hypothesis was based on the non-random relative frequencies, within Australopithecine-bearing deposits, of bones from different parts of the skeleton, as well as certain patterns in the way these bones were broken. However, careful taphonomic analysis, primarily by Brain (1967a, b, 1976, 1981) demonstrated that these patterns were typical of those produced by carnivore activity. Thus the apparent osteodontokeratic culture was simply the product of natural causes. This does not mean that such material was never used as tools, but simply that there is no evidence for systematic, habitual modification of them for use as tools.

[2] From a neurolobiological point of view "learning" is perhaps an imprecise or even misleading term. Although "plasticity" might be a better term, "learning" is more familiar to archaeologists, and I shall continue to use it here. What I am putting in this category are all the information processing mechanisms that reside within an individual's nervous system that can change independently of the underlying genetic code. The exact process by which such change takes place is not important in the current context.

[3] The boundaries between the levels of information processing I am describing are not always sharp. This is most obvious in the case of developmental learning. For example, all humans, given a normal physique and not deprived of the opportunity to do so, learn to walk, and they all walk in more or less the same manner. Walking must be learned, but learning to walk is both prompted and tightly controlled by the genetic code. Developmental learning straddles the borderline between genetically determined and learned. Such ambiguous cases should not detract from the point I am trying to make here, however.

[4] There is considerable disagreement as to whether the brain structures that make language possible are highly specific and localized or generalized(e.g., Geschwind 1970, Gibson 1990, 1994, Jerison 1988, Lieberman 1984, Holloway1983a, b, Wilkins and Wakefield 1995). In either case, however, the inability of non-human primates to learn more than a rudimentary form of human language indicates that the ability to learn language like a human is the product of genetic evolution in the hominid line.

[5] Such signs are what Peirce (1932/1960) called symbols and what de Saussure (1959) called linguistic signs. See Chase (1991) for a more complete discussion.

[6] Vervet monkeys present an intriguing case (Cheney and Seyfarth 1990). They use calls to warn of different kinds of predators, calls that appear to be genetically determined in that they are the same in all groups. However, the vervets must learn what constitutes a given kind of predator. For example, young vervets have been observed giving an avian-predator alarm to a falling leaf, and one captive group uses the terrestrial predator alarm for vet-

erinarians. While there is thus a certain freedom in the relationship between sign and referent, the freedom is very limited by comparison with true symbolic reference. The signs are genetically determined, and each can be used to refer to only a narrow range of referents. Vervets are not free, for example, to use the snake alarm call to refer to a particularly tasty food.

[7] I am assuming for the sake of argument something that is not necessarily clear from the data, that the chimpanzees are in fact cooperating, that an individual's purpose is not to enhance the probability that he will catch the prey but to enhance the probability that some individual in the group will catch the prey.

[8] It is possible that Upper Paleolithic stone tools (at least in Europe) were more standardized than in the Middle Paleolithic (Mellars 1989:365, 1996:405, Stringer and Gamble 1993:208-9). This is perhaps less demonstrable for the early than for the later Upper Paleolithic. I know of no systematic study designed to demonstrate such standardization beyond the constraints imposed by blade technology.

References Cited

Axelrod, R.
   1984 *The Evolution of Cooperation*. Basic Books, New York.

Bechtel, W.
   1988 *Philosophy of Science: an Overview for Cognitive Science*. Lawrence Erlbaum, Hillsdale, N.J.

Bickerton, D.
   1990 *Language and Species*. University of Chicago Press, Chicago.

Boesch, C.
   1993 Aspects of transmission of tool-use in wild chimpanzees. In *Tools, Language and Cognition in Human Evolution*, edited by K. Gibson and T. Ingold, pp. 171-184. Cambridge University Press, Cambridge.

Brain, C.
   1967a Hottentot food remains and their meaning in the interpretation of fossil bone assemblages. *Scientific Papers of the Namib Desert Research Station* 32:1-11.
   1967b Procedures and some results in the study of Quaternary cave fillings. In *Background to Evolution in Africa*, edited by W. Bishop and J. D. Clark, pp. 285-301. The University of Chicago Press, Chicago.
   1976 Some principles in the interpretation of bone accumulations associated with man. In *Human Origins*, edited by G. Isaac and E. McCown, pp. 97-116. Benjamin, Menlo Park.
   1981 *The Hunters or the Hunted?* The University of Chicago Press, Chicago.

Bronowski, J. and U. Bellugi
   1970 Language, Name and Concept. *Science* 163:669-673.

Byers, M.
   1994 Symboling and the Middle-Upper Palaeolithic transition: a theoretical and methodological critique. *Current Anthropology* 35:369-400.

Churchland, P. S.
   1986 *Neurophilosophy: toward a Unified Science of Mind/Brain*. Bradford Books. MIT Press, Cambridge, MA.
   1988 *Matter and Consciousness: a Contemporary Introduction to the Philosophy of the Mind. (Revised edition)*. MIT Press, Cambridge.

Cheney, D. L. and R. M. Seyfarth
   1990 *How Monkeys see the World*. University of Chicago Press, Chicago.

Chase, P. G.
   1991 Symbols and Paleolithic Artifacts: Style, Standardization, and the Imposition of Arbitrary Form. *Journal of Anthropological Archaeology* 10:193-214.

Chase, P. G. and H. L. Dibble
   1992 Scientific Archaeology and the Origins of Symbolism: A reply to Bednarik. *Cambridge Archaeological Review* 2:43-51.

Dart, R. A.
   1949 The predatory implemental technique of Australopithecus. *American Journal of Physical Anthropology* 7:1-38.
   1957 *The osteodontokeratic culture of Australopithecus prometheus*. Transvaal Museum Memoirs, no. 10.
   1958 The minimal bone-breccia content of Makapansgat and the australopithecine predatory habit. *American Anthropologist* 60:923-931.
   1960 The bone tool-manufacturing ability of Australopithecus prometheus. *American Anthropologist* 62:134-143.

Dawkins, R.
   1976 *The Selfish Gene*. Oxford University Press, Oxford.

Dennett, D. C.
    1987   *The Intentional Stance.* Bradford Books. MIT Press, Cambridge.

Dibble, H. L. and A. Pelcin
    1995   The Effect of Hammer Mass and Velocity on Flake Mass. *Journal of Archaeological Science* 22:429-439.

Dibble, H. L. and J. Whittaker
    1981   New Experimental Evidence on the Relation Between Percussion Flaking and Flake Variation. *Journal of Archaeological Science* 6:283-296.

Fouts, R. S. and R. L. Rigby
    1977   Man-chimpanzee communication (reprinted 1980]. In *How Animals Communicate*, edited by T. A. Sebeok, pp. 1034-54. Indiana University Press, Bloomington.

Gallistel, C.R.
    1980   *The Organization of Action: A New Synthesis.* Erlbaum, Hillsdale, N.J.

Gamble, C.
    1982   Interaction and alliance in palaeolithic society. *Man* 17:92-107.

Gardner, B. T. and R. A. Gardner
    1969   Teaching sign language to a chimpanzee. *Science* 165:664-672.
    1978   Comparative psychology and language acquisition [reprinted Sebeok and Sebeok 80]. In *Psychology: the State of the Art*, edited by K. Salzinger and F. L. Denmark. Annals of the New York Academy of Sciences 309.

Geschwind, N.
    1970   Organization of language and the brain. *Science* 170:940-944.

Gibson, K. R.
    1990   New perspectives on instincts and intelligence: Brain size and the emergence of hierarchical mental constructional skills. In *"Language" and Intelligence in Monkeys and Apes*, edited by S. Parker and K. Gibson, pp. 97-128. Cambridge University Press, Cambridge.
    1994   Continuity theories of human language origins versus the Lieberman model. *Language and Communication* 14:37-58.

Goodall, J.
    1986   *The Chimpanzees of Gombe: Patterns of Behavior.* The Belknap Press of Harvard University Press, Cambridge.

Gould, R., D. Koster and A. Sontz
    1971   The Lithic Assemblage of the Western Desert Aborigines of Australia. *American Antiquity* 36:149-69.

Griffin, D. R.
    1992   *Animal Minds.* University of Chicago Press, Chicago.

Hamilton, W. D.
    1964   The genetical evolution of social behavior I, II. *Journal of Theoretical Biology* 7:1-52.

Hockett, C. and R. Ascher
    1964   The human revolution. *Current Anthropology* 5:135-168.

Holloway, R. L.
    1983a   Human Brain Evolution: A Search for Units, Models and Synthesis. *Canadian Journal of Anthropology* 3:215-230.

Holloway, R. L.
    1983b   Human paleontological evidence relevant to language behavior. *Human Neurobiology* 2:105-114.

Howe, M. J. A.
    1988   Intelligence as explanation. *British Journal of Psychology* 79:349-360.

Jerison, H. J.
    1988   Evolutionary neurobiology and the origin of language as a cognitive adaptation. In *The Genesis of Language*, edited by M. E. Landberg, pp. 3-10. Mouton de Gruyter, Berlin.

Kail, R. and J. W. Pellegrino
    1985   *Human Intelligence: Perspectives and Prospects.* W.H. Freeman and Company, New York.

Kelsall, J. P.
    1968   *The Migratory Barren-Ground Caribou of Canada.* Department of Indian Affairs and Northern Development, Canadian Wildlife Service, Ottawa.

Lawick-Goodall, Jane van
    1970   Tool-using in primates and other vertebrates. In *Advances in the Study of Behavior*, edited by D.S. Lehrman, R.A. Hinde, and E. Shaw, vol. 3, pp. 195-249. Academic Press, New York.

Lieberman, P.
    1984   *The Biology and Evolution of Language.* Harvard University Press, Cambridge.
    1991   *Uniquely Human: The evolution of Speech, Thought, and Selfless Behavior.* Harvard University Press, Cambridge.
    1994   Human thought and human uniqueness. *Language and Communication* 14:86-96.

Martinet, A.
    1960   *Eléments de liguistique générale.* Armand Colin, Paris.

Mech, L.
  1970 *The Wolf*. Natural History Press, New York.

Mellars, P.
  1989 Major issues in the emergence of modern humans. *Current Anthropology* 30:349-385.
  1996 *The Neanderthal Legacy. An Archaeological Perspective from Western Europe*. Princeton University Press, Princeton.

Miles, H. and S. Harper
  1994 "Ape Language" studies and the study of human language origins. In *Hominid Culture in Primate Perspective*, edited by D. Quiatt and J. Itani, pp. 253-276. University Press of Colorado, Niwot.

Parker, S. T. and B. Baars
  1990 How scientific usages reflect implicit theories: Adaptation, development, instinct, learning, cognition, and intelligence. In *"Language" and Intelligence in Monkeys and Apes*, edited by S. T. Parker and K. R. Gibson, pp. 65-96. Cambridge University Press, Cambridge.

Peirce, C. S.
  1932/1960 The icon, index, and symbol. In *Collected Papers of Charles Sanders Pierce, vol. II*, edited by C. Hartshorne and P. Weiss, pp. 156-73. Harvard University Press, Cambridge.

Plotkin, H. C.
  1988 Intelligence and evolutionary epistemology. *Human Evolution* 3:437-448.

Premack, D. and G. Woodruff
  1978 Does the chimpanzee have a theory of mind? *Behavioral and Brain Sciences* 1:515-26.

Ristau, C. A. and D. Robbins
  1982 Language in the great apes: a critical review. *Advances in the Study of Behavior* 12:141-255.

Roebroeks, W., J. Kolen and E. Rensink
  1988 Planning depth, anticipation and the organization of Middle Paleolithic technology: the "archaic natives" meet Eve's descendants. *Helinium* 28:17-34.

Roetzheim, W.
  1994 *Enter the Complexity Lab*. Sams Publishing, Indianapolis.

Roitblat, H. L.
  1985 *Introduction to Comparative Cognition*. W.H. Freeman, New York.

Sanders, R. J.
  1985 Teaching apes to ape language: explaining the imitative and nonimitative signing of a chimpanzee (*Pan troglodytes*). *Journal of Comparative Psychology* 99:197-210.

Saussure, F. d.
  1959 *Course in General Linguistics*. Edited by Charles Bally and Albert Sechehaye in collaboration with Albert Riedlinger. Translated by Wade Baskin. Philosophical Library, New York.

Savage-Rumbaugh, E. S., D. M. Rumbaugh and S. Boysen
  1978 Linguistically-mediated tool use and exchange by chimpanzees (*Pan troglodytes*) [reprinted in Sebeok and Sebeok 1980 - see Sebeok 1980]. *Behavioral and Brain Science* 4:539-554.
  1980 Do apes use language? *American Scientist* 68(January-February):49-61.

Schaller, G. B.
  1972 *The Serengeti Lion*. University of Chicago Press, Chicago.

Sebeok, T. A.
  1980 Looking in the destination for what should have been sought in the source. In *Speaking of Apes: a Critical Anthology of Two-Way Communication with Man*, edited by T. A. Sebeok and J. Umiker-Sebeok, pp. 407-428. Plenum Press, New York.

Sevcik, R. A. and E. S. Savage-Rumbaugh
  1994 Language comprehension and use by great apes. *Language and Communication* 14:37-58.

Sternberg, R. J. and W. Salter
  1982 Conceptions of intelligence. In *Handbook of Human Intelligence*, edited by R. J. Sternberg, pp. 3-28. Cambridge University Press, Cambridge.

Stringer, C. and C. Gamble
  1993 *In Search of the Neanderthals: Solving the Puzzle of Human Origins*. Thames and Hudson, London.

Teleki, G.
  1973 *The predatory behaviour of wild chimpanzees*. Bucknell University Press, Lewisburg, PA.

Terrace, H. S.
  1979 *Nim*. Knopf, New York.

Terrace, H. S., L. Petitto and T. G. Bever
  1979 Can an ape create a sentence? *Science* 206:891-900.

Toth, N., K. D. Schick, E. S. Savage-Rumbaugh, R. A. Sevcik and D. M. Rumbaugh
  1993 Pan the Tool-Maker: Investigations into

the stone tool-making and tool-using capabilities of a Bonobo (*Pan paniscus*). *Journal of Archaeological Science* 20:81-91.

Trivers, R. L.
1971 The evolution of reciprocal altruism. *Quarterly Review of Biology* 46:35-57.

Waal, F. de
1989 *Chimpanzee Politics: Power and Sex among Apes*. Johns Hopkins University Press, Baltimore.

Waldrop, M. M.
1992 *Complexity: the Emerging Science at the Edge of Order and Chaos*. Simon and Schuster, New York.

Whallon, R.
1989 Elements of Cultural Change in the Later Palaeolithic. In *The Human Revolution*, edited by P. Mellars and C. Stringer, pp. 433-454. Edinburgh University Press, Edinburgh.

Wilkins, W. K. and J. Wakefield
1995 Brain evolution and neurolinguistic preconditions (with comments). *Behavioral and Brain Sciences* 18:161-226.

10. Behavioral Response to Variable Pleistocene Landscapes

Richard Potts

Abstract

Data from Olduvai, Olorgesailie, Kanjera (eastern Africa) and Bose (southern China) illustrate that Pleistocene toolmakers confronted diverse rock sources, vegetation and water distributions, and unstable ecological conditions. While stone artifacts seem to imply highly standardized behavior, land use suggests that mid-Pleistocene hominids were responsive to new contingencies. This responsiveness was honed by habitat variability, which increased over the past 700,000 years. This process of response to variable landscapes (rather than refined adaptation to a single habitat type) incited cognitive change in *Homo*, including greater information processing, neuronal plasticity, and open programs of behavior enabling mental and social flexibility.

The Boundary Conditions of Hominid Adaptation

Over the past 20 years, I have been engaged in comparing the environmental settings and behavioral traces of late Pliocene to mid-Pleistocene hominids *sensu stricto* (or hominins, referring to bipedal descendents after the human-chimp divergence). This research has two objectives. The first is to uncover how the activities of toolmakers covaried with habitat and how hominids used resources across past landscapes. This objective requires sampling, through excavation, single stratigraphic layers that represent relatively brief periods (ten to one thousand years long) and are exposed over broad areas (Potts 1994; Potts et al., n.d.). The second objective is to document sequences of environmental change over hundreds of thousands of years, and to assess whether hominids and their behavior persisted or changed in relation to their surroundings (Potts 1998b). Both of these approaches offer data about the adaptive contexts of hominid social groups, from which we may develop and test hypotheses about the cognitive challenges they faced.

The union of paleoenvironmental and archeological data enables one to address the ecological boundary conditions of early hominid adaptation. By *boundary conditions* I mean the climatic, physical geographic, biotic, and resource settings in which hominids persisted within a basin or spread across a region during particular time spans. Hominid evolution has a dramatic ecological and geographic aspect—the oldest hominids appear to have been tropical African in origin, while the single remaining species is worldwide. It is, therefore, edifying to ascertain whether hominids of a particular time and place tracked certain types of climate or vegetation, and under what conditions they proved capable of breaking particular environmental or geographic barriers. The only way to determine this is by correlation, using the finest possible resolution, between hominid fossils/artifacts and environmental data in a variety of stratigraphic sequences. With this approach, archeological data become evermore pertinent to the study of hominid adaptation and evolutionary history.

Archeologists involved in early hominid research are asking increasingly difficult questions that go beyond the traditional focus on artifact typology and stone technology. They seek to connect their research to paleoecological issues and to discover the adaptive context of hominid evolution. So, for example, stone tools can be studied as resources (i.e., raw materials) obtained from known places (e.g., outcrops) and carried over measurable distances (Toth 1985; Potts 1988, 1994). Artifact discard and reduction strategies can thus be examined in terms of alternative source outcrops, transport distances, and mechanical properties of the rocks. Archeological faunal remains can be assessed from a functional morphological perspective to infer the habitat breadth of hominid bone collectors (Plummer and Bishop 1994). The range and abundance of faunal body parts, furthermore, offer important clues about foraging strategy, diversity, and competition (Blumenschine 1991; Monahan 1996; Potts 1988). By comparing archeological fauna across regions, researchers may, moreover, determine whether certain species were "fellow travelers"—animals associated with hominids, possibly part of an ecological community or susceptible to similar ecological constraints and opportunities (Turner 1984; Turner and Wood 1993). This type of research helps to assess one of the central questions of human evolution research —to what extent were hominid toolmakers unique in their environmental adaptations, or governed by general constraints and opportunities similar to those experienced by other large mammals? Was the spread of early Pleistocene hominids from Africa to Eurasia, for example, unique among mammals, thus indicating some peculiar form of adaptation related perhaps to stone technology, dietary breadth, or social flexibility? In answering these questions, the study of other mammals is as important as the study of hominids.

These new research directions represent a stunning development for archeologists. What were once distinct boundaries between archeology and such diverse fields as evolutionary biology, paleoecology, geology, primatology, and evolutionary psychology can now be viewed as a complex intersection, the rigorous synthetic field of human origins research, which appears to promise a more penetrating scrutiny and complete understanding of hominid behavioral ecology and evolution. Critical to this endeavor is the comparison of early hominid localities where the actual evidence of hominids, the traces of their activities, the landscapes they inhabited, the organisms they lived with, and the adaptive contexts vital to their evolutionary history, all can be investigated.

In this chapter I will present evidence about the diversity of landscapes confronted by early hominids. In doing so, I will briefly compare four late Pliocene and early Pleistocene basins. Data from these localities illustrate that hominid toolmakers, over a period of ~2.3 to 0.7 million years ago (Ma), confronted diverse rock sources and changing ecological conditions. While stone artifacts seem to imply highly standardized behaviors, an investigation of land use shows that Pleistocene toolmakers exhibited an increasing degree of responsiveness over time to the varied terrains and shifting habitats represented in these regions. Accordingly, I propose, the unique characteristics of human cognition and cultural capacity evolved as a means of adapting to variable landscapes, resource unpredictability, and dramatic habitat fluctuation.

Hypotheses of Hominid Cognitive Evolution

There are two main types of hypothesis concerning hominid cognitive and brain evolution. The first posits how a particular constraint on brain size — such as available metabolic energy (Aiello and Wheeler 1995) or the need to dissipate heat (Falk 1990) — was alleviated during hominid evolution, thereby allowing the brain to increase in complexity and function (see also Weaver et al., this volume and Lieberman, this volume). The second deals directly with the adaptive issue of brain function and, by implication, the factors of natural selection that contributed to, or drove, the evolution of brain size, complexity, and cognition. This type of hypothesis concentrates on what the brain is good for, how certain of its operations enhanced reproduction and survival among early hominids, and why the brain attained a remarkably large size in several hominid lineages. A complete understanding of hominid brain evolution must ultimately bridge the two types of hypothesis — i.e., it must consider the costs, constraints, and benefits of encephalization, reorganization, and the evolution of specific cognitive functions.

In the realm of adaptive hypotheses, two competing views have dominated efforts to identify the driving forces of cognitive evolution in primates and other mammals. Ecological hypotheses draw attention to the relative complexity of various foraging strategies. Brain evolution, accordingly, is related to food distribution and the mental representation required to obtain food, or to the demands of extracting and processing food once it is found (e.g., Milton 1981; Parker and Gibson 1977). By contrast, the social brain hypothesis relates brain evolution to social complexity — i.e., group size or the intricacy of intragroup encounters (e.g., Jolly 1966; Humphrey 1976; Byrne and Whiten 1988; Dunbar 1998). Advocates of this hypothesis have argued that neocortical size in higher primates is poorly correlated with foraging area or extractive foraging methods (Dunbar 1995; but see Byrne [1998]).

Much thinking about human brain evolution, in particular, has been devoted to function and adaptive factors — e.g., how brains helped hominids in finding food, flaking tools, foiling predators, and forging social alliances. But these considerations have usually been too general, conducted in a vacuum impoverished of information about the time, place, or settings in which a particular behavior or cognitive operation might have been advantageous, or a particular constraint could have been lifted. The tenet of evolution by natural selection implies that costs, benefits, and adaptive causes are tied to particular conditions, resource distributions, evolutionary opportunities, and adaptive constraints. Most hypotheses about hominid brain evolution are, however, devoid of context, which is as vital to adaptive evolution as genetic variation itself.

It is here where paleoenvironmental information may productively lead to another type of hypothesis by adding contextual detail about the specific adaptive settings encountered by early hominid populations. Natural selection does not work in relation to vague, contextless stretches of time. Rather, its effects are realized in relation to specific resource distributions, predation risks, social group compositions, and densities of intra- and inter-specific competitors, with the results compounded over time according to both stability and variability in these factors. As we will see, certain of these variables may be appreciated by examining the actual basins and environments where hominids lived.

Environmental hypotheses of hominid evolution are typically habitat-specific. They identify a particular type of habitat or environmental trend that provided the most challenging or likely setting for the evolution of human adaptive characteristics. In recent years, habitat-specific hypotheses of hominid evolution have focused on the importance of environmental drying in Africa (Vrba 1988; Vrba *et al.* 1989; deMenocal 1995), open savanna habitat (Klein 1989), riparian woodland (Blumenschine 1987), a stable mosaic of grassy woodland (Kingston *et al.* 1994), or forest (Rayner *et al.* 1993; Boesch-Achermann and Boesch 1994).

An alternative, the variability selection hypothesis, is suggested by the dramatic increase in environmental fluctuation apparent in deep-sea and regional records of continental vegetation, temperature, and moisture over the span of hominid evolution (Potts 1996a,b; 1998a,b). According to this hypothesis, the escalation of environmental remodeling from seasonal to orbital time scales presented a crucial adaptive problem. Episodic revamping of landscapes led to shifts in resource distributions and disparities in adaptive settings. Most species were either sufficiently mobile or dispersed to track favored habitats or they became extinct. The idea of variability selection applies to other species, notably certain hominids and other large mammal lineages (Potts 1998a): Different individuals of the same lineage at different times faced diverse conditions of reproductive success and survival; large, repeated shifting of selective conditions favored genes and developmental paths that happened to build adaptive versatility.

A visual model of this process (Potts 1996b; 1998a) suggests how inconsistencies in natural selection may bias against habitat-specific adaptations and favor a type of adaptive versatility that essentially decouples an organism from narrow habitat conditions. The result, depending on genetic variation, is the evolution of complex anatomical, cognitive, and social functions capable of processing and responding to intricate and variable environmental information. These new functions, which evolve as a lineage faces an escalating series of habitat extremes, are considered to be adapted to environmental novelty. In this sense, the variability selection hypothesis runs counter to the canonical aim of paleoanthropology to discover the single niche, habitat, or environmental trend that most "challenged" hominid populations, thus to which hominid evolution was a response.

Traces of Early Hominid Artifacts and Environments in Ancient Basins

In this section, I will characterize the geomorphology and environmental settings of four sedimentary basins in which hominids deposited stone tools. The basins, which span the late Pliocene to early middle Pleistocene, include the African localities of Omo-Turkana, Olduvai Beds I and II, Olorgesailie, and, in southern China, the Bose basin (Figure 1). The objective of examining these disparate localities is to begin to develop the comparative basis needed to uncover the range of environmental variation encountered by early hominids and to infer the responses by toolmakers to those varied settings. First, however, we must turn to a problem that has substantially been resolved in human evolutionary studies but, oddly, still looms large in the archeology of human origin.

The Single Species Problem

The problem posed by the single species hypothesis — the idea that tools reflect the broad cultural niche of a single hominid species at any given time (Wolpoff 1971) — has not been satisfactorily resolved in early hominid archeology. Per-

TIME (Ma)	LOCALITY
~ 0.78	BOSE BASIN (southern China)
1.20 - 0.78	OLORGESAILIE (southern Kenya)
1.7 - 1.0	KANJERA (western Kenya)
1.85 - 1.20	OLDUVAI, BEDS I AND II (northern Tanzania)

Fig. 1. Study localities discussed in the text and their temporal ranges.

haps it is the only subfield of paleoanthropology where this hypothesis persists. There are several reasons to reject it, including:

(1) compelling evidence that tool manipulation and manufacture occur widely among higher primates, though tool behavior is not a shared primitive trait of African apes, including hominids. (Free-ranging bonobos and gorillas are not known to be toolmakers, thus toolmaking is not necessarily a character at the root of the hominid clade *sensu stricto*);

(2) the growing consensus that multiple hominid species co-existed for much of the past 4 million years (thus the availability of several possible stone-toolmaking species of bipeds);

(3) evidence that multiple Plio-Pleistocene taxa (e.g., *Australopithecus (P.) robustus* and early *Homo*) possessed a subset of morphological traits indicative of stone toolmaking (Susman 1994), though Marzke (1997) presents evidence that certain Plio-Pleistocene hominids (e.g., early *Homo*) evolved a wider complex of such traits, making them more effective in stone tool behavior than other hominid species; and

(4) a growing consensus that early stone tools cannot be equated with "culture" in a modern human sense.

The last point represents a key paradigm shift in early hominid archeology and begs further comment. Paleolithic stone artifacts are an outcome of several interacting factors such as stone procurement, rock mechanics, intended tool function, and skill in core reduction. It has become apparent that these flaking behaviors and factors affecting stone transport and tool activity may cut across species boundaries, in which case they do not reflect relatively narrow rule-bound systems of symbolic behavior in the same sense as distinct cultures do in modern humans. The mere presence of a distinct lithic industry or tool type (e.g., an Acheulean handaxe, a Châtelperronian point) evidently does not signal a behavioral adaptation unique to one hominid species much less a particular cultural group within a species.

If indeed stone tools and archeological sites cannot safely be ascribed to one hominid species or another, what contribution can early hominid archeology make to the study of human cognitive evolution? Stone tools, their spatial and temporal distribution, and the range of habitats used by toolmakers can be considered traits, like cranial capacity or dental measures, which can be compared and plotted over time in order to discern trends within the hominid clade without explicitly assigning those traits to particular taxa. The goal of this exercise is to try to discover any pattern or trend in the activities and environmental interactions of hominid toolmakers, especially to ascertain the responsiveness of hominid toolmakers to the range of environmental conditions they encountered. Did they (all stone toolmakers) confine their tool behavior to certain features of the terrain (outcrops, water sources) independent of the variation between areas? Did Pleistocene hominids, moreover, tend to track certain favored habitats despite environmental fluctuations over time?

Paleolithic tool manufacture is often considered to have been highly conservative; thus many researchers conclude that early and middle Pleistocene hominids were not especially "smart" with regard to the variation around them. While there was more than one way to make an Acheulean biface (Belfer-Cohen and Goren-Inbar 1994), it is striking how persistent handaxe/cleaver/pick morphologies were over broad regions, independent of the diverse properties of rocks, from 1.5 to 0.2 Ma.

If, as I will propose, responsiveness to surrounding contingencies is a useful proxy for cognitive functioning (or "intelligence"), it would be helpful to expand the range of prehistoric data for assessing behavioral conservatism versus flexibility in hominids over time. The point here is that hominid toolmakers did leave behind signals of their responsiveness to Pliocene and Pleistocene landscapes, and the discovery of these signals on a landscape and interbasin scale is quickly becoming the pre-eminent charge of early hominid archeological research.

This still leaves us with a "grade-oriented" analysis in which we must be content with finding overall trends in hominid behavior — and accept our continued ignorance of how stone tools and habitat use corresponded with hominid species where more than one stone-toolmaking taxon might have existed.

The Omo-Turkana Basin (2.3 and 1.6 million years ago)

In present-day northern Kenya and southern Ethiopia, sites of late Pliocene age (ca. 2.3 Ma) have been found on both the west and east sides of the ancient Omo river system. The artifact occurrences were formed between the channels of alluvial fans near their confluence with the Omo axial drainage (Feibel et al. 1991; Rogers et al. 1994). While volcanic highlands lay far to the west of the Lokalalei site of West Turkana (Kibunjia 1994), dispersed clasts used in making stone tools occurred locally, where minor alluvial channels approximated the major river axis. In this time range, only low density clusters of artifacts are known, and they are located very close to the original stone sources, usually within several tens of meters. To the north, sites found on the eastern side of the Omo river system (Member F, Shungura Formation) also occurred in the toes of the alluvial fans, though the latter originated in metamorphic (rather than volcanic) highlands. Ephemeral streams coming off the marginal drainage introduced metamorphic clasts close to the meandering axial system. It was near this intersection that hominids found rocks and created artifact clusters. According to Rogers et al. (1994:149), "proximity to both the axial and marginal systems does seem to be critical [for the toolmakers] at this early stage, for in other localities of the same temporal interval and with a proximity to only one or the other system, artifact occurrences have not been recorded."

Rogers et al. (1994) provide an excellent comparison of this late Pliocene pattern with that of later times in the same basin. In the interval 1.9-1.8 Ma, hominid toolmakers are recorded in a wider diversity of settings, including a lake margin setting and within the axial drainage itself, where hominids first appear to have collected stone from the ancestral Omo. This apparent diversification of archeological settings took place during a period of local climatic instability.

In the Lower Okote Member of the Koobi Fora Formation (East Turkana) approximately 1.6 Ma, we find that new, distinctive tool types occur and that the artifact clusters are more varied. Research organized by G.Ll. Isaac and J.W.K. Harris (Isaac 1997) and recent work by M.J. Rogers (1997) document that artifact sites were less tied to stone sources. Toth (1985), for example, shows that the size of cores declines with distance from the lava highlands, while Bunn's (1994) study of faunal remains indicates that hominids venturing farthest from the highlands used stone tools for butchery without discarding them. Thus, although the toolmakers were limited by the size of available stream cobbles, stone toolmaking by one or more species of early Pleistocene hominids involved considerable transport of implements away from the stone sources.

Feibel's paleogeographic work points to the increased complexity of settings encountered by hominids during this span (Rogers et al. 1994). Reconstructions of the period 1.7 to 1.5 Ma suggest an intricate response by toolmakers to a dynamic landscape that was dominated on occasion by a meandering fluvial regime similar to late Pliocene settings and, at other times, by a broad system of shallow, braided channels. The latter created a highly varied spatial and temporal array of environments, modified as new channels were cut, easily flooded, and rapidly filled by sediment. Toolmakers left lithic debris in nearly the entire range of depositional settings—channels, gravel bars, the banks of major river beds and minor channels, and in proximal and distal floodplains. Artifact clusters ranged from extremely dense concentrations to sparse scatters.

In summary, Pliocene toolmakers in the Omo-Turkana basin (1) created low-level artifact concentrations that were (2) very specifically located within the fluvial system and (3) very close to the channel sources of rocks used in making tools. By the early Pleistocene, the toolmakers' activities were far less tethered to particular stone sources or specific geomorphological settings. Hominids applied their toolmaking and stone-carrying skills over wider and wider distances and types of habitat.

Olduvai (1.85 - 1.20 million years ago)

Based on work by M.D. Leakey (1971), R. Hay (1976), and others, the excavations and geology of Beds I and II, Olduvai Gorge, are very well known. Archeological sites in Bed I (1.85 - 1.76 Ma) are largely if not completely confined to the ancient lake margin facies (i.e., a zone of periodic lake flooding). Stone sources were mainly within 3 km of sites, though transport distances of up to 10 km occurred from a source (Kelogi gneiss) within the lake margin (Hay 1976:184). The spatial distribution of sites suggests that the toolmakers were confined to this lakeside zone. The main localities of Bed I, moreover, exhibit a unique pattern of vertical stacking of stone artifact concentrations (and, typically, associated faunal materials) in superjacent stratigraphic layers (Leakey 1971; Potts 1988, 1994). At the FLK North locality, for example, Leakey was able to uncover abundant stone artifacts and faunal remains within relatively small excavations simply by digging downwards. This approach does not usually yield large concentrations of remains in Turkana, Olorgesailie, or other early Pleistocene basins. In the latter, the densest artifact clusters are found in separate spatial locations, not stacked vertically. Olduvai toolmakers certainly did not limit their discard of stone artifacts to these dense concentrations (Leakey 1971; Blumenschine and Masao 1991). The vertical stacking of archeological clusters at certain localities does, however, imply a strong and unique tethering of stone-tool behavior not only to the lake margin but also to specific places within that zone, cutting across several stratigraphic levels.

In addition to flaked pieces, the toolmakers of Bed I also carried unmodified stone pieces, manuports, composed of the same rock types as the flaked pieces. Manuports may be interpreted as raw materials distributed by hominids within the lake margin as a means of solving the problem of assuring the presence of suitable stone in the lakeside foraging range (Potts 1988). This may reflect a behavior not unlike the mental mapping of hammerstones relative to food trees apparent in some chimpanzees (Boesch and Boesch 1984). The fact that hominids carried at least two types of resources (stones and carcass parts) over some distance to the same delimited places, however, signifies a foraging technique distinct from that of chimps (Potts 1988:288-289).

In Bed II Olduvai (1.76 to ~1.2 Ma), we see the continued concentration of implements, manuports, and other archeological debris in separate strata at the same confined localities (e.g., MNK Main). But, by 1.6 Ma, the archeological localities themselves have a wider distribution, including areas outside the lake margin zone. Hominid toolmakers began to exploit new stone sources, such as Engelosin phonolite, that were located well beyond the lake margin, and also fauna that probably ranged well beyond the lakeside (e.g., *Equus oldowayensis,* whose abundance greatly increased in the archeological fauna by middle Bed II times).

In contrast with the Omo-Turkana region, Olduvai was a lake-dominated basin during the span of Beds I and II. Despite this different setting, similar changes in hominid land use occurred in both regions. During the early Pleistocene at Olduvai, the toolmakers' activities became more widely distributed, both within and beyond the lake margin zone, and tool-assisted behaviors occurred in lakeside, channel, and alluvial floodplain contexts. Furthermore, several shifts in climate and vegetation are documented in Beds I and II at Olduvai, including highly arid spans (upper Bed I, Bed II Lemuta Member) and wetter, wooded periods (lower Bed I, base of Bed II). Evidence of the toolmakers (e.g., dense artifact concentrations) is found throughout the sequence, suggesting the ability of these hominids to accommodate to periodic revisions of climate and vegetation.

Olorgesailie (1.2-0.5 million years ago)

Building on the work of Robert Shackleton (1978), Glynn Isaac (1977, 1978), and Louis Leakey (1952), our research at Olorgesailie has established a precise chronology, developed a paleolandscape-scale methodology of excavation, and mapped the rock sources used by hominids (Potts 1989; 1994; Deino and Potts 1990; Tauxe *et al.* 1992; Potts *et al.* 1996; Potts *et al.* n.d.). Hominids from about 1 Ma to 490 thousand years ago (ka) found a variety of source rocks for making tools on the slopes of Mt. Olorgesailie and surrounding highlands. These rocks consist of lava materials that vary in composition and flaking qualities from east to west, paralleling the erosional face of the lake basin sediments below the mountain (Figure 2A).

Stone handaxes occur in large accumulations almost exclusively on the eastern margin of the Legemunge sub-basin. These concentrations are directly in line with several distinctive lava rock sources also located on that edge of the sub-basin. If raw material was the main determinant of where hominids left handaxes, however, we would expect those made of other raw materials to be left close to their sources. That is not the case. Handaxes made from materials obtainable further west were also deposited in the eastern area, suggesting that handaxe deposition was not tied to the source outcrops but to some other feature of the landscape (Potts 1994). At the time of the largest accumulations of handaxes, ca. 900 ka, a braided stream system occurred in this eastern area; handaxes made from all types of stone are associated with the channels of this system (although it is impossible as yet to say what other factors, such as shelter, water, or food, may have led the toolmakers to drop their handaxes in this vicinity). By contrast, in the contemporaneous floodplain paleosols and sands, which stretch away from the main stream axis, handaxes are uncommon; the assemblages are dominated by utilized flakes, scrapers, discoids, and broken pieces of handaxes (Isaac 1977; Potts 1994). The overall spatial pattern indicates, therefore, that diverse artifacts were deposited by the toolmakers contingent on specific features of the landscape. In other basins where different features prevailed, we may expect to find a different pattern of artifact discard.

The lake and entire hydrological regime of Olorgesailie were subject to dramatic, episodic changes during the early and middle Pleistocene (Figure 2). During Member 1 of the Olorgesailie Formation (ca. 992 ka), the lake was confined to the northern and eastern areas (Figure 2B) but then expanded (Member 2) to the lava boundaries of the basin (Figure 2C). After a series of fluctuations, the lake became much smaller (Members 4 and 5, ca. 974 ka), and its center moved back and forth between the Legemunge and Oltepesi sub-basins. Then, on two occasions — Member 7 (900 ka) and Member 10 (662 ka) - the lake shifted to the southwestern part of the basin (Figure 2D). At times during this sequence, the lake disappeared altogether (Owen and Renaut 1981; Potts 1994). Since hominid toolmakers left signs of their activity throughout the sequence, they were evidently able to adapt to the climatic shifts, tectonic disruptions of the landscape, changes caused by nearby eruptive events, and fluctuations in lake-fluvial hydrology and resource distributions (Potts 1998b).

Bose (~0.78 million years ago)

A rather different setting is presented by the Bose basin in Guangxi Autonomous Region, southern China, where I have collaborated with Huang Weiwen of the IVPP, Beijing (Huang *et al.* 1990). In contrast with other localities considered here, there was no large mountain, no series of distinct rock outcrops, no major lake or lake margin zone, and no single river or stream channel axis. A lateritic soil characterizes the ~800 km2 of Pleistocene exposures in the Bose basin. Stone artifacts are preserved at a single level, ~40 cm thick, situated ~1 m beneath the top of the 8-meter-thick laterite. Such soils typify much of southern China and Southeast Asia, and are notoriously difficult to date, but the artifacts of Bose have a direct *in situ* association with tektites (part of the Australasian tektite field) dated between 730 and 780 ka (Guo *et al.* 1996).

The stone industry of Bose is characterized by powerful percussion, the production of large flakes and flake scars, followed by retouch, and also the standardization of final form — all of which are characteristic of the Acheulian. Indeed, the Bose basin does preserve handaxes, bifacially flaked and with a tear-drop shape typical of handaxes of western Eurasia and Africa. For the most part, however, the Bose stone industry is characterized by unifacial flaking. What initially look like well-flaked handaxes and picks when viewed from one side, when turned over, reveal the unflaked surfaces (cortex) of large cobbles. This pattern is very repetitive, as repetitive as the bifacial handaxe is at Olorgesailie. In Bose, the large handaxes, picks, and their unifacial equivalents are almost entirely confined to the central part of the basin, where the very largest sizes of stone cobbles were available. In the eastern part of the basin, where smaller cobble sources prevailed, smaller unifacial cores predominate.

As at Olorgesailie, we see in southern China that hominid toolmakers about 780 ka made and left behind their main standardized artifact form in a confined part of the basin. But their response was to a rather different setting from that of

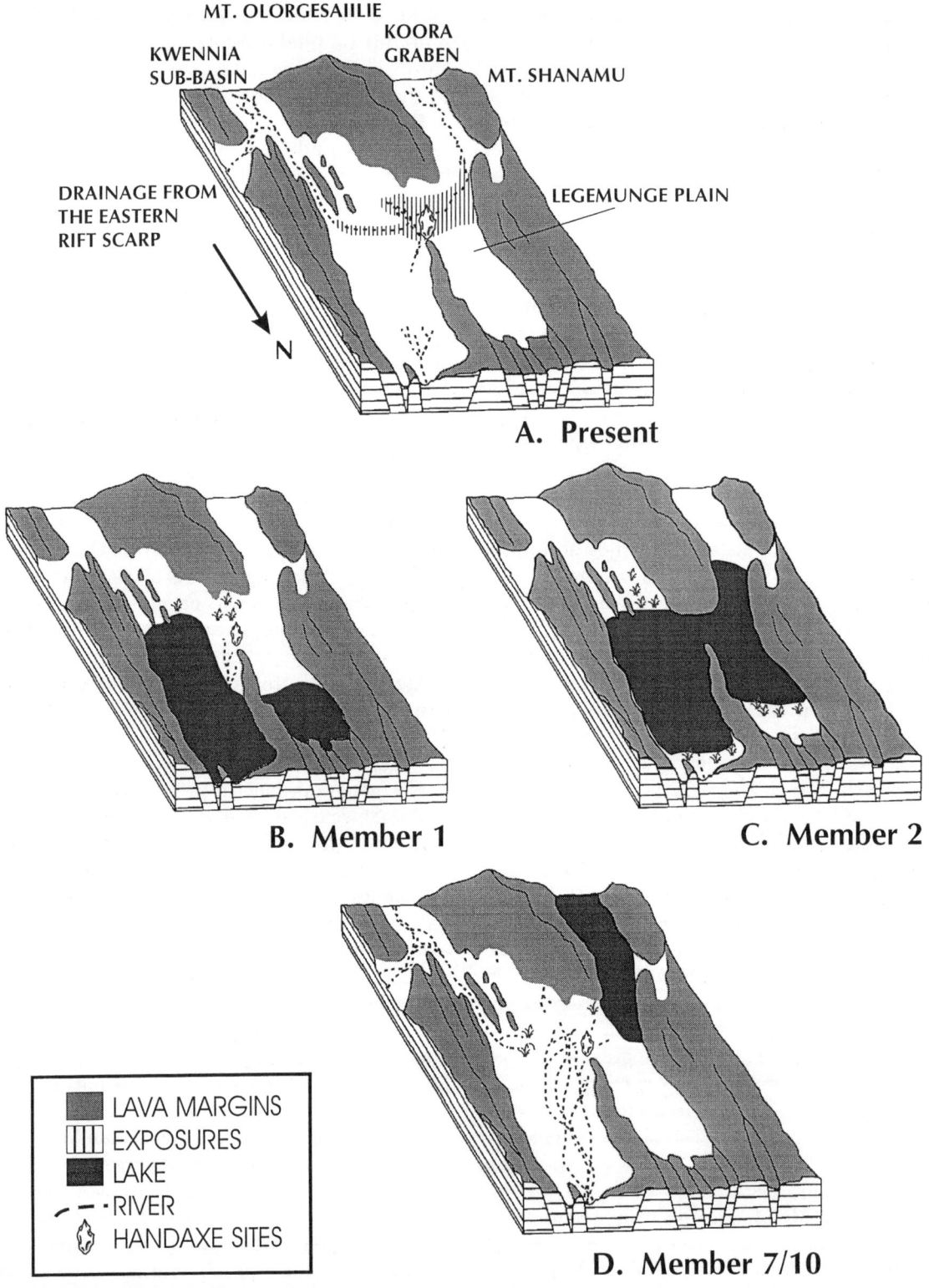

Fig. 2. Extensive hydrological changes in the Olorgesailie basin during the Pleistocene. (A) The present basin and the exposed artifact-rich sediments lying north of Mt. Olorgesailie. (B-D) Paleogeographic reconstructions of Member 1 (~992 ka), Member 2 (~980 ka), and Members 7 and 10 (900-780 ka and 662 ka, respectively). See text for further explanation.

Olorgesailie. From what we can discern at this point in our research, the response was largely to the size of cobbles deposited by Pliocene streams and exposed to Pleistocene toolmakers.

Human Cognitive Evolution: Responsiveness to Environmental Variability

The four basins considered here exhibit diverse hydrological regimes, a variety of geomorphological settings, and important differences in the types and distributions of rocks available to hominids. Several of the interbasinal differences and changes over time are summarized in Figure 3. Between 2.3 to 0.7 Ma, the empirical clues regarding hominid land use suggest a greater responsiveness over time to contextual variables. This is evidenced by (1) longer distances of stone transport, (2) more mobile (less tethered) site production, (3) the presence of sites in a wider diversity of depositional settings, and (4) the production and concentration of different artifact forms in different areas of the landscape, correlated in some regions (e.g., Bose) with the size and form of available rocks and, in other basins (e.g., Olorgesailie), with streams and possibly related resources such as food, shade, and water.

Archeological assessments of hominid behavior used to focus on typological form, a convenience afforded by the stone tools that have traditionally defined the discipline. The repetitive manufacture

DIVERSE PATTERNS OF HOMINID LAND USE

TIME (Ma)	LOCALITY	LAND USE
~ 0.78	Bose Basin, S. China	Manufacture of handaxe-like tools dependent on raw material size
1.20 - 0.78	Olorgesailie, Kenya	Handaxes carried but clustered near a central resource (water?)
1.50	**STANDARDIZED TOOL FORMS, INCLUDING HANDAXES**	
1.85 - 1.20	Olduvai, Tanzania	Transport of manuports; Bed II: sites less tied to margin; Bed I: sites strongly tethered to place within lake margin
1.60	Turkana, Kenya	Tools & clusters more varied; sites less tied to stone sources
2.30	Turkana, Kenya	Low density clusters of tools; sited tethered to stone sources

Fig. 3. Comparison of several basins that preserve evidence of hominid toolmakers. Over time, site production becomes less tethered to specific depositional environments, rock sources, and places on the landscape. By 1.5 Ma, the broadly standardized artifact forms of Mode 2 industries — including bifacial handaxes, cleavers, picks, and analogous unifacial forms — were added to the more fluid and continuous typological variation of Mode 1 industries. Landscape-scale comparisons illustrate more clearly than tools an enhanced capacity by early and middle Pleistocene hominids to accommodate to diverse environments.

of certain broadly-standardized forms at certain sites—e.g., thousands of handaxes at Olorgesailie—and the persistence of these forms for over one million years, makes it easy to be impressed by the overall homogeneity of hominid tool behavior, within wide limits. It is conceivable, though, that these tool forms represent resilient aspects of hominid behavior, and that they facilitated many different tasks as implements, cores, and conveniently portable pieces of stone. Yet this implies that no matter how many *chaînes opératoires* can be read into a tool assemblage, typology and tool manufacture are poor indicators of the breadth and variation of hominid response to diverse adaptive settings. I suggest, instead, that the comparative study of hominid land use and habitat offers a better way to evaluate how responsive hominids were to the contingencies of their environments. Response variation may be considered, moreover, a useful proxy of cognition—the application of mental processes (at both the individual and social levels) to solving complex problems of periodic change in resource distributions and abundance.

With the spread of *Homo*, the range of variation experienced by hominids increased. Within a given region, significant temporal shifts also occurred. At Olorgesailie, for example, remodeling of the landscape due to tectonic activity and shifts in climate (Isaac 1978; Owen and Renaut 1981; Potts 1996a, 1998b) caused large revisions in water, plants, and animal resources on which hominid foraging, social strategies, survival, and reproductive success depended.

Olorgesailie was not unusual in this regard. The marine oxygen isotope records indicate wide global fluctuations in temperature and water distribution, especially strong over the past 700 thousand years. Numerous local pollen and sedimentary records support the point that moisture and vegetation underwent dramatic oscillation during this span (for a review, see Potts 1996a; Potts 1998a).

The key to cognitive change in *Homo*, I propose, is this disparity in environmental conditions met by hominids. As the conditions of Darwinian selection were altered periodically, genetic changes were favored that proved effective in processing highly variable environmental information, creating versatile responses, and mediating diverse strategies of behavior. The archeological record offers a strong hint of this process, as tool behaviors that were once tethered to specific resources and places were ultimately replaced by those responsive to diverse landscapes.

Various aspects of human brain evolution provide further evidence for this view. If we examine cranial capacity data in fossil hominids (e.g., Aiello and Wheeler 1995) (see Figure 4), we discover that the largest increase in hominid brain size independent of body size took place over the past 700,000 years. Encephalization largely coincided, therefore, with the widest recorded fluctuations of global and local habitat of the late Cenozoic (Potts 1998a).

I believe that this temporal correlation indicates why certain aspects of brain evolution occurred (Figure 5)—i.e., as part of a process of adaptation to the variable aspects of environments. This process (i.e., variability selection) was implemented by the disparate selective regimes that certain hominid populations confronted.

Although the cerebellum and other parts of the brain underwent a similar increase in relative size, the cerebral neocortex contributed disproportionately to the expansion of absolute brain size. The neocortex is functionally involved with analysis of sensory input, mental representation, and integration and secondary processing of external data initially processed by various cortical and subcortical regions. Elaboration of these key cognitive functions apparently coincided with several hundred thousand years of extreme environmental oscillation. These functions, moreover, are well *designed* to handle variable data input and to mediate varied and sometimes novel response.

Furthermore, neocortical expansion in mammals has typically involved a decrease in the packing of neurons, enabling a higher degree of interconnection among the neurons with increasing cortical size (Holloway 1968; Jerison 1973). The increase in connectivity appears to correspond with better integration of sensory modes and the possibility of more complex processing of environmental data. The connectivity among neocortical neurons is, furthermore, highly plastic. While certain connections are fixed early in post-natal life, these are broadly tuned to environmental input. Other connections bring neurons together into assemblies throughout much of life. These associative connections depend entirely on environmental input.

Enlargement of the neocortex over the past 1 million years involved, therefore, an expanding sheet of connections sensitive to external data and a generator of variable response. Open programs of behavior, which require detailed environmental input for proper functioning, arose out of both the fixed connections and the changeable assemblies of neurons found in the neocortex and other brain areas. Those programs that deal with mental rep-

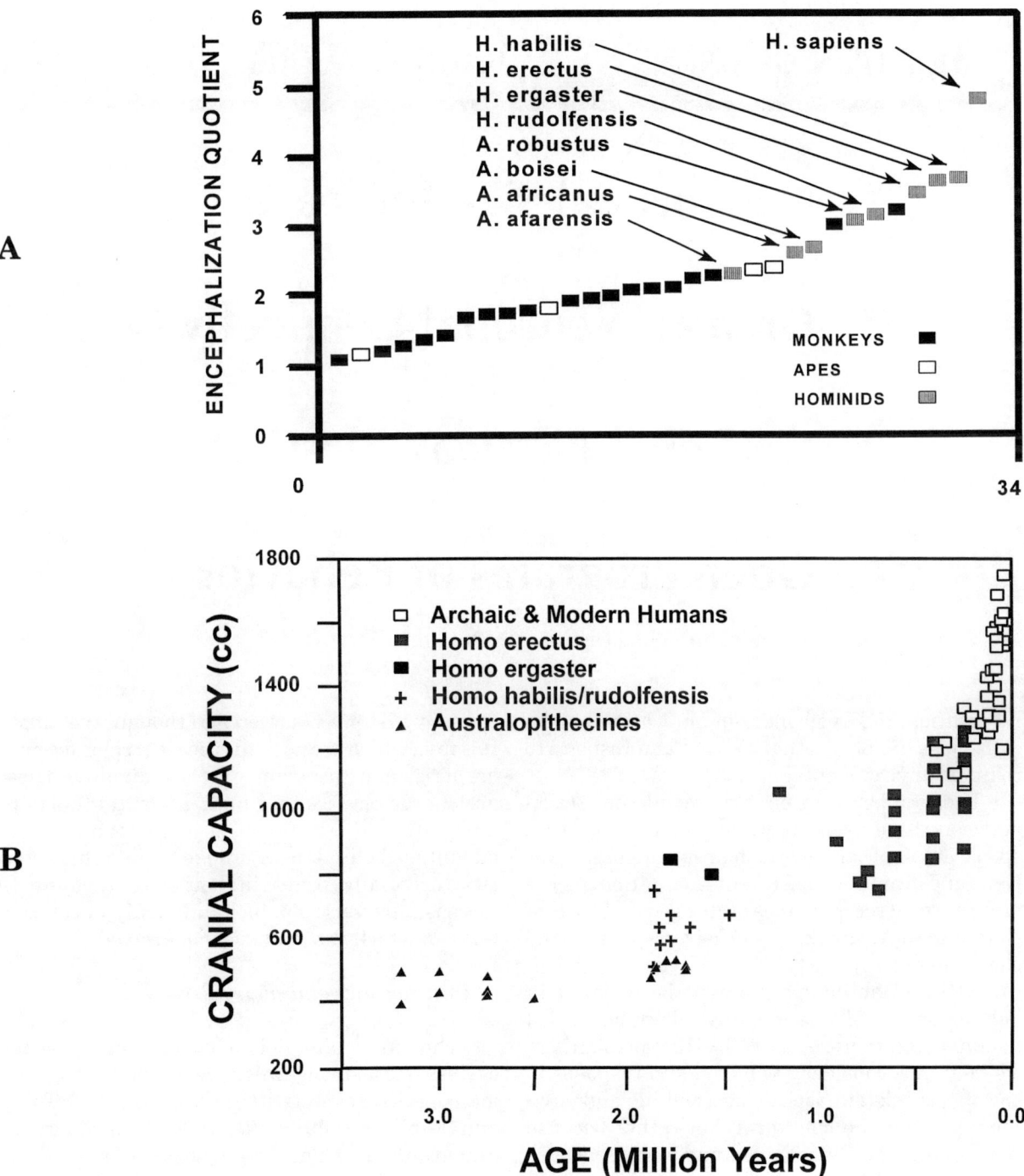

Fig. 4. Hominid brain size increase, based on Aiello and Wheeler (1995). (4A) A ranking of encephalization quotient from low to high for cercopithecoid monkeys and hominoids, including estimates for fossil hominids. Encephalization in hominids prior to *Homo sapiens* falls along a gradient with other higher primates; relative brain size in *H. sapiens* is considerably above the gradient. (4B) A plot of hominid fossil cranial capacity over time, which indicates brain size increases around 1.8-2.0 million years ago, and especially over the past 700 thousand years. The two graphs combined show that most of the encephalization in humans occurred over the past 700 ka, correlated with the largest recorded oscillations of climate and habitat during the late Cenozoic.

HUMAN BRAIN & COGNITIVE EVOLUTION

1. Neocortical Size Increase

2. Greater Neuronal Connectivity

3. Neuronal Pathway Plasticity

4. Open Programs of Behavior

Fig. 5. Key aspects of human brain and cognitive evolution.

resentation and symbolic communication are localized, as are others that process and respond to complex environmental data.

There appears to be a powerful connection between these open programs, the enhanced responsiveness of hominids to their surroundings over time, and dramatic Pleistocene fluctuation. As the archeological record makes quite clear, behavioral change in the toolmakers was not a response to the challenges of any one type of landscape or ecological setting. Habitat-specific hypotheses falter in helping us to understand many salient aspects of hominid evolution (Potts 1998b). Human cognition, we may thus consider, evolved not in response to specific, consistent aspects of social life and environment over the long term, but rather because individuals, at different times and places, encountered highly diverse situations to which mental and social skills were applied.

Episodic change in adaptive settings placed a premium on the social mechanisms that enhanced buffering of environmental uncertainty. In modern *Homo*, these mechanisms include (Figure 6): (1) intricate social networks by which individuals share resources; (2) complex symbolic language, which people use to create abstractions and transfer their mental and temporal maps to each other; (3) technology and the capacity to innovate; and (4) complex systems of action and thought (i.e., institutions), which are perhaps the most rigorous buffers of risk and uncertainty to the individual. These are characteristics that human mental functions deal with daily. But I suggest that the selective advantage behind these unique human characteristics originally resided in their power to cope with increasingly variable, including truly novel, contexts in which the genus *Homo* evolved.

At Odds with Evolutionary Psychology

The burgeoning field of evolutionary psychology asserts a strong influence on the study of human cognitive evolution (Barkow *et al.* 1993; Tooby and Cosmides 1989; Pinker 1994). It considers the human mind to comprise a series of specialized mental modules that are universal to *Homo sapiens* and that evolved to solve the specific adaptive problems faced by Pleistocene hunter-gatherers. Researchers in this field emphasize that cognitive evolution—indeed all adaptive change—occurs in relation to the "statistical consistencies of environments" or the "regularities of adaptive conditions" over long time spans. According to Tooby and Cosmides (1990:375): "All adaptations evolved in response to the repeating elements of past environments, and their structure [including that of

EVOLVED CHARACTERISTIC	BUFFERING OF ENVIRONMENTAL CHANGE/RISK
Intricate Social Networks	Reciprocity (e.g., food sharing)
Symbolic Language	Ability to refer to distant resources & habitat features; complex mental maps
Technology	Complex & creative modification of immediate surroundings
Cultural Institutions	Complex social activity powered by accepted symbols & meanings

Fig. 6. Four ways in which *Homo sapiens* moderates the effects of environmental uncertainty.

the human mind] reflects in detail the recurrent structure of ancestral environments."

We may note here a few of the many recurrent elements in hominid environments:

1. *Predation risks*: It seems likely that virtually all hominid habitats had large predators. Avoiding them must have required clever use of time and space by hominids. Predator diversity would have multiplied the behavioral complexity of the problem, fostering the retention and simultaneous use of multiple strategies of predator defense.

2. *Habitat and resource complexity*: Habitat patchiness posed challenges of spatial memory and foraging strategy. Seasonal variations also fostered the cognitive and social means for switching to alternative foods at specific times, thus expanding the range of hominid resource-acquiring strategies employed throughout the annual cycle.

3. *Tools for the toolmakers*: For hominids dependent on stone-tool behavior, tool manufacture (especially standardized forms such as handaxes) required special memory of flaking sequences, mental imaging of tool characteristics, and experimental testing of rocks — all of which were factors on which cognition impinged.

4. *Social group interaction*: Certain aspects of social life, especially the presence of kin and the dynamics of mate selection, must have been consistent in all hominid groups. As alluded earlier, there are many proponents of the idea that cognitive evolution has largely been a response to the recurrent problems of social manipulation, i.e., outwitting competitors and forming intricate alliances within the social group (Humphrey 1976; Byrne and Whiten 1988).

These aspects of hominid environments probably had strong influences on hominid brain and cognitive evolution. The variability selection hypothesis I have considered here adds an important dimension—i.e., the remodeling of local habitats had a powerful modifying effect on each of these factors. Thus the interactions between hominids and potential predators, food and water sources, microhabitats, and stone sources were periodically recast. Each major alteration of climate and resources would have elicited changes in hominid group size and composition, population densities, and thus possibly the success of particular mate-searching strategies.

Evolutionary psychologists assert that the design properties of the human brain could only have arisen by interaction with the long-term, uniform properties of Pleistocene environments. I have argued here, by contrast, that design properties may also arise by an organism's interaction with the highly dynamic, inconsistent properties of environments. Accordingly, hominid information processing, including mental means of computation and response to surrounding conditions, evolved in relation to large environmental extremes and inconsistencies in adaptive milieu.

In one sense, we might consider the wide disparities in Pleistocene landscapes and adaptive conditions as a repetitive element in the lives of hominids. Tooby and Cosmides (1990) have, in fact, argued that such *inconsistent* environmental properties were part of the *regularities* of the past. But this argument jibes poorly with the evolutionary psychologists' emphasis on the consistency of selection over long time periods (e.g., the Pleistocene). Furthermore, it clouds the distinction between the changeable and the more uniform qualities of past environments — a very important distinction in the analysis of adaptation.

The functions of the brain that are *unique* to humans (as opposed to those shared with many other organisms) appear to facilitate responsiveness to diverse contexts. These universal functions are equivalent to what the evolutionary psychologists call "adaptive design properties of an organism." But in humans, these design properties evolved, I have argued, in response to the uncertainty and mutability of past environments, not merely their predictable qualities.

Since present conditions are not like the past, according to evolutionary psychologists, the structure of human minds cannot be considered "adapted to" the present. It is important to note, however, that temporal disparity in adaptive conditions typified the entire span of human evolution. The effects of distinct adaptive regimes have been integrated over time. Consequently, cognitive mechanisms are found universally in humans that from birth are primed to deal effectively with novel contexts, including many aspects of the present. They are also responsible for creating ever-new conditions, well beyond past boundaries, that further challenge the abilities of people to accommodate. In this sense, the human mind is indeed a reflection of a past environmental *dynamic*; but it is not limited to some hard-and-fast collection of Pleistocene problems typically envisioned by evolutionary psychologists (e.g., Ridley 1993). By implication, the cognitive mechanisms unique to modern *Homo sapiens* offer more than the sum of innate modules believed by evolutionary psychologists to operate solely in relation to past adaptive problems, and that are said to tie human thought and activity directly to specific, repetitive elements of ancestral environments (Tooby and Cosmides 1990, 1993; Pinker 1994).

Conclusion

We have considered the environments inhabited by early hominid toolmakers and the conditions of natural selection that may account for human brain evolution. The time-space comparison of hominid sites indicates one of the most fundamental aspects of hominid behavioral evolution during the Pleistocene was an enhanced responsiveness to increasingly variable settings.

I further suggest that this enhanced responsiveness directly reflects on key elements of brain evolution, particularly the high rate of encephalization during the Pleistocene; the strong absolute size contribution made by cortical association areas, responsible for integrating information received and processed by other brain areas; enhancement of neuronal connectivity and ontogenetic plasticity in humans; and the importance of open programs of behavior and cognition, which require extensive environmental input for appropriate functioning.

The conventional dichotomy between social and ecological hypotheses of brain evolution has been based, in part, on underestimating the environmental dynamics and real habitat complexity with which populations have had to contend (e.g., Humphrey 1976). Environmental novelty and complexity were strongly escalated over time spans in which lineages of hominids lived. Landscapes and resource distributions were revamped, in effect changing the rules of Darwinian optimality by which individuals at different times and places may have benefited from particular social or foraging strategies. An important implication of the variability selection idea, then, is that a new basis exists for bridging social and ecological explanations of human cognitive evolution, weakening the traditionally strict division that exists between these types of hypotheses.

Paleoanthropologists, neuroscientists, and psychologists have long sought *the* key factor, or prime mover, that selected for brain and cognitive evolution. Among paleoanthropologists, the favored factors have included toolmaking, tool using, foraging, and the benefits of complex sociality. According to the idea forwarded here, all of these aspects of behavior were involved, but largely as proximate causes.

At a more fundamental level, these aspects of behavior and social life were exhibited in an increasing range of basinal settings over time. Within the basins, hominids experienced a series of shifting habitats in which the contingencies of reproductive success and survival were subject to episodic change. This context of environmental instability (on local, regional, and global scales) and the associated disparity in selective conditions, I propose, were largely responsible for the evolution of unique aspects of human cognition and the emergence of social diversity and adaptability in modern *Homo sapiens*.

Acknowledgments

Research at Olorgesailie, Olduvai, and Bose has been funded by the Smithsonian Institution and the National Science Foundation. Support and permission of the National Museums of Kenya (Nairobi) and the Institute for Vertebrate Paleontology and Paleoanthropology (Beijing) are gratefully acknowledged, as are the contributions of A.K. Behrensmeyer, T. Plummer, A. Deino, Huang Weiwen, Hou Yamei, and other members of our research teams. My thanks to Jennifer Clark for her assistance with the manuscript, and to April Nowell for her convincing invitation to contribute to this volume. This is a publication of the Smithsonian's Human Origins Program.

References Cited

Aiello, L.C. and P. Wheeler
   1995 The expensive-tissue hypothesis: The brain and the digestive system in human and primate evolution. *Current Anthropology* 36:199-221.

Barkow, J. H., L. Cosmides and J. Tooby (eds.)
   1993 *The Adapted Mind: Evolutionary Psychology and the Generation of Culture*. Oxford University Press, New York.

Belfer-Cohen, A. and N. Goren-Inbar
   1994 Cognition and communication in the Levantine Lower Paleolithic. *World Archaeology* 26:144-157.

Blumenschine, R. J.
   1987 Characteristics of an early hominid scavenging niche. *Current Anthropology* 28:383-407.
   1991 Hominid carnivory and foraging strategies and the socio-economic function of early archaeological sites. *Phil. Trans. R. Soc. London* B 334:211-219.

Blumenschine, R. J. and F. T. Masao
   1991 Living sites at Olduvai Gorge, Tanzania? Preliminary landscape archaeology results in the basal Bed II lake margin zone. *Journal of Human Evolution* 21:451-462.

Boesch, C. and H. Boesch
   1984 Mental map in wild chimpanzees: An analysis of hammer transports for nut cracking. *Primates* 25:160-170.

Boesch-Achermann, H. and C. Boesch
   1994 Hominization in the rainforest: The chimpanzee's piece of the puzzle. *Evolutionary Anthropology* 3:9-16.

Bunn, H. T.
   1994 Early Pleistocene hominid foraging strategies along the ancestral Omo River at Koobi Fora, Kenya. *Journal of Human Evolution* 27:247-266.

Byrne, R. W.
   1998 The Technical intelligence hypothesis: An additional evolutionary stimulus to intelligence? In *Machiavellian Intelligence II*, edited by A. Whiten and R. W. Byrne, pp.289-311. Cambridge University Press, Cambridge.

Byrne, R. and A. Whiten (eds.)
   1988 *Machiavellian intelligence*. Oxford University Press, Oxford.

Deino, A. and R. Potts
   1990 Single crystal 40Ar/39Ar dating of the Olorgesailie Formation, southern Kenya rift. *Journal of Geophysical Research* 95 (B6):8453-8470.

deMenocal, P. B.
   1995 Plio-Pleistocene African climate. *Science* 270:53-59.

Dunbar, R. I. M.
   1995 Neocortex size and group size in primates: a test of the hypothesis. *Journal of Human Evolution* 28:287-296.
   1998 The social brain hypothesis. *Evolutionary Anthropology* 6:178-190.

Falk, D
   1990 Brain evolution in *Homo*: The 'radiator' theory. *Behavioral and Brain Sciences* 13:333-381.

Fiebel, C. S., J. M. Harris and F. H. Brown
   1991 Neogene paleoenvironments of the Turkana Basin. In *Koobi Fora Research Project*, Volume 3, edited by J. M. Harris, pp.321-370. Clarendon Press, Oxford.

Guo Shilun, Hao Xiuhong, Chen Baoliu and Huang Weiwen
   1996 Fission track dating of Paleolithic site at Bose in Guangxi, South China. *Acta Anthropologica Sinica* 15:347-350.

Hay, R. L.
   1976 *Geology of the Olduvai Gorge*. University of California Press, Berkeley.

Holloway, R. L.
   1968 The evolution of the primate brain: Some aspects of quantitative relations. *Brain Research* 7:121-172.

Huang Weiwen, Leng Jian, Yuan Xiaofeng and Xie Guangmao
   1990 Advanced opinions on the stratigraphy

and chronology of Baise stone industry. *Acta Anthropologica Sinica* 9:105-112.

Humphrey, N. K.
- 1976 The social function of intellect. In *Growing Points in Ethology*, edited by P. P. G. Bateson and R.A. Hinde, pp.303-317, Cambridge University Press, Cambridge.

Isaac, G. Ll.
- 1977 *Olorgesailie*. University of Chicago Press, Chicago.
- 1978 The Olorgesailie Formation: Stratigraphy, tectonics, and the palaeogeographic context of the Middle Pleistocene archaeological sites. In *Geological Background of Fossil Man*, edited by W.W. Bishop, pp.173-206. Scottish Academic Press, Edinburgh.

Isaac, G. Ll. (ed.)
- 1997 *Koobi Fora Research Project, Volume 5: Pleistocene archaeology*. Clarendon Press, Oxford.

Jerison, H. J.
- 1973 *Evolution of the brain and intelligence*. Academic Press, New York.

Jolly, A.
- 1966 Lemur social behavior and primate intelligence. *Science* 153:501-506.

Kibunjia, M.
- 1994 Pliocene archaeological occurrences in the Lake Turkana basin. *Journal of Human Evolution* 27:159-172.

Kingston, J. D., B. D. Marino and A. Hill
- 1994 Isotopic evidence for Neogene hominid paleoenvironments in the Kenya rift valley. *Science* 264:955-959.

Klein, R. G.
- 1989 *The Human Career: Human Biological and Cultural Origins*. University of Chicago Press, Chicago.

Leakey, L. S. B.
- 1952 The Olorgesailie prehistoric site. In *Proceedings of the First Pan-African Congress on Prehistory, 1947, Nairobi*, edited by L. S. B. Leakey and S. Cole, p.209. Philosophical Library, New York.

Leakey, M. D.
- 1971 *Olduvai Gorge*, Volume 3. Cambridge University Press, London.

Marzke, M. W.
- 1997 Precision grips, hand morphology, and tools. *American Journal of Physical Anthropology* 102:91-110.

Milton, K.
- 1981 Distribution patterns of tropical plant foods as a stimulus to primate mental development. *American Anthropologist* 83:534-548.

Monahan, C. M.
- 1996 New zooarchaeological data from Bed II, Olduvai Gorge, Tanzania: Implications for hominid behavior in the Early Pleistocene. *Journal of Human Evolution* 31:93-128.

Owen, R. B. and R. Renaut
- 1981 Paleoenvironments and sedimentology of the Middle Pleistocene Olorgesailie Formation, southern Kenya rift valley. In *Palaeoecology of Africa*, edited by J. Coetzee and E. M. van Zinderen Bakker, pp. 147-174. Balkema, Rotterdam.

Parker, S. T. and K. R. Gibson
- 1977 Object manipulation, tool use, and sensorimotor intelligence as feeding adaptations in early hominids. *Journal of Human Evolution* 6:623-641.

Pinker, S
- 1994 *The Language Instinct*. Morrow, New York.

Plummer, T. W. and L.C. Bishop
- 1994 Hominid paleoecology at Olduvai Gorge, Tanzania as indicated by antelope remains. *Journal of Human Evolution* 27:47-76.

Potts, R.
- 1988 *Early Hominid Activities at Olduvai*. Aldine de Gruyter, New York.
- 1989 Olorgesailie: New excavations and findings in Early and Middle Pleistocene contexts, southern Kenya rift valley. *Journal of Human Evolution* 18:477-484.
- 1994 Variables vs. models of early Pleistocene hominid land use. *Journal of Human Evolution* 27:7-24.
- 1996a. *Humanity's Descent: The Consequences of Ecological Instability*. Morrow, New York.
- 1996b. Climate and human evolution. *Science* 273:922-923.
- 1998a. Variability selection in hominid evolution. *Evolutionary Anthropology* 7:81-96.
- 1998b. Environmental hypotheses of hominin evolution. *Yearbook of Physical Anthropology* 41:93-136.

Potts, R., T. Jorstad and D. Cole
- 1996 The role of GIS in interdisciplinary investigations at Olorgesailie, Kenya, a Pleistocene archeological locality. In *Anthropology, Space, and Geographic*

Information Systems, edited by M. Aldenderfer and H. D. G. Maschner, pp. 202-213. Oxford University Press, New York.

Potts, R., A. K. Behrensmeyer and P. Ditchfield
n.d. Paleolandscape variation in early Pleistocene hominid activities: Members 1 and 7, Olorgesailie Formation, Kenya. *Journal of Human Evolution.*

Rayner, R. J., B. P. Moon and J. C. Masters
1993 The Makapansgat australopithecine environment. *Journal of Human Evolution* 24:219-232.

Ridley, M.
1993 *The Red Queen*. Macmillan, New York.

Rogers, M. J
1997 *A Landscape Archaeological Study at East Turkana, Kenya*. Ph.D. dissertation, Rutgers University.

Rogers, M. J., C. S. Feibel and J. W. K. Harris
1994 Changing patterns of land use by Plio-Pleistocene hominids in the Lake Turkana Basin. *Journal of Human Evolution* 27:139-158.

Shackleton, R. M.
1978 A Geological map of the Olorgesailie Formation. In *Geological Background of Fossil Man*, edited by W. W. Bishop. Scottish Academic Press, Edinburgh.

Susman, R. L.
1994 Fossil evidence for early hominid tool use. *Science* 265:1570-1573.

Tauxe, L., A. D. Deino, A. K. Behrensmeyer and R. Potts
1992 Pinning down the Brunhes/Matuyama and upper Jaramillo boundaries: A reconciliation of orbital and isotopic time scales. *Earth and Planetary Science Letters* 109:561-572.

Tooby, J. and L. Cosmides
1989 Evolutionary psychology and the generation of culture, part I. *Ethology and Sociobiology* 10:29-49.

1990 The past explains the present: Emotional adaptations and the structure of ancestral environments. *Ethology and Sociobiology* 11:375-424.

1993 The psychological foundations of culture. In *The Adapted Mind: Evolutionary Psychology and the Generation of Culture*, edited by J. H. Barkow, pp.19-136. Oxford University Press, New York.

Toth, N.
1985 The Oldowan reassessed: A close look at early stone artifacts. *Journal of Archaeological Science* 12:101-120.

Turner, A.
1984 Hominids and fellow-travelers: Human migration into high latitudes as part of a large mammal community. In *Hominid Evolution and Community Ecology*, edited by R. Foley, pp.193-217. Academic Press, London.

Turner, A. and B. Wood
1993 Taxonomic and geographic diversity in robust australopithecines and other African Plio-Pleistocene larger mammals. *Journal of Human Evolution* 24:147-168.

Vrba, E. S.
1988 Late Pliocene climatic events and human evolution. In *Evolutionary History of the "Robust" Australopithecines*, edited by F. E. Grine, pp. 405-426. Aldine de Gruyter, New York.

Vrba, E. S., G. H. Denton and M. L. Prentice
1989 Climatic influences on early hominid behavior. *Ossa* 14:127-156.

Wolpoff, M. H.
1971 Competitive exclusion among Lower Pleistocene hominids: the single species hypothesis. *Man* 6:601-614.

11. The Fossil Evidence for the Evolution of Human Intelligences in Pleistocene *Homo*

Anne H. Weaver, Trenton W. Holliday, Christopher B. Ruff and Erik Trinkaus

Abstract

One of the most striking evolutionary patterns of Pleistocene hominids revealed by the fossil record is a dramatic increase in brain size relative to body size. This evolutionary increase in overall encephalization presumably reflects changes in information-processing capacity, or "intelligence." However, the brain is a costly organ to grow and to maintain. In order to understand the fitness advantage of larger brains, we need to consider what aspects of information processing would have provided benefits outweighing their costs. Four specific kinds of information processing ("intelligence") were elaborated in the Pleistocene:

1. object oriented intelligences (including conceptual and motor aspects);
2. conceptual intelligences (landscape use, long-term planning, and complex patterns of resource exploitation;
3. social intelligences (empathy, cooperation, anticipation of others' actions, communication);
4. linguistic intelligences (motor anatomy and conceptual processing).

A better understanding of the temporal pattern of both endocranial and postcranial evidence, in conjunction with the archeological behavioral record, will shed light on when and under what circumstances each of these aspects of intelligence emerged. We examine the changing pattern of Pleistocene hominid overall encephalization reflecting overall information processing capacity. We also present evidence related to two specific aspects of "object-oriented" intelligences:

1. cervical neural canal dimensions in WT-15000 in comparison with Neandertals and modern humans;
2. asymmetry of the upper limb in Pleistocene hominids reflecting differential mechanical loading of the arms related to handedness.

Finally, we suggest areas of further research to address the evolutionary emergence of specific "intelligences" rather than generalized information-processing:

1. fine-grained analysis of patterns of encephalization;
2. more extensive analysis of anatomical asymmetries (endocranial and postcranial);
3. quantitative investigation of specific brain regions, especially the cerebellum;
4. analyses of patterns of brain growth related to secondary altriciality.

Introduction

To consider the evolution of intelligence in the hominid lineage, we must include evidence from the hominid paleontological record, as a baseline at the very least. For more than a century, brain size, as a gross biological measure of human cognitive abilities, has figured strongly in expositions on hominid evolution, and there have been multiple attempts to read the differential evolution of particular portions of the brain from endocranial casts. Yet these human biological reflections of past hominid intelligence have frequently existed in a vacuum.

There has been little systematic consideration of what particular aspects of intelligence might be relevant, or of how selection might operate upon human cognitive anatomy. Moreover, there is usually only general reference as to how the aspects of human neuroanatomy reflected in the human fossil record might actually relate to cognitive abilities in general or to any specific aspect of intelligence.

We begin to address these questions here with data recently derived from the hominid paleontological evidence of the past million and a half years. During this period, beginning with early *Homo erectus* and continuing through "archaic" and "modern" *H. sapiens*,[1] hominids built upon evolutionary changes evident in *Australopithecus* and early members of the genus *Homo*. However, it was a period of substantial changes in both human brain size and archeological reflections of behavioral complexity. Moreover, basic modern human functional anatomy had become fully established, providing an essentially modern human biological context in which these neurological and behavioral changes took place. In addition, most of the archeological evidence for the evolution of human intelligence is derived from this interval. It is therefore an important period for a paleontological discussion of the evolution of intelligence.

Intelligence

To a human paleontologist, "intelligence" looks like a black box, or better yet, a series of nested black boxes, in the shape of the endocranium. The outermost box encompasses the relation of neuroanatomy to its tracings upon the endocranial bone. The next deepest box encloses the relation of the gross anatomical features to actual brain function. The next deepest box surrounds the dynamics of brain function and behavior that may be inferred from the paleoanthropological record, including hominid facial and postcranial skeletal morphology as well as associated archeological traces. The innermost box conceals the ways in which behavior, brain structure, and their paleoanthropological traces may reflect "intelligence."

To learn more about the evolution of intelligence from the fossil evidence, we must look for homologous aspects of neuropsychology, neuroanatomy, and endocranial morphology. Despite differences in methodology and materials, the fundamental historical controversy in each of these fields of research has revolved around a similar dichotomy: are intelligence and its underlying anatomy better modeled as unitary phenomena, or are they better modeled as aggregations of more or less independent functions or domains? Paleoanthropologists have used both unitary and domain-oriented models to look at the fossil evidence.

Unitary Models:

Unitary models of intelligence are appealing to human paleontologists because endocranial capacity is (relatively) easily observed in fossils. If the information-processing capacity of the brain depends upon its size, then bigger brains can process more information. Given the close anatomical relationship between the brain and its meninges and the endocranial surface, endocranial capacity and brain size are highly correlated (Martin 1990); hence larger endocranial capacity can be taken as a proxy for larger brain size. Once a basic proportion of the brain devoted to body maintenance, scaled to body size, is accounted for, any "extra neurons" in a proportionately larger brain will reflect an increased, if unspecified, capacity to process extra information. Departures from the "expected" degree of encephalization for a given taxon are calculated as an index or "encephalization quotient," or "EQ" (Jerison 1973, this volume, see also Potts, this volume). The evidence for a dramatic increase in EQ within the genus *Homo* is unequivocal. As we learn more about the temporal and geographical patterning of this increase, we can relate it to specific environmental and cultural contexts. If intelligence is indeed unitary, then it is reasonable to assume that an increase in EQ within the genus *Homo* bears directly on the modern human pattern of information processing.

There are two potential pitfalls of an approach to the evolution of intelligence based on brain size alone. Intelligence and its neuroanatomical basis may not be a singular, generalized, global capacity. Moreover, the cognitive evolution that took place within the genus *Homo* was probably a subtle, complex and mosaic process. A simple size-based analysis alone will not be fine-grained enough to discriminate such changes.

What is the theoretical support for a unitary view of intelligence? The early work of Piaget (Piaget and Inhelder 1969) provides a classic neuropsychological model for unitary, globally organized intelligence. In Piaget's model, a single tightly organized set of mental operations that develops with increasing age controls the nature and transformation of many different abilities, content domains and symbolic faculties. In his later work, Piaget himself began to question the unitary model (Piaget 1972; Campbell 1993), but the concept of a "global" intelligence remains embedded in the work of many contemporary neuropsychologists.

What might be the anatomical basis for unitary cognitive processing? In unitary models (Jackson 1864; Lashley 1929, see also Kosslyn and Koenig 1992) the brain works as a single system, in which all parts are equally capable of all tasks. Jackson (1864) provided a neuroanatomical model that is in many ways structurally similar to Piaget's neuropsychological model. Jackson believed that various mental abilities are hierarchically organized. Less complex activities are dependent upon "low-level" neural structures such as the spinal cord and brain stem. In adition, more complex activities recruit motor and sensory processes. Very complex "higher" functions also involve "higher" cortical regions of the frontal lobes.

By the 1960s more refined neuroanatomical experimentation had largely refuted the extreme unitary viewpoints represented by Jackson and Lashley. It became apparent that IQ is not a simple function of EQ.

Domain Specific Models:

Technological advances in neuro-imaging techniques [computerized tomography (CT), magnetic resonance imaging (MRI) and positron emission tomography (PET)] have been important in understanding the modular nature of cognition. In addition to traditional studies of neural pathology and invasive surgical methods, non-invasive neuro-imaging is permitting many specific neural functional regions to be mapped. Finer and finer neurofunctional distinctions are being made (e.g., Fox et al.,1985; Grasby et al. 1993; Charles et al. 1994; Kim et al. 1994; Geyer et al. 1996). On the strength of the evidence provided by such neuro-imaging analyses, localization of many neurological functions is no longer in doubt (Kosslyn and Koenig 1992; Hirschfeld and Gelman 1994; Cosmides and Tooby 1994).

We are just beginning to understand the delineation, interaction and organization of cognitive functions as they are manifested in behavior. As a result, the field of cognitive neuroscience currently encompasses a diversity of models, methods, experimental goals, and terminology. As a simplifying conceptual framework, we will adopt the useful concept of "graded localization" of functions (Luria 1966). In this model, comprehensive but neurological "functions," (e.g., language) comprise arrays of "subfunctions" (e.g., syntax, grammar, lexical memory, and retrieval, or motor processing). Subfunctions, in turn, may be aggregations of sub-subfunctions (e.g., storage of verbs or different categories of nouns). As we use it here, the term "function" is roughly equivalent to Fodor's "domain" (Fodor 1983) and Chomsky's "module" (Chomsky 1988, see Hirschfeld and Gelman 1994; also see Lieberman, this volume, for a discussion of complex functional systems).

While much of cognitive neuroscience is devoted to detailed mapping of specifically localized subfunctions and their modular interaction (the trees), some researchers, including cognitive psychologists and archaeologists (Gardner 1983; Donald 1991; Demetriou et al. 1993; Mithen 1996) are providing models of overarching functions (the forest). Gardner's domain-based model, for example, postulates a minimum of seven characteristic, independent "intelligences" of modern humans. These include verbal, mathematical and analytical, musical, emotional, kinesthetic and visuo-spatial functions, which he calls "intelligences." Similarly, Demetriou et al. propose five major cognitive domains: the qualitative-analytic, the quantitative-relational, the causal-experimental, the verbal-propositional, and the spatial-imaginal while Mithen hypothesizes that various aspects of cognition arose mosaically during the Pleistocene, but were only fully integrated in modern humans.

Intelligence in the Hominid Past:

If intelligence is a domain-specific rather than a unitary, additive property, and if we cannot directly translate EQ into IQ, how then can we interpret the dramatic increase in endocranial capacity in Pleistocene hominid evolution? What specific functions or functional systems might have been supported by localized neurological expansion? Unfortunately, the cognitive abilities which interest most neuroscientists are those that are evident in modern humans. Moreover, models of cognitive function are constructed from a modern, western, and academic human viewpoint. Paleoanthropologists must approach the problem of cognitive domains from a slightly different perspective. We assume the pattern of uniquely human intelligences evolved either as direct adaptations or as exaptations under Pleistocene ecological conditions. Domains of intelligence that may have been elaborated as Pleistocene adaptations include:

1. *Object-oriented intelligences* (including conceptual and motor aspects). Traces of material culture provided by the archeological record provide data for evaluating object-oriented intelligences. Certain features of neurological and postcranial anatomy (particularly those related to object manipulation with the upper limb as well as the anterior dentition) may also suggest adaptations related to the production and use of objects. "Spatial" functions and subfunctions involving hand-eye coordination, proprioception and trajectory control, as well as visualization of internal object templates are relevant here.

2. *Conceptual Intelligences* (probably closely related to object-oriented intelligences). Again, archeological data may offer clues, for example, related to differentiation of conceptual domains. "Spatial" functions and subfunctions related to long-term planning, landscape use and large-scale conceptual mapping of resources, as well as the multiple steps (*chaînes opératoires*) involved in technology production and use, reflect conceptual intelligence.

3. *Social/"Machiavellian" Intelligences* (Essock Vitale and Seyfarth 1986; Byrne and Whiten 1988). We may be able to draw inferences about social intelligence in the genus *Homo* from the size and spacing of archeological sites as well as by evidence of burial practices, personal adornment, artifact "styles," and art.

4. *Linguistic Intelligences* (Deacon 1996; Noble and Davidson 1996, this volume). Language may subserve, enable, enhance, be involved in, be exapted for, be generated from or be catalyzed by behaviors related to all three of the above categories: conceptual intelligences (in its grammatical and lexical aspects), object-oriented intelligences (in its lexical aspects) and social intelligences (in its communicative aspects).

Paleontological Evidence for the Evolution of Hominid Intelligences

The archeological record of changing human ecological relationships and technology, as well as relevant aspects of past human biology, furnish the currently available paleoanthropological evidence reflecting the emergence of modern human cognitive abilities. The data presented in this paper concentrate on three aspects of the Pleistocene fossil evidence for central nervous system evolution: 1) overall encephalization; 2) spinal canal dimensions reflecting upper limb innervation; and 3) postcranial asymmetries suggesting neural reorganization for handedness.

Patterns of Encephalization in Pleistocene Homo:

1. The Paleontological Evidence

As the hominid fossil record has filled in for early members of *Homo erectus* (ca. 1.5 ma BP) and as chronological control has increased for the late Early and Middle Pleistocene, several aspects of genus *Homo* brain size evolution have emerged (Figure 1).

First, there was a modest increase in average brain size (measured from endocranial capacity) in early *H. erectus* relative to that of preceding members of the genus *Homo* (Tobias 1991; Begun and Walker 1993). Given uncertain cranial-postcranial associations in the diverse species of early *Homo*, the significance of this increase is difficult to assess (but see Begun and Walker 1993). Second, there was no significant change in the genus *Homo* mean endocranial capacity through the Early to the early Middle Pleistocene (Trinkaus and Wolpoff 1992; Begun and Walker 1993; Trinkaus 1995). Third, there was a major increase in hominid brain size through the Middle Pleistocene, peaking in the early Late Pleistocene among late archaic and early modern humans (Trinkaus and Wolpoff 1992; Trinkaus 1995). Fourth, there appears to be little change in brain size from the late Middle through the Late Pleistocene. And fifth, there has been a gradual decrease in overall cranial capacity through the late Late Pleistocene and Holocene (Henneberg 1988).

These inferences have been based almost exclusively on assessments of absolute endocranial capacity (and by inference brain size). However, a significant proportion of the variation observed may be driven by overall trends in body size through the genus *Homo* (Dubois 1921; Holloway 1981a; Henneberg 1988). Consequently, we have used appropriate estimates of body mass for Pleistocene *Homo* (Ruff et al. 1997), to compute new encephalization quotients for these fossil hominids.

Body mass was estimated using two models (Ruff et al. 1997). First, we employed a biomechanical model, in which a lower limb weight-bearing articulation (the femoral head) is taken to be proportional to the habitual loads upon it. Such articulations are primarily loaded by body mass in bipedal hominids and are less sensitive to activity-induced mechanical loads than other aspects of the lower limb (Ruff et al. 1991; Trinkaus et al. 1994). Femoral head size has a number of advantages: it is frequently preserved, easily measured (either directly or from the associated acetabulum), and its relationship to body mass in modern humans has been determined (Ruff et al. 1991; McHenry 1992; Grine et al. 1995). Second, we estimated body mass using a non-biomechanical geometric model (Ruff 1994). In this model, body mass is proportional to stature and bi-iliac breadth. Stature can be estimated from long bone lengths, using the ecogeographical and sex appropriate formula (Trotter and Gleser 1952; Ruff and Walker 1993). Body breadth can be measured directly as

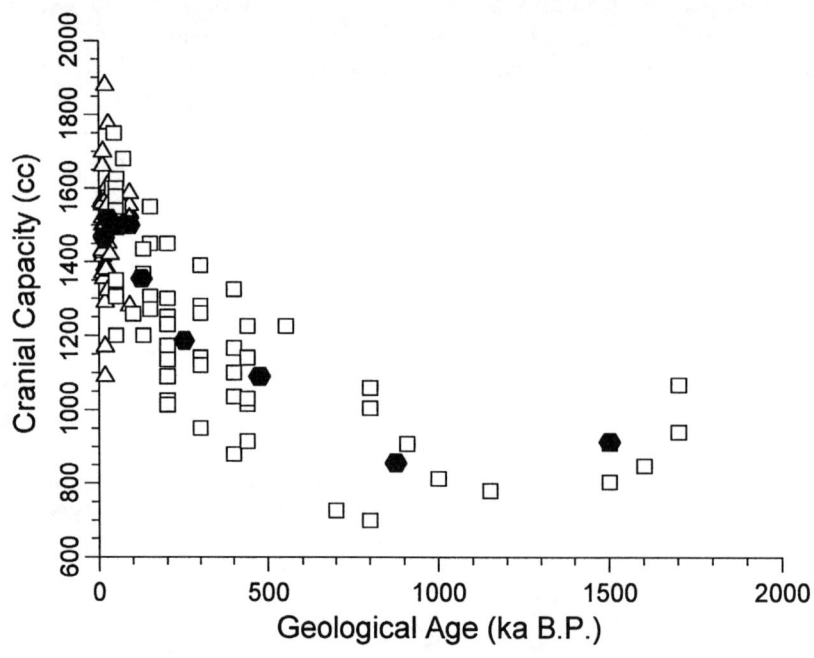

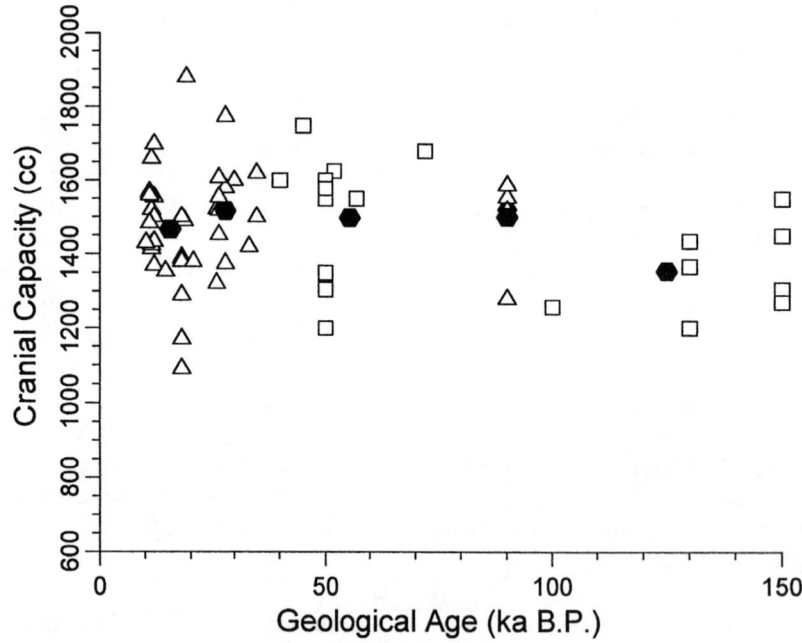

Fig. 1. Bivariate plots of encodranial capacity versus geological time (0 = present) for specimens of archaic (open squares) and modern (open triangles) Pleistocene *Homo*, plus sample means for temporal/taxonomic groups (solid hexagons). Above: the 1800 to 10 ka BP distribution. Below: expansion of the 150 to 10 ka BP distribution to show Late Pleistocene patterns. The samples are Early Pleistocene (1200-1800 ka BP), late Early-early Middle Pleistocene (600-1150 ka BP), middle Middle Pleistocene (400-550 ka BP), late Middle Pleistocene (200-300 ka BP), early Late Pleistocene (100-150 ka BP), Qafzeh-Skhul (ca.90 ka BP), late archaic human (36-75 ka BP), early Upper Paleolithic (21-35 ka BP), and late Upper Paleolithic (10-20 ka BP).

bi-iliac or maximum pelvic-breadth, or it can be estimated using within population and global ecogeographic patterns in relative pelvic breadth (Ruff 1991; Holliday 1995; Trinkaus 1996). These two body mass estimation techniques, when they can be applied to the same specimens, yield non-significant differences, with a mean absolute difference of <5 kg.

Martin's (1981) mammalian formula was then used to compute encephalization quotients (EQs) for individuals with associated brain and body mass estimates. Because Early and Middle Pleistocene hominid remains are scarce, we also used mean brain and body mass estimates for designated chronological periods (see caption to Figure 2). A similarly derived mean EQ estimate for a representative late Holocene sample (Pecos Pueblo Amerindians) is included for reference.

The plot of EQ versus geological age (Figure 2) for the complete sample largely reiterates the pattern observed with absolute cranial capacities. There is little change between ca.1.5 ma BP and ca. 800 ka BP, followed by a dramatic increase through the Middle Pleistocene, reaching modern human values sometime during the Late Pleistocene. The EQ for KNM-WT 15000, the only individual who predates the late Middle Pleistocene, falls close to the mean EQ for the early Early Pleistocene sample.

If the plot is changed to expand the last 150 ka years (Figure 2), however, an interesting pattern emerges. Individual late archaic humans (mostly European and Near Eastern Neandertals) fall largely below the Holocene human reference line. Last interglacial and early last glacial late archaic human EQs based on means fall clearly below that recent human line as well. In contrast, all of the Pleistocene early modern human remains cluster around the Holocene line, with their three means falling essentially on that Holocene EQ line. These early modern human samples include the early Last Glacial Near Eastern Qafzeh-Skhul sample, and pan-Old World early and late Upper Paleolithic samples.

Apparently EQ, like absolute brain size, remained static through Early Pleistocene and early Middle Pleistocene *Homo*, then increased dramatically through the later Middle and into the early Late Pleistocene. Moreover, this trend continued well into the Late Pleistocene, with EQs showing a modest but significant increase across the late archaic to early modern human transition. Since the emergence of early modern humans, there has been no significant change in human EQs through time.

2. Costs of Encephalization

There is a common impression that the large brains (or high EQs) of later Pleistocene and especially modern humans were an inevitable consequence of hominid cultural and biological evolution. Many authors, however, have pointed out that encephalization did not exist in an ecological or selective vacuum (e.g., Abitbol 1990; Falk 1990; Foley 1990). Even when we consider the possible interactions between culture and the genome, we must still view increased encephalization as the consequence of directional selection (presumably for augmented neurological functions) acting against ongoing stabilizing selection. Since large brains are costly in a number of ways, any localized neurological expansion which contributed to encephalization in Pleistocene hominids would have had to provide benefits which outweighed stabilizing selection to limit brain size. Costs of larger brains would have included:

A. *Obstetrical difficulties from fetal-maternal cephalo-pelvic disproportion* (see Hogan and Gallup, this volume, for a detailed discussion of this topic). Maternal pelvic dimensions in large bodied hominids are relatively large, allometrically scaled to their large body size, and birth is relatively rapid and uncomplicated (Leutenegger 1982). Hominoid neonates are relatively well developed neurologically (Jolly 1985; Trevathan 1987). Modern human neonates, who are large and large-brained relative to maternal body size, are born with much more difficulty, because the biomechanical demands of bipedalism limit the size of the human pelvic outlet. A number of perinatal complications affecting future maternal reproductive success and newborn viability are attributable to the small size of the pelvic outlet relative to overall infant size (Trevathan 1987). In one cross-cultural study of modern humans, cephalo-pelvic disproportion *sensu stricto*, (diagnosed by failure of the infant's head to engage at the level of the ischial spines) occurred in 0.9% of cases (Trevathan 1987). While the ultimate outcome of cephalo-pelvic disproportion will vary with both its degree and the efficacy of medical intervention, selection on neonatal brain size is clearly an active force even in modern humans.

B. *Secondary Altriciality*. In non-human primates, brain size approximately doubles from birth to adulthood; in humans, it more than triples (Martin 1990). That is, the human brain at birth is smaller than would be predicted by adult brain size; and gestation length, which is a function of

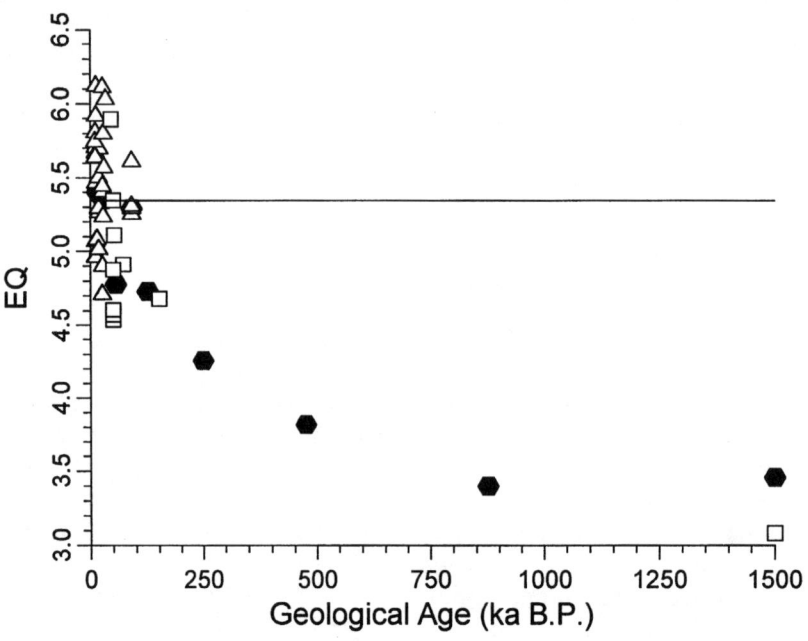

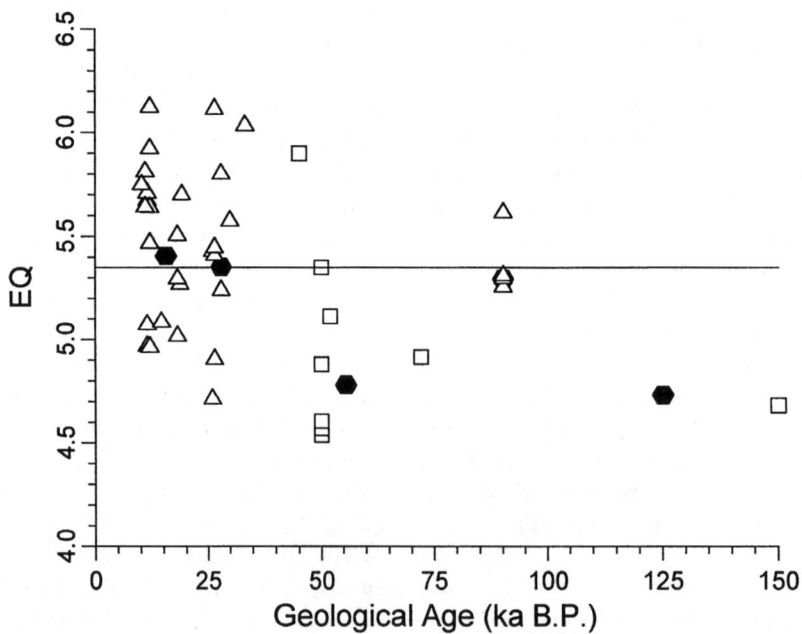

Fig. 2. Bivariate plots of encephalization quotients (EQs) versus geological time (0 = present) for specimens of Pleistocene *Homo* (archaic: open squares; early modern: open triangles) preserving associated brain and body mass estimations, plus EQs based on sample means of brain and body mass for temporal/taxonomic groups (solid hexagons). In addition, the mean EQ for a recent human sample is provided as a horizontal reference line. Above: the complete distribution. Below: the past 150 ka to show archaic versus modern differences in mean EQs. Samples as in Figure 1.

brain size in other primates, is shorter in humans than would be expected based on adult brain size (Trinkaus and Tompkins 1990). As a result, human neonates are in some respects more helpless (i.e., more "altricial") than the newborns of most other primates, and for 6 to 9 months after birth, human infants continue to function in many ways more like a fetus than an infant (Gould 1977). During this period of secondary altriciality, the brains of human infants continue to grow at prenatal rates, dramatically outpacing somatic growth for the first twelve months of infancy (Martin 1990). This rapid growth is primarily attributable to axonal proliferation and myelination, simultaneously with the extensive selective elimination of synaptic connections (Racik 1995; Reinis and Goldman 1980). During the period of rapid brain growth associated with secondary altriciality, human infants are especially vulnerable, with little margin for environmental stress (Reinis and Goldman 1980). This places high demands upon the mother and the social group. Human altriciality may have arisen as a secondary adaptation to permit greater encephalization in hominids despite biomechanical limits imposed by bipedalism on the size of the pelvic outlet. The actual timing of the emergence of this altriciality during Pleistocene *Homo* remains uncertain, given apparently different trajectories in pelvic aperture and brain size evolution during this period (Ruff 1995:Figures 1 and 2).

C. *High energetic demands of an enlarged brain.* Large brains continue to be energetically demanding throughout life. The interaction between the need for high quality food and the cognitive ability required to obtain it may be somewhat self-perpetuating. High quality nutritional resources (fruit, meat) tend to be scattered or difficult to obtain, requiring increased cognitive ability (Milton 1988, 1993). At the same time, the metabolic demands of larger brains may result in a compensatory reduction in gut size (Aiello and Wheeler 1995). Reduced gut size may, in its turn, increase the need to ingest high quality food (e.g. animal protein) (Aiello and Wheeler 1995).

D. *A prolonged developmental period.* The large size of the recent hominid brain carries with it an allometric general increase in the lengths of life history periods, especially the period before biological maturity. Populations that support juveniles who are not able to protect or support themselves face an accumulated risk of mortality before reproductive maturity. Population growth rates may also be slowed as generation length increases.

E. *Thermoregulation.* The brain tolerates little temperature variation, requiring extensive biological and cultural adaptations for thermoregulation (Falk 1990; Wheeler 1988, 1992), especially in a primate that dispersed, primarily during the Middle Pleistocene, into higher and higher latitude climates.

Despite these costs, it is apparent that both encephalization and neurological reorganization did occur, presumably involving the cognitive architecture of conceptual, object-related, social and linguistic functions. Such neurological evolution would have involved a number of components of the central nervous system:

1. Areas of the cerebral motor cortex and cerebellum, as well as afferent and efferent spinal nerves involved with fine manipulation and eye-hand coordination;

2. Vocal tract motor and sensory innervation for speech, including at least the spinal tract, cerebral motor cortex and cerebellum.

3. Language association regions of the cerebral cortex for lexical memory, syntactical coordination, and auditory interpretation of speech.

4. General and specific cortical and subcortical areas for cognitive abilities, relating to social complexity, increased territory sizes, more complex provisioning strategies in time and space, and more elaborate technology.

Cervical Vertebral Canal Size:

Vertebrate spinal canal measurements have shown that any given species has its own unique pattern of vertebral canal dimensions, with swellings in the spinal cord in the lower cervical and upper lumbar regions (Giffen 1990, 1992). These patterns reflect the relative innervation of the torso, tail, forelimbs and hindlimbs. In most vertebrates, the cervical and lumbar enlargements are associated with an increase in gray matter (cell bodies) for the innervation of the upper and lower limbs, respectively.

Modern humans exhibit larger cervical canal cross-sectional areas relative to body size than other mammals, including non-human primates. The increase is attributable to a relative increase in white matter (axons) rather than gray matter in the spinal cord (MacLarnon 1996). There is evidence that this relative increase in white matter is related to increased cortico-spinal inter-communication and more complex, higher level cerebral motor processing. For instance, a larger area of the cerebral cortex is devoted to processing for the right upper limb than to the left (White et al. 1994). In addition, the lateral cortico-spinal tracts have been found to be larger on the right side of the cervical enlargement in humans (Yakolev and Rakic 1966; Nathan et al. 1990; see also White et al. 1994).

Thus, fine manual dexterity appears to involve enhanced integration of muscular function, mediated by complex brain-cord communication, rather than an increase in motor neuron number or density in the peripheral nervous system. Given the archeological evidence of increased manual dexterity during hominid evolution, and a postulated common neurological basis for ballistic aspects of manipulation and language (Calvin 1983; Davidson and Noble 1996), we might look to the fossil record to see when this derived anatomy in the cervical vertebrae first appeared.

The original description of the KNM-WT 15000 Nariokotome *H. erectus* skeleton (Brown et al. 1985), which preserves the C7 vertebra, suggested that in this Early Pleistocene specimen the cervical canal was smaller than would be expected in a modern human. The Nariokotome skeleton belongs to a child of about 11 years (Smith 1993). In modern humans of this age, the spinal cord has already attained adult or near-adult dimensions (MacLarnon 1993), permitting reasonably direct comparison of this specimen with modern human adults.

In a further study, however, MacLarnon (1993) found that the Nariokotome child falls within the low end of the modern human range of variation in terms of cross-sectional area, and in terms of spinal canal shape (i.e., a relatively greater transverse canal diameter) he resembles modern humans. We have extended MacLarnon's comparison of the Nariokotome cervical vertebral dimensions to include Late Pleistocene humans, including Neandertals and early modern humans, as well as recent humans.

Our evidence supports MacLarnon's conclusion that the Nariokotome skeleton, at 1.5 ma B.P., shows relative cervical canal breadth enlargement, as do modern humans. A plot of C7 canal dorsoventral diameter versus transverse diameter (Figure 3) places KNM-WT 15000 well in line with a recent human distribution as well as most of the known Late Pleistocene specimens; only a couple of Neandertals appear to have relatively broader C7 canals. In contrast, data for a few C7 vertebrae of *Pan troglodytes* of generally similar body size to these hominids have canal dimensions which are similar to KNM-WT 15000 in dorso-ventral diameter but are much lower than any of the *Homo* specimens in canal transverse diameter.

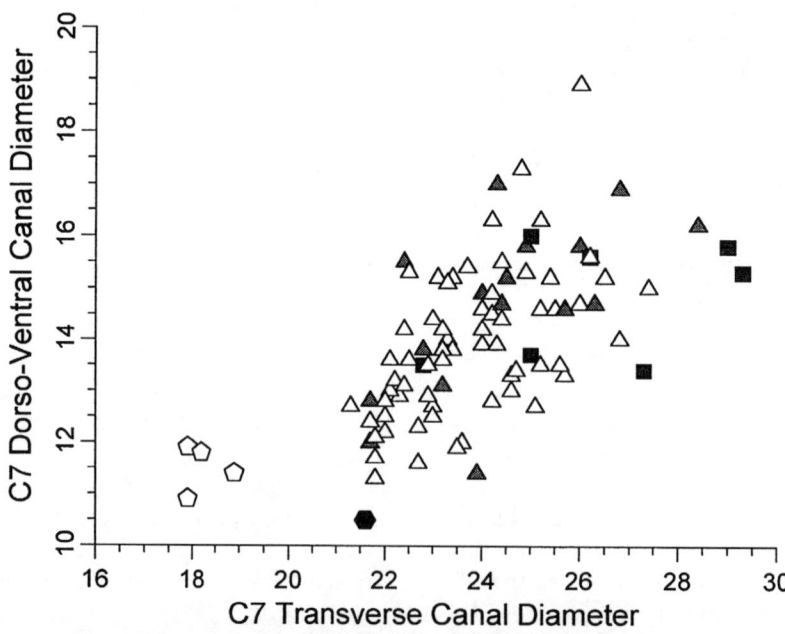

Fig. 3. Bivariate plot of C7 spinal canal dorsoventral diameter versus transverse diameter, for *Pan troglodytes* specimens [data from MacLarnon (1993; figure 15.9)] (open pentagons), KNM-WT 15000 (solid hexagon), late archaic humans (shaded squares), early modern humans (shaded triangles), and recent humans (open triangles).

In absolute C7 canal "area" (dorso-ventral times transverse canal diameters), the Late Pleistocene late archaic and early modern human distributions largely overlap (Figure 4), even though the Late Pleistocene specimens tend to be toward the larger half of the recent human distribution. It is likely that this modest shift, as with the brain size versus EQ trends, reflects body size decreases.

In this, KNM-WT 15000 falls at the low end of the more recent human range of variation in terms of absolute C7 canal area, being at the lower limit of the Late Pleistocene and Holocene distributions. This same pattern is indicated by comparisons of its C7 canal diameters to summary statistics for recent human samples (Hasebe 1912; Lanier 1939; MacLarnon 1993), in which the KNM-WT 15000 values fall below the means of all of the samples and at or below the minimum values for those recent human canal dimensions.

This pattern is reinforced when C7 canal "area" is plotted against femoral length (as a body size indicator) for both Late Pleistocene and recent humans and against body mass estimates for Late Pleistocene humans (Figure 4). In the first, KNM-WT 15000 falls at the low end of the more recent human distribution. However, the difference between KNM-WT 15000 and the more recent human remains, and especially the Neandertals, would be increased if the C7 canal "area" were plotted against estimated body mass rather than femoral length, given the tropical linear proportions of that early specimen and the warm temperate to arctic proportions of the more recent remains (Ruff and Walker 1993; Holliday 1995). The resultant plot (Figure 4) completely separates KNM-WT 15000 from the Late Pleistocene sample. Even if the canal "area" of KNM-WT 15000 were increased by ca.10%, to adjust for possible additional growth through adolescence, it would still remain well below those of the Late Pleistocene hominids and most recent humans.

Consequently, a relatively larger lower cervical vertebral canal, compared to other primates, may well have evolved in the hominid lineage to some extent prior to the emergence of *H. erectus*. Yet, if KNM-WT 15000 is representative of the middle of its population range in relative C7 canal size, there appears to have been additional enlargement of the canal through Pleistocene *Homo*, perhaps in association with the accompanying encephalization. This would imply, among other things, that a significant portion of the encephalization which took place during this human evolutionary period was related to aspects of upper limb innervation.

Postcranial Asymmetry:

Modern humans are consistently and strongly right handed at a species level (Schultz 1937; Annett 1972; Corballis 1983). By contrast, in non-human primates, functional asymmetry varies with the kind of task performed, and handedness is neither as pronounced nor as consistent as it is in modern humans (MacNeilage et al. 1987; Fagot and Vauclair 1988; but see Marchant and McGrew 1991). Modern human upper limbs provide evidence of differential mechanical loading of the arms, presumably as a result of handedness. Given that diaphyseal bone responds to habitual biomechanical load levels through the deposition and/or resorption of cortical bone (see Trinkaus et al. 1994 and references therein), it is possible to measure the relative structural strengths of upper limb long bones and assess patterns of differential loading of the right versus left arms during an individual's lifetime. The best measure of this strength is the polar moment of area, which combines the quantity of bone in a cross section with its mechanically relevant distribution relative to the mid-axis of the shaft.

Measurements of the mid-distal humeral shaft polar moments of area indicate that approximately three-quarters of modern humans possess stronger right humeri (Table 1). This percentage corresponds reasonably well with observed frequencies of right handedness in modern human populations (Annett 1972; Corballis 1983; Plato et al. 1984). Among Pleistocene humans, extending back to KNM-WT 15000, virtually every individual exhibits greater robusticity on the right side (Table 1), suggesting that the modern human right-sided handedness pattern was present, possibly in even higher frequencies, in Pleistocene *Homo*.

The degree as well as the level of humeral asymmetry shows diachronic patterning within the genus *Homo*. Modern humans have a median asymmetry level of about 10%. Late archaic and early modern Late Pleistocene humans have even more pronounced median levels of asymmetry, of about 28% and 50% (Figure 5). These changes probably reflect a similar baseline pattern of handedness, with differential levels of habitual loading in more robust Pleistocene humans.

A number of statistical correlations have been established for modern humans between cerebral asymmetries, handedness and linguistic functions (Geschwind and Levitsky 1968; Galaburda et al. 1978; Habib et al. 1995). Endocranial asymmetries, similar in degree and direction to those observed

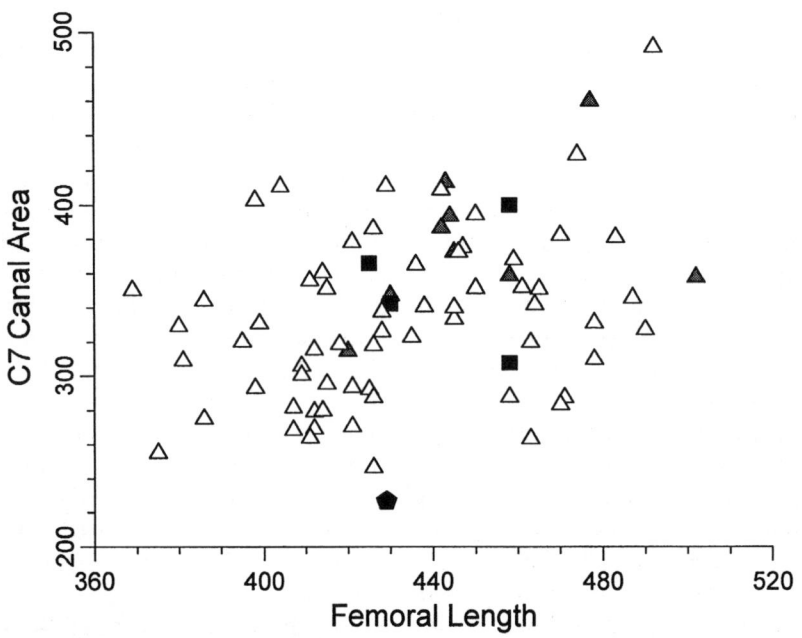
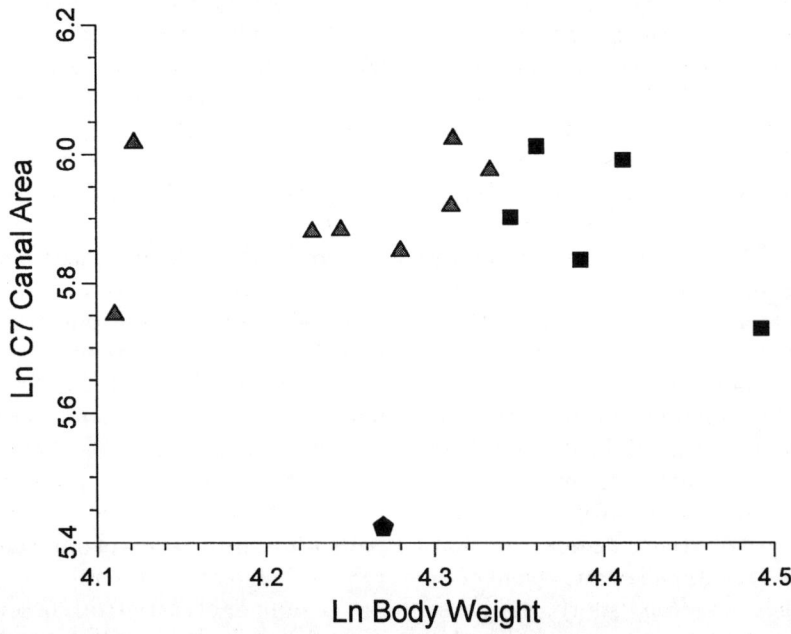

Fig. 4. Bivariate plots of C7 canal "area" (AP x ML) against measures of body size. Above: relative to femoral length, as a measure of stature. Below: relative to estimated body mass for Pleistocene *Homo*. Symbols as in Figure 3.

Table 1. Percent of individuals with the right humerus having a greater diaphyseal strength (as measured by the mid-distal diaphyseal polar moment of area) than the left one, presumably reflecting right-side dominant handedness. Since small amounts of random variation in bone strength may influence which side is more robust for specimens with little asymmetry, the recent human values were computed using the total available sample and only those specimens with asymmetry >5%.

	Percent Right > Left	N
H. erectus (KNM-WT 15000)	100%	1
Late archaic humans	100%	7
Early modern humans	93.3%	15
Recent humans (asym. >5%)	78.9%	114
Recent humans (all)	74.6%	126

in modern humans, have been noted for Pleistocene hominids (LeMay 1976, 1985; LeMay et al. 1982; Holloway 1981b; Holloway and De La Coste-Lareymondie 1982). Innervation for right hand dominance is asymmetrical, with additional neural space given over to areas of the left cerebral cortex which govern right-hand motor function (White et al. 1994). Given these correlations, coupled with archeological evidence for right-handed tool manufacture (Toth 1985) and dental-wear asymmetry related to use of the front teeth as a vice or "third hand" (Bermúdez de Castro et al. 1988), a consistent right-biased upper limb asymmetry in archaic humans provides important evidence that elements of a modern human pattern of manipulative function were in place by the Early Pleistocene.

Humans are also predominantly left-hemisphere dominant for most language functions. However, although we are relatively secure in inferring left hemisphere involvement in right-handedness for Pleistocene *Homo*, we must be very cautious in jumping to conclusions about linguistic functions on the basis of statistical correlations between handedness and language dominance. Statistical correlation is not evidence of a causal relationship, and to date no satisfactory biological causal explanation exists for the correlation between left-hemisphere dominance for language and right-handedness. One promising hypothesis postulates that language and tool use depend on a common cognitive ability to generate sequences (Corballis 1991; Calvin 1993). However, this hypothesis has not yet been thoroughly tested.

Just as spatial and linguistic functions may not be functionally "complementary," but simply statistically correlated (Bryden et al. 1983), so handedness and linguistic functions may also be functionally independent. In evaluating functional morphological asymmetries, a number of issues need to be addressed. While left-hemispheric dominance and right handedness reflect a species-level pattern, handedness and language dominance are frequently disassociated in individuals. Developmentally plastic effects and gender differences further complicate the picture, rendering asymmetry inferences for fossils problematic (Damasio and Damasio 1993; Falk 1993; Charles et al. 1994).

Ongoing Issues in Pleistocene *Homo* Paleoneurology

If we wish to pursue the question of the evolution of intelligence, we must continue to pursue research strategies which both confine and define human intelligences in ways that can be addressed by the hominid fossil record. These might include:

1. Additional fine-grained analysis of patterns of encephalization. This requires both continued appropriate estimates of body mass and brain size, plus their integration with other forms of paleoanthropological (archeological and paleoecological) evidence. Moreover, what patterns of stabilizing and directional selection might have been operating during periods of stasis or change in hominid encephalizaton?

2. Further analysis of anatomical asymmetries. Correlations of patterns of postcranial and endocranial asymmetry will help us to tease apart the evolutionary trajectories of object-oriented and conceptual intelligences. Can we further refine at least Late Pleistocene patterns and correlate them with overall encephalization, endocranial morphology, ecological factors, and archeological evidence?

3. Studies of non-cerebral regional endocranial morphology. One promising and hitherto largely neglected region of the brain is the cerebellum. Cerebellar asymmetry related to handedness has been observed in modern humans (Snyder et al. 1995). Unlike most "sub-cortical" structures, the cerebellum leaves a clearly delineated impression on endocasts, and underwent notable enlargement in the Plio-Pleistocene (Dean 1988). The cerebellum is implicated in a number of important subfunctions, including adaptive motor learning and memory, fine manipulation, and the control of repetitive rhythmic manual behaviors (Leiner et al. 1986; Ito 1993; Thach et al. 1992). Cerebellar function may also be related to storage of internalized ballistic trajectories (permitting very rapid motor sequences without recourse to time-consuming feedback mechanisms) or inhibition of motor movement while ballistic trajectories are being compared and evaluated (cf. Calvin 1993). Further studies of cerebellar asymmetry, shape and size relative to the cerebral hemispheres, may provide insight into both motor and cognitive aspects of object-related intelligences.

4. Analysis of patterns of brain growth. While secondary altriciality is costly in certain respects, heterochrony related to neonatal neurological maturity may have an important selectively advantageous influence on cognitive architecture during the critical period of neuronal "pruning" and axon formation.

Conclusion

The timing, pattern, and degree of central nervous system evolution in Pleistocene *Homo* is becoming clearer. The marked increase in encephalization during the Middle and Late Pleistocene

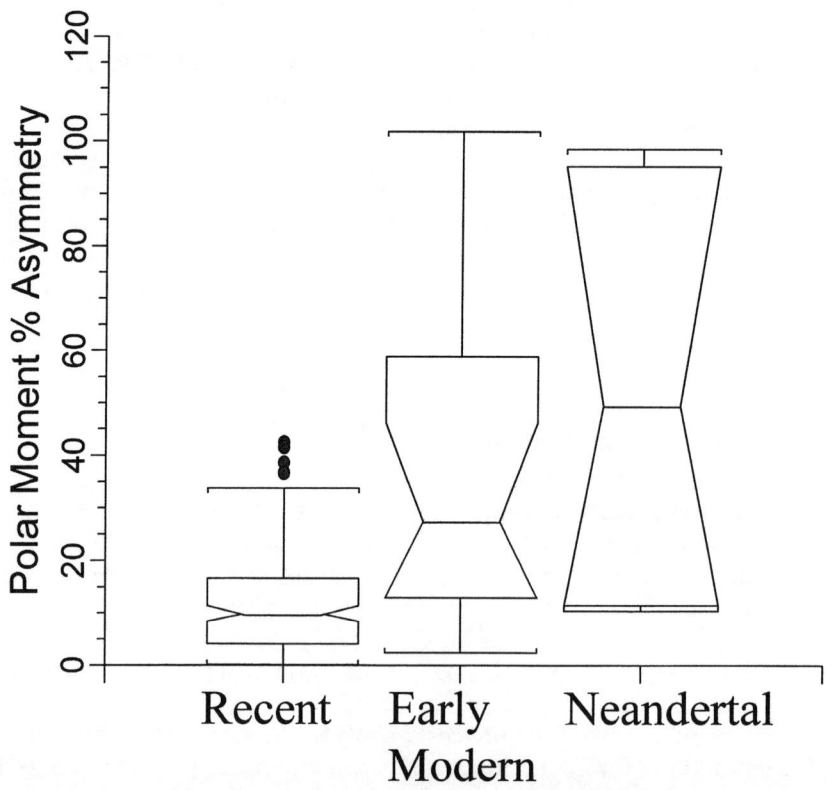

Fig. 5. Boxplots of asymmetry [((max − min)/ max) x 100] in humeral polar moment of area (a measure of overall diaphyseal rigidity) for mid-distal humeral shafts. Samples include diverse Holocene human groups (Recent, N = 126), European and Near Eastern early modern humans (Early Modern, N = 15) and Near Eastern and European late archaic humans (Neandertal, N = 7). Data from Trinkaus et al. (1994), Churchill (1994) and Trinkaus (unpub. data).

probably reflects the mosaic development of a number of independent modern human cognitive functions. Changes in relative spinal cord size and a persistence of bilateral asymmetries, reflecting at least handedness, suggest patterns of evolution for object-related intelligences. To understand these paleoneurological changes, we must consider the nature of human "intelligences" as well as the selective context in which these changes took place.

Tobias (1971:p. 2) has said that "it is an amazing yet inescapable fact that a great deal remains to be said [about the overall size of the brain], for much that has already been asserted about the brain size of both fossil man and the living races does not stand up to the cold light of scientific scrutiny." As our empirical understanding of both Pleistocene hominid neurological anatomy and modern primate brain function grows, we look forward optimistically to a more thorough grounding of theory in fact as the implications of the fossil evidence come into clearer focus.

Note

1. We are aware of the current trends regarding the number of species within the genus *Homo*, including the subdivision of these traditional species of the genus *Homo* into multiple species [such as dividing *H. erectus* into *H. erectus* and "*H. ergaster*" and *H. sapiens* into *H. sapiens*, "*H. neanderthalensis*" and possibly "*H. heidelbergensis*" (e.g., Wood 1992; Stringer 1996)] and the lumping of them together (Wolpoff 1996). Whatever the theoretical, morphological and/or phylogenetic arguments for these and other taxonomic schemes, we feel that the identification of boundaries between groups of Pleistocene *Homo* are frequently arbitrary and typological, and that emphasizing such distinctions may well obscure important continuities in the Pleistocene evolution of the genus *Homo*. At the same time, we recognize that there may well be legitimate speciation boundaries within Pleistocene *Homo* and that the magnitude of variation and directional change between early Early Pleistocene non-habiline *Homo* and modern humans exceeds that normally included even within variable polytypic species. We will therefore either refer to this period of hominid evolution simply as "Pleistocene *Homo*" or use the currently accepted, however inaccurate, divisions of *H. erectus* and archaic vs. modern *H. sapiens*.

References Cited

Abitol, M. M.
 1990 The multiple obstacles to encephalization. *Behavioral and Brain Sciences* 13:349.
Aiello, L. C. and P. Wheeler
 1995 The expensive-tissue hypothesis. *Current Anthropology* 36:199-221.
Annett M
 1972 The distribution of manual asymmetry. *British Journal of. Psychology* 63:343-358.
Begun, D. and A. Walker
 1993 The endocast. In *The Nariokotome Homo erectus Skeleton*, edited by Alan Walker and Richard Leakey, pp. 327-358. Harvard University Press, Cambridge.
Bermúdez de Castro, J. M., T. G. Bromage and Y. Fernández Jalvo
 1988 Buccal striations on fossil human anterior teeth: evidence of handedness in the middle and early Upper Pleistocene. *Journal of Human Evolution* 17:403-412.
Brown, F., J. Harris, R. Leakey and A. Walker
 1985 Early *Homo erectus* skeleton from west Lake Turkana, Kenya. *Nature*. 316:788-792.
Bryden, M. P., H. Hécaen and M. DeAgostini
 1983 Patterns of cerebral organization. *Brain and Language* 20:249-262.
Byrne, R. W. and A, Whiten, Editors
 1988 *Machiavellian intelligence*. Clarendon Press, Oxford.
Calvin, W. H.
 1983 A stone's throw and its launch window: timing precision and its implications for language and hominid brains. *Journal of Theoretical Biology* 104:121-135.
 1993 The Unitary Hypothesis: a common neural circuitry for novel manipulations, language, plan-ahead, and throwing? In *Tools, Language and Cognition in Human Evolution*, edited by K. R. Gibson and T. Ingold, pp. 230-250. Cambridge University Press, Cambridge.
Campbell, R. L.
 1993 Epistemological problems for neo-Piagetians. In *The Architecture and Dynamics of Developing Mind: Monographs of the Society for Research in Child Development*. Serial No. 234. Vol. 58(5-6):168-191.

Charles, P. D., R. Abou-Khalil, B. Abou-Khalil, R. T. Wertz, D. H. Ashmead, L. Welch and H. S. Kirschner
    1994 MRI asymmetries and language dominance. *Neurology* 44:2050-2054.

Chomsky, N.
    1988 *Language and Problems of Knowledge*. MIT Press, Cambridge, MA.

Churchill, S. E.
    1994 *Human Upper Body Evolution in the Eurasian Later Pleistocene*. Ph. D. Thesis, University of New Mexico.

Corballis, M. C.
    1983 *Human Laterality*. Academic Press, New York.
    1991 *The Lopsided Ape*. Oxford University Press, New York.

Cosmides, L. and Tooby, J.
    1994 Origins of domain specificity: the evolution of functional organization. In *Mapping the Mind: Domain specificity in Cognition and Culture*, edited by L. Hirschfeld and S. Gelman, pp. 85-116. Cambridge University Press, Cambridge.

Damasio, A. R. and H. Damasio
    1993 Brain and Language. In *Mind and Brain: Readings from Scientific American Magazine*. W. H. Freeman & Co., New York.

Deacon, T.
    1996 The Symbolic Species: The Co-evolution of Language and the Brain. W. W. Norton and Company, New York.

Dean, M. C.
    1988 Growth processes in the cranial base of hominoids and their bearing on morphological similarities that exist in the cranial base of *Homo* and *Paranthropus*. In *Evolutionary History of the Robust Australopithecines*, edited by F. E. Grine, pp. 107-112. Aldine de Gruyter, New York.

Demetriou, A., A. Efklides and M. Platsidou
    1993 *The Architecture and Dynamics of Developing Mind: Monographs of the society for research in child development*. Serial No. 234. Vol. 38(5-6).

Donald, M.
    1991 *Origins of the Modern Mind*. Harvard University Press, Cambridge, MA.

Dubois, E.
    1921 On the significance of the large cranial capacity of *Homo Neandertalensis*. *Proc. Kon. Akad. Wetenschappen* 23:1271-1288.

Essock-Vitale, S. and R. M. Seyfarth
    1986 Intelligence and Social Cognition. In *Primate Societies*, edited by B. Smuts, D. L. Cheney, R. M. Seyfarth, R. W. Wrangham and T. T. Struhsaker, pp. 452-461. University of Chicago Press, Chicago.

Fagot, J. and J. Vauclair
    1988 Handedness and bimanual coordination in the lowland gorilla. *Brain Behavior Evolution* 32:89-95.

Falk, D.
    1990 Brain evolution in *Homo*: the radiator theory. *Behavioral and Brain Sciences* 13:333-381.
    1993 Sex differences in visuospatial skills: implications for hominid evolution. In *Tools, Language and Cognition in Human Evolution*, edited by K. R. Gibson and T. Ingold, pp. 216-229. Cambridge University Press, Cambridge.

Fodor, J.
    1983 *Modularity of Mind*. The MIT Press, Cambridge, MA.

Foley, R. A.
    1990 The causes of brain enlargement in human evolution. *Behavioral and Brain Sciences* 13:354-356.

Fox, P. T., M. E. Raichle, and T. Thach
    1985 Functional mapping of the human cerebellum with positron emission tomography. *Proceedings of the National Academy of Science* 82:7462-7466.

Galaburda, A. M., M. LeMay, T. L. Kemper, and N. Geschwind
    1978 Right-left asymmetries in the brain. *Science* 199:852-856.

Gardner, H.
    1983 *Frames of Mind: The Theory of Multiple Intelligences*. Basic Books, New York.

Geschwind, N. and W. Levitsky
    1968 Human brain: left-right asymmetries in the temporal speech region. *Science* 161:186-187.

Geyer, S., A. Ledberg, A. Schleicher, S. Kinomura, T. Schormann, U. Bürgel, T. Klingberg, J. Larsson, K. Zilles and P. E. Roland
    1996 Two different areas within the primary motor cortex of man. *Nature* 382:805-807.

Giffen, E. B.
    1990 Gross spinal anatomy and limb use in living and fossil reptiles. *Paleobiology* 16:448-458.

1992 Functional implications of neural canal anatomy in recent and fossil marine carnivores. *Journal of Morphology* 214:357-374.

Gould, S. J.
1987 *Ontogeny and Phylogeny.* Belknap Press, Cambridge, MA.

Grasby, P. M., C. D. Frith, K. J. Friston, C. Bench, R. S. J. Frackowiak and R. J. Dolan
1993 Functional mapping of brain areas implicated in auditory-verbal memory function. *Brain* 116:1-20.

Grine, F. E., W. L. Jungers, P. V. Tobias, and O. M. Pearson
1995 Fossil *Homo* femur from Berg Aukas, northern Namibia. *American Journal of Physical Anthropology* 97:151-185.

Habib, M., F. Robichon, O. Lévrier, R. Khalil and G. Salamon
1995 Diverging asymmetries of temporo-parietal cortical areas: a reappraisal of Geschwind/Galaburda theory. *Brain and Language* 48:238-258.

Hasebe, K.
1912 Die Wirbersäule der Japaner. *Zeitschrift für Morphologie und Anthropologie* 15:259-380.

Henneberg, M.
1988 Decrease in human skull size in the Holocene. *Human Biology* 60:395-405.

Hirschfeld, L. A. and S. A. Gelman
1994 Towards a topography of mind: an introduction to domain specificity. In *Mapping the Mind: Domain Specificity in Cognition and Culture*, edited by L. Hirschfeld and S. Gelman, pp. 3-36. Cambridge University Press, New York.

Holliday, T. W.
1995 *Body Size and Proportions in the Late Pleistocene Western Old World and the Origins of Modern Humans.* Ph. D. Thesis, University of New Mexico

Holloway, R. L.
1981a The Indonesian *Homo erectus* brain endocasts revisited. *American Journal of Physical Anthropology* 55:503-521.
1981b Volumetric and asymmetry determinations on recent hominid endocasts: Spy I and II, Djebel Irhoud I, and the Sale *Homo erectus* specimens, with some notes on Neandertal brain size. *American Journal of Physical Anthropology* 55:385-393.

Holloway, R. L. And M. C. De La Coste-Lareymondie
1982 Brain endocast asymmetry in pongids and hominids: some preliminary findings on the paleontology of cerebral dominance. *American Journal of Physical Anthropology* 58:101-110.

Ito, M
1993 New Concepts in cerebellar function. *Rev. Neurol.* (Paris) 149:596-599.

Jackson, H. H.
1864 Clinical remarks on defects of expression (by words, writing, signs, etc.) in diseases of the nervous system. *Lancet* 1:604-605.

Jerison, H. J.
1973 *Evolution of the Brain and Intelligence.* Academic Press, New York.

Jolly, A
1985 *The Evolution of Primate Behavior.* Macmillan, New York.

Kim, S.-G., K. Ugurbil and P. L. Strick
1994 Activation of a cerebellar output nucleus during cognitive processing. *Science* 265:949-951.

Kosslyn, S. M. And O. Koenig
1992 *Wet Mind: The New Cognitive Neuroscience.* New York, Macmillan.

Lanier, R. R.
1939 The presacral vertebrae of American white and negro males. *American Journal of Physical Anthropology* 25:341-419.

Leiner, H. C., A. L. Leiner and R. S. Dow
1986 Does the cerebellum contribute to mental skills? *Behavioral Neuroscience* 100:443-454.

LeMay, M.
1976 Morphological cerebral asymmetries of modern man, fossil man, and nonhuman Primates. *Annals of the New York Academy of Sciences* 280:349-366.

LeMay, M., M. S. Billig, and N. Geschwind
1982 Asymmetries of the Brains and Skulls of Nonhuman Primates. In *Primate Brain Evolution: Methods and Concepts*, edited by Este Armstrong and Dean Falk, pp. 263-278. New York, Plenum Press.

Leutenegger, W.
1982 Neonatal brain size and neurocranial dimensions in Pliocene hominids: implications for obstetrics. *Journal of Human Evolution* 16:291-296.

Luria, A. R
1966 *Higher Cortical Functions in Man.* New York, Basic Books.

MacLarnon, A.
- 1993 The vertebral canal. In *The Nariokotome Homo erectus Skeleton*, edited by Alan Walker and Richard Leakey, pp. 359-390. Harvard University Press, Cambridge, MA.
- 1996 The scaling of gross dimensions of the spinal cord in primates and other species. *Journal of Human Evolution* 30:71-87.

MacNeilage, P. F., M. G. Studdert-Kennedy and B. Lindbolm
- 1987 Primate handedness reconsidered. *Behavioral and Brain Sciences* 10:247-303.

McHenry, H. M.
- 1992 Body size and proportions in early hominids. *American Journal of Physical Anthropology* 87, 407-431.

Marchant, L. F. and W. C. McGrew
- 1991 Laterality of function in apes: a meta-analysis of methods. *Journal of Human Evolution*. 21:425-438.

Martin, R. D.
- 1981 Relative brain size and basal metabolic rate in terrestrial vertebrates. *Nature* 293:57-60.
- 1990 *Primate Origins and Evolution: A Phylogenetic Reconstruction*. Princeton University Press, Princeton.

Milton, K.
- 1988 Foraging behavior and the evolution of primate intelligence. In *Machiavellian Intelligence*, edited by Richard W. Byrne and Andrew Whiten, pp. 285-306. Clarendon Press, Oxford.
- 1993 Diet and primate evolution. *Scientific American* 169:86-93.

Mithen, S.
- 1996 *The Prehistory of the Mind*. Thames and Hudson, New York.

Noble, W. and I. Davidson
- 1996 *Human Evolution, Language and Mind*. Cambridge University Press, Cambridge.

Piaget, J
- 1972 Intellectual evolution from adolescence to adulthood. *Human Development* 15:1-12.

Piaget, J., and B. Inhelder
- 1969 *The Psychology of the Child*. Translated from the French by H. Weaver. Basic Books, New York.

Plato C. C., K. M. Fox and R. M. Garruto
- 1984 Measures of lateral functional dominance: hand dominance. *Human Biology* 56:259-275.

Racik, P.
- 1995 Corticogenesis in Human and Nonhuman Primates. In *The Cognitive Neurosciences*, edited by M. S. Gazzaniga, pp. 127-146.. The MIT Press, Cambridge, MA.

Reinis, S. and J. Goldman
- 1980 *The Development of the Brain*. Charles C. Thomas, Springfield, IL.

Ruff, C. B.
- 1991 Climate, body size and body shape in hominid evolution. *Journal of Human Evolution*. 21:81-105.
- 1994 Morphological adaptation to climate in modern and fossil hominids. *Yearbook of Physical Anthropology* 37:65-107.
- 1995 Biomechanics of the hip and birth in early *Homo*. *American Journal of Physical Anthropology* 98:527-574.

Ruff, C. B., W. W. Scott and A. Y. C. Liu
- 1991 Articular and diaphyseal remodeling of the proximal femur with changes in body mass in adults. *American Journal of Physical Anthropology* 86:397-413.

Ruff, C. B., E. Trinkaus and T. W. Holliday
- 1997 Increased body mass and encephalization in Pleistocene *Homo*. *Nature* 387: 173.

Ruff, C. B. and A. Walker
- 1993 Body size and body shape. In *The Nariokotome Homo erectus Skeleton*, edited by Alan Walker and Richard Leakey, pp. 234-265. Harvard University Press, Cambridge.

Snyder, P. J., R. M. Bilder, H. Wu, B. Bogerts and J. A. Lieberman
- 1995 Cerebellar volume asymmetries are related to handedness: a quantitative MRI study. *Neuropsychologia* 33:407-419.

Thach, W. T., H. P. Goodkin and J. G. Keating
- 1992 The cerebellum and the adaptive coordination of movement. *Annual Review Neuroscience* 15:403-42.

Schultz A. H.
- 1937 Proportions, variability and asymmetries of the long bones of the limbs and the clavicles in man and apes. *Human Biology* 9:281-328.

Smith, B. H.
- 1993 The physiological age of KNM-WT 15000. In *The Nariokotome Homo erectus Skeleton*, edited by Alan Walker and Richard Leakey, pp. 195-220. Harvard University Press, Cambridge, MA.

Stringer, C. B.
　1996　Current issues in modern human origins. In *Contemporary Issues in Human Evolution*, edited by W. E. Meikle, F. C. Howell and N. G. Jablonski, pp. 115-134. *California Academy of Sciences Memoir* 21.

Tobias, P. V.
　1971　*The Brain in Hominid Evolution.* Columbia University Press, New York.
　1991　*Olduvai Gorge 4: The Skulls, Endocasts and Teeth of Homo habilis.* Cambridge University Press, Cambridge.

Toth, N.
　1985　Archaeological evidence for preferential right-handedness in the Lower and Middle Pleistocene, and its possible implications. *Journal of Human Evolution* 14:607-614.

Trevathan, W. R.
　1987　*Human Birth: An Evolutionary Perspective.* Aldine de Gruyter, New York.

Trinkaus, E.
　1995　Brains and bodies: mosaic trends in Middle Pleistocene archaic *Homo* morphology (Cerebros y cuerpos: tendencias evolutivas en mosaico de la morfología del hombre arcaico durante el Pleistoceno Medio). In *Evolución Humana en Europa y los Yacimientos de la Sierra de Atapuerca*, edited by J. M. Bermúdez de Castro, J. L. Arsuaga and E. Carbonell, pp. 1:205-228. Consejería de Cultura y Turismo, Junta de Castilla y León,. Valladolid.
　1996　The humerus versus the femur: changing patterns of diaphyseal robusticity across the late archaic to early modern human transition. *Anthropologie (Brno)* 34:101-110.

Trinkaus, E., Churchill, S. E., and Ruff, C. B.
　1994　Postcranial robusticity in *Homo*, II: humeral bilateral asymmetry and bone plasticity. *American Journal of Physical Anthropology* 93:1-34.

Trinkaus, E., and Tompkins, R. L.
　1990　The Neandertal life cycle: the possibility, probability, and perceptibility of contrasts with recent humans. In *Primate Life History and Evolution. Monographs in Primatology*, edited by C. J. DeRousseau, pp. 153-180. Monographs in Primatology Vol. 14. Wiley-Liss, New York.

Trinkaus, E., and Wolpoff, M. H.
　1992　Brain size in post-habiline archaic *Homo* (abstract). *American Journal of Phyical Anthropology, Supplement* 14:163.

Trotter, M. and Gleser, G. C.
　1952　Estimation of stature from long bones of American whites and Negroes. *American Journal of Physical Anthropology* 10:463-514.

Wheeler, P. E.
　1988　The thermoregulatory advantages of hominid bipedalism in open equatorial environments: the contribution of increased convective heat loss and cutaneous evaporative cooling. *Journal of Human Evolution* 21:107-115.
　1991　The influence of bipedalism on the energy and water budgets of early hominids. *Journal of Human Evolution* 21:117-136.

White, L. E., G. Lucas, A. Richards, D. and Purves
　1994　Cerebral asymmetry and handedness. *Nature* 368:197-198.

Wolpoff, M. H.
　1996　*Human Evolution*, 1996-1997 Edition. McGraw-Hill, New York.

Wood, B. A.
　1992　Origin and evolution of the genus *Homo*. *Nature.* 355:783-790.

12. On the Neural Bases of Spoken Language

Philip Lieberman

Abstract

One of the derived properties that differentiates us from other living primates appears to be a neural network that constitutes the "functional language system" (FLS) of the human brain. The FLS regulates the production and comprehension of spoken language which allows humans to rapidly transmit and integrate information with knowledge stored in the brain's dictionary. Although neither the neuroanatomical structures or neural circuits implicated in this system can be specified with certainty given the present state of knowledge, converging behavioral, neuroanatomical, and neurophysiogic data point to the FLS being a distributed neural network that includes subcortical structures traditionally associated with motor control as well as the traditional cortical areas associated with language (Broca's and Wernicke's areas), and other regions of neocortex. The FLS may have evolved from functional neural systems that enhance biological fitness by achieving timely motor responses to environmental challenges and opportunities.

The FLS regulates the motor commands necessary for voluntary speech, integrates "knowledge" of speech production in the perception of speech, accesses words from the brain's "dictionary," and comprehends distinctions in meaning conveyed by syntax. It is not possible to completely fractionate human linguistic and cognitive ability since many of the neuroanatomical structures that support the circuits of the FLS also are implicated in other aspects of cognition, including our ability to take account of and appropriately act on new information. Early anatomically modern *Homo sapiens* (AMHS) fossils had fully human speech producing anatomy and most likely had a modern human FLS. Given the involvement of neuroanatomical structures of the FLS in cognition, they also most likely had fully modern cognitive ability. A reappraisal of Neanderthal speech-producing anatomy indicates that they had superior speech abilities than newborn human infants. However, Neanderthals nonetheless lacked fully human speech-producing anatomy and would have been incapable of producing the full range of speech-sounds of any human language. Since recent genetic studies show that distinctions in dialects serve as reproductive isolating mechanisms in modern human populations, Neanderthal speech may have served as a genetic isolating mechanism accounting for their replacement over an extended period.

Introduction

One of the derived features that differentiates human beings from other primates is spoken language. Human beings can learn to use manual sign languages to communicate similar referential information. However, the speech is the "default" medium for human language and thought. Recent studies suggest that the plasticity of the brain allows neuroanatomical structures that would otherwise have regulated speech to instead regulate sign language (Elman et al. 1997). I will outline a theory concerning the brain bases of human linguistic ability. I propose that a specialized "functional language system" (FLS) evolved to regulate the production and comprehension of spoken language. Spoken language allows us to communicate information rapidly, transcending the temporal limitations imposed by the human auditory system. The FLS thereby enhances biological fitness, it allows human beings to respond rapidly to environmental challenges and opportunities, integrating sensory information with linguistic knowledge (Lieberman 1984, 1991, 1998, 2000).

Although neither the neuroanatomy or neurophysiology of the FLS can be specified with certainty given the present state of knowledge, converging behavioral, neuroanatomical, and neurophysiogic data show that it is a distributed neural network (Mesulam 1990) that includes subcortical structures traditionally associated with motor control as well as the traditional cortical areas associated with language (Broca's and Wernicke's areas), and other regions of neocortex. The traditional view of the neural bases of complex behaviors derives from early 19th phrenology (Gall 1809). Phrenologists claimed that different parts of the cranium were the "seats" of various aspects of behavior or character. Neo-phrenological theories do not claim that the bumps on a person's skull can tell you whether he is honest or has mathematical skills. However, neo-phrenological theories map complex behaviors to localized regions of the brain; they explicitly or implicitly claim that a particular part of the brain regulates some aspect of behavior. Perhaps the best known example is the Broca-Wernicke model of the brain bases of human language. The traditional interpretation of Paul Broca's 1861 study of a patient who had suffered a series of strokes, was that damage limited to "Broca's area," a frontal region of neocortex, will result in expressive language deficits—word finding difficulties, impediments in speech production, simplified syntax, as well as difficulties in the comprehension of distinctions in meaning conveyed by syntax. Damage to a posterior area of cortex, Wernicke's area (1874), were correlated with "receptive" linguistic deficits shortly after Broca's paper. The models first proposed by Lichtheim (1885), claimed that the neurological basis of human language was a system linking Wernicke's area with Broca's area. This model which was revived by Geschwind (1970), stated that Wernicke's area processed incoming speech signals; information was then transmitted via a hypothetical cortical pathway to Broca's area which served as the "expressive" language output device. The Lichtheim-Geschwind theory is taken by linguists such as Chomsky (1986) and Pinker (1994) to be a valid model of the neural architecture underlying human linguistic ability. Pinker (1994) specifically identifies Broca's area as the site of the "syntax organ."

Although this model has the virtue of being simple, current neurophysiogic data show that it is wrong. Some regions of neocortex are specialized to process particular stimuli, visual or auditory, while others are specialized to regulate motor control, emotional responses, or hold information in long-term or short-term (working) memory, etc. But complex behaviors, such as the way that a monkey protects his eyes from intruding objects, derive from neural networks formed by circuits that link populations of neurons in many different neuroanatomical structures. It is impossible to localize the "seat" of a complex behavior because a particular neuroanatomical structure can support different neuronal populations that form components of circuits regulating other aspects of behavior. In other words, while specific operations may be performed in particular parts of the brain, these operations are integrated into circuits or "networks" that regulate an observable aspect of behavior. A particular aspect of behavior thus involves the activity of neuroanatomical structures distributed throughout the brain. As Mesulam (1990) notes,

> complex behavior is mapped at the level of multifocal neural systems rather than specific anatomical sites, giving rise to brain-behavior relationships that are both localized and distributed.

In short, it is necessary to differentiate between *neuroanatomical structures*, i. e., various areas of cortex and subcortical structures which have been discerned using traditional anatomical procedures, and *neural circuits* which only can be mapped by means of "tracer" and electrophysiologic techniques at a microscopic level. Particular neuroanatomical structures appear to perform specific "computations," but a given neuroanatomical structure can support distinct neuronal populations which project to neuronal populations in different neuroanatomical structures, forming neural circuits. Circuits and groups of circuits, "networks" or "systems," regulate particular aspects of behavior. Moreover, the architecture of the circuit or network that regulates a particular aspect of behavior is usually neither "logical" or "parsimonious." As Mayr (1982) notes, the logic of biology inherently reflects the opportunistic nature of biological evolution. Neuroanatomical structures that were adapted to carry out one function have been modified to regulate "new" behaviors that contributed to biological "fitness," the survival of progeny. Thus the neural model presented here is that:

(1) Human language is regulated by a neural "functional language system" (FLS) similar in principle to the neural functional systems that regulate other aspects of behavior in human beings and other animals. Many of these systems evolved to allow timely responses to environmental challenges and opportunities that enhanced biological fitness.

(2) The FLS evolved by means of Darwinian mechanisms to regulate speech, which allows information to be transferred vocally at a rate faster than that of other auditory signals (Liberman et al. 1967). Speech is a central element of human language. The FLS regulates the motor commands

that underly speech, it perceives speech by bringing a listener's "knowledge" of the articulatory constraints of speech production to bear on the interpretation of the acoustic signal, it accesses words from the brain's dictionary through their sound pattern, and maintains words in verbal working memory by means of a rehearsal mechanism (silent speech) in which words are internally modeled by the neural mechanisms that regulate speech production.

(3) At least some of the neuroanatomical structures that are implicated in the circuitry of the FLS play a part in regulating "lower" aspects of behavior such as manual motor control. Other neuroanatomical structures implicated in the FLS also play a part in aspects of cognition outside the domain of language. Their involvement in these tasks reflects the general pattern of evolution—organs adapted to one function took on new tasks.

(4) Natural selection that enhanced the computational power of a neuroanatomical structure that functions in different aspects of behavior may enhance its role in all of these behaviors. Thus, enhanced linguistic ability probably cannot be morphologically or functionally differentiated from enhanced cognitive ability and certain aspects of motor control.

(5) Finally many aspects of human language are not unique to anatomically modern *Homo sapiens* (AMHS). Lexical and syntactic ability exist to a degree in present day apes. We can also infer that archaic hominids such as the Neanderthals and *Erectus* grade hominids must have possessed vocal language since there would have been no basis for adaptations that enhanced the efficiency and saliency of speech production in AMHS.

Neural Functional Systems

Recent neurophysiologic studies show that neural functional systems have evolved in many species that generate timely responses to environmental challenges and opportunities. These neural systems link sensory inputs to neuroanatmical structures that execute appropriate motor responses. A brief example of a neural functional system that has been studied in monkeys may be useful. Although the neurophysiologic techniques used to explore the monkey system cannot be used to study human beings, we undoubtedly have a similar neural functional system.

The traditional view, now known to be incorrect, of the neural basis of the representation of visual space was that some part of the visual cortex contains a "map" of space analogous to the representation of a scene in a photograph. Many aspects of behavior depend on visual information, such as moving one's hand towards an object, identifying a particular animal, or avoiding walking over a cliff. All of these activities hypothetically could involve reference to a single common topographic "map" that codes the location of everything that you see at a given moment. If we were to follow the principles of logic and parsimony that structure mind-brain theories such as that proposed by Fodor (1983) it would be reasonable to propose that a single map existed in some part of the brain, which other neural modules that regulated hand movements, identifying objects, walking, etc. could reference. However, neurophysiologists who study vision conclude that no single, unified, visual "map" exists in the primate brain. Instead, neural functional systems exist that are adapted to achieving particular goals, and that make use of particular aspects of vision.

One system which has been explored by Charles Gross and his colleagues appears to be adapted to grasping or deflecting objects that are moving towards a monkey's face. Their studies of the macaque monkey brain show that cells in the putamen, a subcortical basal ganglia structure, respond vigorously when small objects approach towards the monkey's face and eyes. These neurons do not respond to images, stationary objects, or to moving objects more than a meter or so from the monkey. They instead respond only to moving objects that are close to the monkey. About 25 percent of these sites also respond to tactile sensations on the monkey's face. These sites in the putamen, in turn, communicate to neurons in the monkey's premotor cortex. (Graziano et al. 1994). The visual responses of these neurons are arm-centered; they provide a representation of space near the body that is useful for grasping objects moving towards the monkey and direct appropriate arm movements. This functional close-object-intercept system links inputs from visual cortex, tactile inputs, putamen, premotor cortex, motor cortex and the monkey's arms and hands, integrating sight, touch and muscular control to achieve a specific task—intercepting objects. It doesn't require a study of monkeys in their natural habitat to predict that individuals who could better avoid being hit in the head would be more likely to survive and would produce more descendants.

None of these neuroanatomical structures, putamin, premotor cortex, etc., is the "seat" of this "close-intercept-system." The putamin and closely associated basal ganglia structures, for example, support circuits that regulate the timing of manual motor activity (Cunnington et. al. 1995), speech motor control and syntax (Lieberman et al. 1992a; Pickett et al. 1998) bipedal locomotion, emotion, and cognition (Cummings 1993; Marsden and Obeso 1994). The neural circuitry that is the monkey brain's basis for the close-intercept-system consists of neuronal populations in neuroanatomical structures that project to neuronal populations in other neuroanatomical structures. The circuit may be segregated from the neural circuits that are the physiologic bases of other complex behaviors (the question is still open). However, the circuitry of the close-intercept-system is distributed.

The Human Functional Language System

No living species apart from *Homo sapiens* can talk, comprehend distinctions in meaning conveyed by moderately complex syntax, or acquire more than about 150 words. The weight of present evidence indicates that human beings, alone, have a FLS, a neural functional system adapted for spoken language. The constraints of speech production and neural mechanisms adapted for speech structure the hypothetical FLS of the human brain—it is a system for language organized about *speech*. (This issue and other data and questions concerning the FLS are discussed in detail in Lieberman (1998; 2000).

Since only human beings possess spoken language, this rules out many of the invasive techniques that are necessary to chart neural circuits. However, converging evidence from many studies, including the ones that will be briefly noted below, indicate some of the probable architecture and neuroanatomical structures of the FLS. The study of "experiments in nature" in individuals who have suffered strokes, disease, and trauma that damage particular parts of their brains and new, relatively noninvasive, imaging techniques such as Positron Emission Tomography (PET), and Functional Magnetic Resonance Imaging (FMRI) which can assess relative metabolic activity in different parts of the brain while a person performs a task, indicate that human language is regulated by means of a distributed neural network that constitutes the FLS.

A brief and incomplete account of the processes and neuroanatomical structures involved in comprehending a sentence that has moderately complex syntax, for example, the sentence, *"I saw the cat who ate the mouse run."* will perhaps illustrate the operations of the proposed FLS.

Auditory input and speech perception

The first stage must involve sound. Auditory perception must take place in order to perceive the sounds of speech. But there exists no compelling evidence for a "modular" organization (Fodor 1983). Abundant evidence exists for a "speech mode" of perception (Liberman et al. 1967; Lindblom 1996). However, the process does not involve sequential processing of the incoming acoustic signal by an auditory module which then transfers information to a phonetic module which, in turn, transfers an abstract "phonetic" representation to a lexical module or by independent modules operating in parallel. Robust experimental data show that auditory perception and speech perception can not be completely fractionated. For example, the initial stop consonant of *cat* is differentiated from the sound [g] by its "voice onset time" (VOT). The VOT of a stop consonant is the time interval between the "burst" that occurs when the tongue moves away from the roof of the mouth and the onset of phonation that occurs when the vocal cords of the larynx begin to open and close (Lisker and Abramson 1964). A 20 msec time window produces a sharp categorical "linguistic" boundary when human listeners hear these sounds; if burst and phonation occur within 20 msec the sound is heard as a [g]. However, it is clear that the "linguistic," categorical perception of stop sounds by their VOT (short-lag [g] versus long-lag [k] as well as [b] versus [p] and [d] versus [t]) derives from an auditory mechanism that also enters into the process by means of which human beings and other mammals localize sounds (Hirsch and Sherrick 1961). Chinchillas, therefore, categorize the stop sounds of human speech in much the same manner as humans because the constraint derives from the mammalian auditory system (Kuhl 1978).

Links between speech perception and production

Nonetheless, there is something "special" about the perception of speech; the acoustic speech signal is processed with reference to internalized "knowledge" of speech production. A direct link between speech perception and production was explicitly proposed by Liberman and his colleagues (1967) who correctly noted that the individual sounds of speech (the phonetic segments that roughly correspond to the letters of the alphabet) are melded together in the speech signal. It is impossible to isolate acoustic segments that each correspond to a letter of the phonetic alphabet. The acoustic factors that signal individual speech sounds are melded together in syllable-sized units. Liberman et al. (1967) claimed that the process of speech perception involved internally modeling possible articulatory gestures and then comparing the internally generated speech signal against the incoming acoustic signal. They claimed that "invariant" articulatory gestures that each corresponded to an individual sound existed. Subsequent experimental studies ruled out articulatory "invariants" that corresponded to individual phonetic segments. Different speakers, for example, use different articulatory gestures to generate the same speech signal (Nearey 1979). Moreover, an individual speaker produces the same speech signal by means of different articulatory gestures in different phonetic contexts (e.g., the sound [t] has different acoustic and articulatory characteristics in the syllables in [tu] and [ti] (Sereno et al. 1987). However, it is evident that human listeners do interpret speech signals in a "speech mode" (Liberman and Mattingley 1985; Lindblom 1996) in which they take account of the constraints imposed by human speech-producing anatomy and physiology.

Lip reading, for example, reflects this process; under normal circumstances the accuracy of speech perception is enhanced when we view the face of the person who is talking. McGurk and MacDonald (1976) showed that a listener viewing someone saying [ba] and hearing [ga] will "hear" the syllable [da] because the visual information signaling the production of the sound [ba] is integrated with the acoustic signal of the sound [ga]. The constraints of human speech anatomy yield "quantal" consonantal distinctions. Although the human tongue can form a constriction at any point in the mouth, at certain locations acoustic signals are generated that are perceptually salient and resistant to small errors in tongue placement (Stevens 1972). These locations correspond to the "places of articulation"—the particular locations along the roof of the mouth (the hard palate) where consonants are formed in all human languages. In English only [da] can occur between the conflicting visual-articulatory [ba] and the auditory [ga] presented to the listener. A listener, therefore, "hears" [da] when he sees someone saying [ba] while listening to the sound [ga]. The speech sound that would have been produced at the intermediate "place of articulation" is selected.

Words

Other animals, including dogs (Warden and Warner 1928) and chimpanzees (Gardner and Gardner 1984, Savage-Rumbaugh et al. 1985) have limited lexical abilities despite the fact that they cannot talk. This is not surprising in light of recent studies of the manner by which human beings access the meaning of words stored in the brain's "dictionary." These PET and FMRI studies show that the neuroanatomical structures that process the primary visual and motoric sensory information that constitute the real-world "knowledge" coded in a word are activated when we think of a word or specific "semantic" properties of a word. Areas of primary visual cortex that are known to be part of the system implicated in the perception of color, for example, are activated when a person thinks of the color of the word *pencil*. Cortical areas implicated in the perception of shape are activated when a person thinks of the shape of a pencil. Cortical areas that regulate manual motor activity are activated by the word *hammer* (Martin et al. 1995). Other species clearly possess these neural "primary perceptual" systems; therefore, it is not surprising that dogs and chimpanzees can code some of the concepts represented by a word. It is interesting to note that independently motivated linguistic studies (e. g., Croft 1991) also claim that basic linguistic categories such as *noun* and *verb*, code distinctions motivated by real-world distinctions rather than the formal, linguistic distributional constraints that often motivate linguists of the Chomskian school.

These primary neuroanatomical areas appear to project to posterior temporal regions of the brain, topographically organized in terms of semantically-based *linguistic* classes such as nouns and verbs (Damasio et. al 1996). The phonological representations of a word, its *name* "spelled out" by the sounds of speech appear to be represented in these areas of the brain.

Lexical access and the comprehension of syntax

Deficits in lexical access constitute one of the primary language problems noted in virtually all studies of aphasia. The studies that will be briefly noted below indicate that Broca's areas is implicated in lexical access. However, many other neuroanaomical structures support the neuronal circuitry implicated in lexical access. These structures also are involved in the comprehension of distinctions of meaning conveyed by syntax. Indeed, converging evidence shows that syntax and lexical access are interwoven in the "computations" that take place in verbal working memory (Just and Carpenter 1992; MacDonald et al. 1994). I am using the term "computation" metaphorically to signify the neural processing that must occur in working memory; it is doubtful that algorithmic computations in the usual mathematical sense occur in the brain.

Baddeley in 1975 proposed that words are maintained in "verbal working memory" by means of a "rehearsal" mechanism in which they are continually subvocally articulated. Subsequent studies support this theory. People derive the meaning of the sentence that they hear or read by maintaining the words that constitute the individual sentences in verbal working memory through this process of phonetic rehearsal. Many factors can increase the load placed on verbal working memory. Sentences that have more complex syntax or longer words or less common words place a greater task load on verbal working memory. For example, a "center embedded clause, such as "the cat who ate the mouse" must be maintained in verbal working memory in order to comprehend the meaning of the sentence thereby imposing a greater load on verbal working memory. The PET study of Stromswold et al. (1995) monitored activity in Broca's area when subjects were asked to comprehend English sentences; metabolic activity increased when the sentences contained center embedded clauses. The PET study of Awh et al. (1996) shows that as a person attempts to remember the last word of more and more sentences, a task that increases the task load placed on verbal working memory, increased activity occurs in Broca's area and in the neural substrate for speech production (including premotor and motor cortex) which must work harder to rehearse the words of the sentence. These studies show that speech perception, speech production, lexical access, real-world semantic knowledge, and syntactic processing are linked in the FLS.

Other implicated neuroanatomical structures

A close functional and anatomical link between speech production and the comprehension of meaning thus exists in verbal working memory. As we shall see, pathologies that damage the neural mechanisms regulating speech production impair also the comprehension of sentences. These experimental data show that the neuroantomical structures implicated in regulating speech production and the comprehension of syntax include basal ganglia structures. Basal ganglia also support neural circuits to supplementary motor area and premotor and motor cortex that regulate manual motor control as well as circuits to prefrontal cortical areas that are implicated in maintaining cognitive sets and other aspects of cognition (Middleton and Strick 1994). Anatomical and brain-imaging data (Leiner et al. 1993) also point to cerebellum being involved in cognitive and linguistic tasks by means of circuits through basal ganglia.

Some Experimental Studies Explicating the FLS

Aphasia

Recent studies of aphasia with permanent loss of language show that the traditional view of the brain-bases of aphasia is incorrect. In their neurological text Stuss and Benson (1986) note that damage to "the Broca area alone or to its immediate surroundings...is insufficient to produce the full syndrome of Broca's aphasia..." This reappraisal results from many studies of permanent aphasia in patients with brain damage. Damage to the internal capsule, basal ganglia structures including the putamen, and the caudate nucleus as well as other subcortical structures can yield impaired speech production and agrammatism similar to that of the classic aphasias as well other cognitive deficits (e.g., Naeser et al. 1982; Alexander et al. 1987; Mega and Alexander 1994)

Parkinson's disease

Studies of Parkinson's Disease confirm that subcortical basal ganglia brain structures are implicated in controlling both speech production and syntax. Parkinson's Disease (PD) which derives from depletion of the neurotransmitter dopamine, causes major damage to the basal ganglia, sparing cortex until the late stages when cortical receptors may become damaged (Xuerob et al. 1990). Therefore, PD allows us to assess the effects of damage limited to subcortical basal ganglia. The primary deficits of PD are motoric; tremors, rigidity, and repeated movement patterns occur (Jellinger 1990). However, PD can also cause linguistic and cognitive deficits. Independent studies show that PD subjects have higher error rates or longer response times in tasks that involve their ability to comprehend distinctions of meaning conveyed by syntax (e. g., Lieberman et al. 1990, 1992b; Grossman et al. 1991, 1992; Natsopoulos et al. 1994; Lieberman and Tseng 1994, n.d.). The first study demonstrating deficits in the comprehension of syntax (Lieberman et al. 1990) tested PD subjects using The Rhode Island Test of Language Structure (RITLS), which assesses the extent to which a subject is able to use syntax in understanding sentences. Sentence length and vocabulary are balanced in this test which Engen and Engen (1983) originally designed to assess the comprehension of syntax by hearing-impaired children; 8 year-old hearing children perform with virtually no errors. Thirty percent error rates were noted for some PD subjects for sentences with moderately complex syntax.

Several studies (Lieberman et al. 1992b; Lieberman and Tseng 1994, n.d.), moreover, show that the pattern of speech production and syntax comprehension deficits in non-demented Parkinson's Disease subjects is similar in nature to that noted for Broca's aphasia. Acoustic analyses of the speech of Broca's aphasics show that one of their major speech production problems derives from deficits in their ability to regulate voice-onset-time (VOT) (Blumstein et al. 1980; Baum et al. 1990). Lieberman et al. (1992b) studied 40 PD subjects: those having deficits in the regulation of VOT motor timing also tended to have syntax comprehension and cognitive deficits, though none were demented. Nine PD subjects had VOT overlaps of 18.6 % compared to 3.6 % for normal controls speaking at the same rate. The subjects having VOT overlap had significantly higher syntax error rates and longer response times on the RITLS than the VOT non-overlap subjects, VOT-timing errors and syntax errors were significantly correlated. VOT overlap subjects also had significantly higher error rates in cognitive tasks involving abstraction and the ability to maintain a mental set. A subsequent study of Chinese-speaking PD subjects replicated these VOT timing and sentence comprehension deficits using a Chinese adaption of the RITLS syntax test (Lieberman and Tseng 1994, n.d.). Focal lesions of basal ganglia structures resulting from coma yield even higher speech, syntax and cognitive deficits (Pickett et al. 1998).

Hypoxia

Strong correlations between speech motor control deficits and the comprehension of distinctions of meaning conveyed by syntax can occur in otherwise intact individuals due to oxygen deficits. Speech and syntax deficits in mountain climbers ascending Mount Everest similar, but not as extreme, to those in PD subjects, were found as they reached higher altitudes where less oxygen was present (Lieberman et al. 1994, 1995). The ascent of Mount Everest by the "normal" South Col route involves a series of high camps over a three month period. A battery of speech, syntax and cognitive tests similar to those used to assess deficits in PD was administered to five climbers when they first reached each of these progressively higher camps. VHF radio links were used between Everest Base Camp (5,300 meters altitude) and Camps two (6,300 meters altitude) and Three (7,150 meters altitude). The mean VOT "separation width" that differentiated voiced from unvoiced stop consonants decreased from 26.0 to 6.4 ms. The time needed to comprehend spoken English sentences also increased. Response times on the RITLS were 54 percent longer at Camp Three for simple sentences that are readily comprehended by six-year-old children. VOT decrements and increased sentence response were significantly correlated. Hypoxic insult to the globus pallidus, a basal ganglia structure which is sensitive to oxygen deprivation, probably caused these effects. Marked behavioral and cognitive disruption, similar to those noted for frontal lobe lesions was noted by Laplane et al. (1984) for patients with bilateral globus pallidus lesions resulting from anoxia. Similar losses were noted by Strub (1989) for bilateral globus pallidus haemorrages resulting from exposure to extreme-altitude.

The Evolution of the FLS

It is impossible to determine the morphology or the functional structure of an extinct hominid's brain by examining its endocast (Holloway 1995). For that matter, cortical areas similar to those traditionally associated with language (Broca's and Wernicke's areas) can be discerned in chimpanzees. But chimpanzees cannot talk. The cries of chimpanzees are closely linked to affect. Many of the segmental acoustic cues that convey linguistic meaning in human speech can be found in chimpanzee vocalizations (Lieberman 1968) that signify particular emotional states or activities (Goodall 1986). However, chimpanzees can not permute these cues to form "new" cries (Goodall 1986). What may be missing in the chimpanzee brain is the ability to "unpack" the sequential elements that form "bound" responses to external stimuli, and form novel flexible motoric or cognitive outputs (Lieberman 1994a). A qualitative distinction concerning vocal communication thus clearly exists between the human and ape brains. We have more adaptable "computers" that can perform tasks that apes cannot perform at all.

Biological brains clearly don't perform algorithmic computations similar to those performed by digital computers. However, the analogy to computers is useful since it is evident that in the past twenty years quantitative differences in processing speed and memory have produced computers that can perform tasks that are qualitatively different in nature than those performed by the first slow, memory-limited computers. Similar quantitative changes may perhaps account for the evolution of the human brain. If we compare the volumes of some of the neuro-anatomical structures that make up the human FLS with homologous structures in the chimpanzee, it is apparent that major quantitative increases have occurred. The comparative anatomical data of Stephan et al. (1981) are revealing. The total volume of the chimpanzee brain that they measured was 382,103 mm^3; the human brain's volume was 1,281,8473—3.3 times larger. Compared to the chimpanzee brain, the human neocortex is 3.45 times larger, human caudate nucleus and putamin 3.13 times larger, cerebellum 3.15 times larger. In short, an increase of computational power affecting motor control and cognition, deriving from increases in scale may account for the qualitatively different functional power of the human brain. However, since it is, at present, not possible to compare the neural circuitry of human beings with living apes the question remains open; specialized, derived neural circuits may underly human speech ability.

Inferences from the fossil record

Some aspects of the evolution of hominid behavior can be traced in the fossil record through the "derived" anatomical features that differentiate hominids in our presumed line of descent from other extinct hominids and living apes (see Weaver et al., this volume). One frequently discussed derived feature that clearly differentiates the earliest hominids from other primates is the anatomy that yields upright, bipedal locomotion. The human supralaryngeal airway, which generates the formant frequency patterns that yield the high data transmission rate of human speech, is another derived human characteristic. The formant frequency patterns that specify many of the sounds of human speech are determined by the shape of the entire human supralaryngeal airway. The posterior portion of human tongue in a midsaggital view is round and the larynx is positioned low. The resulting human supralaryngeal airway has an almost right angle bend at its midpoint. As we talk, extrinsic tongue muscles can move the tongue upwards, downwards, forwards or backwards yielding abrupt and extreme changes in the cross-sectional area of the human supralaryngeal airway at its midpoint (Fant 1960; Stevens 1972; Nearey 1979). The vowel sounds that occur most often in the languages of the world (Greenberg 1963), [i], [u], and [a]—the vowels of the words *see*, *do*, and *ma*, can only be formed by the extreme midpoint area function discontinuities of the human supralaryngeal vocal tract (Stevens 1972; Beckman et al. 1995; Carre et al. 1994). These vowels have superior qualities for speech perception and production. Their formant frequencies converge, yielding spectral peaks analogous to saturated colors and are inherently resistant to small errors in speech articulation (Stevens 1972; Beckman et al.,1995). The vowel [i] is the "supervowel" of human speech and is less often confused with other vowels (Peterson and Barney 1952); it yields an optimal reference signal from which a human listener can determine the length of the supralaryngeal vocal tract of a speaker's voice (Nearey 1979).

In all other mammals the body of the tongue is long and relatively flat and fills the oral cavity. During swallowing, the nonhuman tongue propels food through the oral cavity. The nonhuman larynx is positioned high and can lock into the nasopharynx, forming an air pathway sealed off from the oral cavity, thereby enabling simultaneous breathing and drinking fluids or swallowing small food particles. As Darwin (1859:191) first noted, the morphology of the human supralaryngeal airway is maladapted for swallowing since solid objects that are forcefully propelled downwards by the pharyngeal constrictor muscles can lodge in the larynx, causing asphyxiation. Other maladaptive features of the human supralaryngeal vocal tract include a reduction in chewing efficiency and crowding of teeth which can result in impaction, infection and death (Lieberman 1975, 1984, 1991, 1998).

One aspect of the debate concerning human vocal anatomy is whether the lower position of the human larynx relative to the mandible and basicranium evolved in order to allow swallowing in upright bipedal hominids Falk (1975), rather than the enhancement of speech. Falk argued that the low position of the human hyoid bone was an adaptation for swallowing which, in turn, was responsible for the unique human supralaryngeal airway. The hyoid bone supports the larynx so a low hyoid bone position would yield the low larynx and the tongue of the human supralaryngeal airway. The human hyoid bone, according to this argument has to move forwards and upwards when swallowing, whereas it only moves forwards in apes. Thus Neanderthals and other upright, bipedal archaic hominids would have also have had low hyoid bones. However, Palmer et al. (1992a) show that the human hyoid bone does not have to move upwards to swallow; it can follow the same pattern of maneuvers observed in other primates. Therefore, Neanderthal and other extinct hominids could have swallowed with apelike supralaryngeal airways. Therefore, it is apparent that adaptations for swallowing did not play a part in the evolution of the human supralaryngeal airway; the human supralaryngeal airway appears to have been modified to enhance vocal communication at the expense of respiratory efficiency and protection from choking (Lieberman and Crelin 1971; Lieberman 1984). We are the only living species with a low larynx that can be obstructed with food, resulting in asphyxiation.

The human vocal tract first appears in anatomically modern human fossils about 100,000 year ago in fossils such as Skhul V and Jabel Qafzeh (Lieberman 1974, 1984). These early modern hominids had cranial structures that could not have supported a nonhuman supralaryngeal airway. The distance between the spinal column and hard palate is too short to accommodate a larynx positioned close to the base of the skull and positioned before the pharynx, as is the case for human newborns and nonhuman primates. Their skulls and mandibles most likely supported human supralaryngeal airways. These fossils are also the first hominids who can be shown to have used symbolic "tools"— grave goods. I think the connection is significant; it would have been impossible to conceive of the need for grave goods without advanced language. Moreover, the modern vocal tracts of these hominids are an index for the presence of the neural mechanisms that are necessary to produce the voluntary, complex muscular movements that underlie the production of speech. Without these neural mechanisms their human vocal tracts would have reduced fitness. If their brains were similar to our brains, the neural mechanisms that would have allowed them to produce the complex sequential maneuvers of human speech also would have yielded the ability to make use of complex syntax. But what can we say about archaic, extinct hominids such as the Neanderthals (D. Lieberman 1995), who were replaced by AMHS?

The Neanderthal enigma

A reappraisal of the probable speech-producing anatomy of classic Neanderthals shows that some of the skeletal features used in past studies to reconstruct their vocal tracts do not seem to be diagnostic (McCarthy and D. Lieberman n.d.). The Lieberman and Crelin (1971) reconstruction of the classic La Chapelle-aux-Saints Neanderthal must be modified; its shape may be closer to that of an adult human than that of a newborn human. However, the La Chapelle-aux-Saints Neanderthal still would not have been able to produce the full range of sounds of human speech. This phonetic deficit follows from the proportions of the pharynx and mouth that are necessary to produce sounds like the vowels [i] and [u]. Human infants begin life with a nonhuman tongue, laryngeal position and supralaryngeal vocal tract (Negus 1949; Lieberman and Crelin 1971). The human supralaryngeal vocal tract reaches an adult-like configuration in which the oral and pharyngeal cavities have equal lengths, permitting quantal speech production, at age six-years. It subsequently maintains those proportions by means of coordinated oral cavity and pharyngeal growth (McCarthy and D. Lieberman n.d.). Therefore, humans after age six-years can produce acoustically salient "quantal" sounds such as [i] and [u] because the lengths of their mouth and pharynx are almost equal (Stevens 1972; Lieberman et al. 1972; Carre, et al. 1994; Beckman et al. 1995).

The basicranial growth pattern that accompanies the descent of the larynx and subsequent maintenance of oral and pharyngeal cavity proportions appears to be absent in certain Neanderthals (D. Lieberman 1998). These Neanderthals retain the primitive hominid face which like those of apes, projects in front of the brain, but this does not indicate with certainty that their larynges occupied the same position as a newborn human's (McCarthy and D. Lieberman n.d.). However, the appraisal of the La Chapelle-aux-Saints fossil's phonetic capabilities does not hinge on the position of the larynx. The longer mouth of the La Chapelle-aux-Saints Neanderthal fossil makes it impossible to achieve the oral-cavity to pharynx proportions that would produce the full range of human speech sounds, even if we adopt the most conservative position and place the Neanderthal hyoid bone and larynx at the position that they would reach in adult humans (Lieberman 1998). The La Chapelle-aux-Saints oral cavity would have had a longer length than a modern human owing to a longer distance between the hard palate and spinal column (Howells 1989; D. Lieberman 1998).) A pharynx that had the equal length necessary to produce quantal sounds (Stevens 1972; Beckman et al. 1995) would place the La Chapelle-aux-Saints Neanderthal larynx in its chest (Lieberman 1984), an impossible position. In this regard, it should be noted that not all of the fossils often classified as Neanderthals would have had this limitation (Lieberman 1984).

The debate concerning the classic Neanderthal supralaryngeal airway has taken several turns. As noted above, Falk (1975) argued that upright, bipedal Neanderthals must have had a low hyoid bone and a modern supralaryngeal airway in order to swallow, but Palmer et al. (1992) show that the hyoid bone maneuvers necessary to swallow are similar in humans and other primates who are not bipedal. Houghton's (1993) claim that the La Chapelle-aux-Saints Neanderthal had a human supralarygeal vocal tract is invalidated by the small tongue of his reconstruction. Houghton's hypothetical Neanderthal tongue would been incapable of propelling food down the length of the oral cavity; the reconstruction is incorrect (Lieberman 1994c). Arensburg et al. (1990) claimed that the Kebara Neanderthal fossil had a human supralaryngeal vocal tract. Their evidence was the fossil's hyoid bone, which bears some resemblance to a human hyoid (though the quantitative measurements noted in Arensburg et al. (1989), but omitted in Arensburg et al. (1990) place it outside the range of modern human hyoid bones). The supralaryngeal airway reconstruction of Arensburg et al. (1990) hinges on the claim that the human hyoid bone does not change its position during the course of human ontogenetic development, so a humanlike hyoid would always signify a human supralaryngeal vocal tract. However, all published data show that the human hyoid descends during the course of ontogenetic development without any systematic change in its shape (Lieberman 1994b). As noted in many independent studies (e. g. Negus 1949; Bosma 1975; Senecail 1979) the hyoid bone is positioned close to the base of the skull in newborn infants and gradually descends relative to the vertebral column and mandible in the course of human development. Therefore, any inferences derived from the Kebara hyoid bone concerning its position and the shape of the fossil's supralaryngeal vocal tract are an act of faith. Schepartz (1993), one of the coauthors of the Arensburg study, claimed that all Neanderthals had modern human supralaryngeal airways on the basis of the Arensburg et al. (1990) and Houghton (1993) papers, but she presents no additional evidence to support this claim. Studies such as Duchin (1990) which claim that Neanderthals and other extinct hominids had full speech capabilities, reflect a basic misunderstanding of the physiology of speech production. Duchin's argument is based on Bell's 1867 theory, that the sounds of speech are determined solely by the position of the tongue in the mouth. Bell (1867), of course, lacked radiographic data and this theory is wrong (c.f. Fant 1960, Nearey 1979; Stevens 1972; Lieberman et al.1992a).

However, we can not conclude that Neanderthals lacked a FLS. Indeed, as Edmund Crelin and I noted in 1971, we can be certain that they must have possessed some form of speech and language. Vocal communication must have been present in *Homo erectus*, ancestral to humans and Neanderthals, else there would have been no selective advantage for the evolution of the human vocal tract, which apart from its contribution to speech intelligibility, reduces biological fitness. Given the participation of the known neuroanatomical components of the FLS in higher cognition, we must conclude that Neanderthals possessed advanced cognitive ability. But the question is still open, was Neanderthal cognitive-linguistic ability as developed as that of contemporary AMHS? The fossil record suggests that it was not, since Neanderthals retained a non-optimal vocal tract. However, it is probable that the cognitive-linguistic distinction between Neanderthals and modern humans was not great. We can account for Neanderthal extinction without invoking any radical cognitive-linguistic distinction if we take account of their speech capabilities.

Recent studies show that distinctions in dialect act as genetic isolating mechanisms in modern human populations (Barbujani and Sokal 1990). Therefore, even slight speech distinctions between contemporary Neandertal and early AMHS populations would have served to genetically isolate them, even if Neandertals and early AMHS populations acted in much the same way manner as modern humans. The Neanderthals would been physically incapable of producing the "universal" vowels of human speech ([a], [i], and [u]. Given the isolating effects of dialect, a small difference in linguistic and/or cognitive ability operating over many generations, could have resulted in Neanderthal extinction (Zubrow 1989). Any cultural or cognitive distinctions that may have differentiated contemporary AMHS and Neanderthal populations would have acted to intensify genetic isolation. However, as is still the case when we attempt to assess the intelligence and linguistic ability of a living human, we must observe behavior. Only archaeological evidence, our window on hominid behavior in the distant past, can resolve this question.

References Cited

Alexander, M. P., M. A. Naeser and C. L. Palumbo
    1987   Correlations of subcortical CT lesion sites and aphasia profiles. *Brain* 110:961-991.

Arensburg, B., A. M. Tiller, B. Vandermeersch, H. Duday, L. A. Schepartz and Y. Rak
    1989   A Middle Paleolithic human hyoid bone. *Nature* 338:758-760.

Arensburg, B., L. A. Schepartz, A. M. Tiller, B. Vandermeersch, H. Duday, and Y. Rak
    1990   A reappraisal of the anatomical basis for speech in Middle Paleolithic hominids. *American Journal of Physical Anthropology* 83:137-146.

Awh, E., J. Jonides, R. E. Smith, E. H. Schumacher, R. A. Koeppe and S. Katz
    1996   Dissociation of storage and rehearsal in working memory: Evidence from Positron Emission Tomography. *Psychological Science* 7:25-31.

Baddeley, A. D.
    1975   *Working Memory*, Clarendon Press, Oxford.

Barbujani, G. and R. R. Sokal
    1990   Zones of sharp genetic change in Europe are also linguistic boundaries. *Proceedings of National Academy of Sciences, USA* 187:1816-1819.

Baum, S. R., S. E. Blumstein, M. A. Naeser, and C. L. Palumbo
    1990   Temporal dimensions of consonant and vowel production: An acoustic and CT scan analysis of aphasic speech. *Brain and Language* 37:327-338.

Beckman, M. E., T.-P. Jung, S.-H. Lee, K. de Jong, A. K. Krishnamurthy, S. C. Ahalt, K. B. Cohen and M. J. Collins
    1995   Variability in the production of quantal vowels revisited. *Journal of the Acoustical Society of America* 97:471-489.

Bell, A.
    1867   *Visible Speech*. Simpkin and Marshall, London.

Blumstein, S. E., W. Cooper, H. Goodglass, H. Statlender and J. Gottleib
    1980   Production deficits in aphasia: a voice-onset time analysis. *Brain and Language* 9:153-170.

Bosma, J. F.
    1975   Anatomic and physiologic development of the speech apparatus. In *Human Communication and its Disorders*, edited by D. B. Towers, pp. 469-481. Raven, New York.

Broca, P.
    1861   Remarques sur le siège de la faculté de la parole articulée, suivies d'une observation d'aphemie (perte de parole). *Bulletin de la Societé d'Anatomie* (Paris) 36:330-357.

Carre, R., B. Lindblom and P. MacNeilage
    1994   Acoustic factors in the evolution of the human vocal tract. *Journal of the Acoustical Society of America* 95:2924

Chomsky, N.
    1986   *Knowledge of Language: Its Nature, Origin and Use*. Prager, New York.

Croft, W.
    1991   *Syntactic Categories and Grammatical Relations*. University of Chicago Press, Chicago.

Cummings, J. L.
    1993   Frontal-Subcortical circuits and human behavior. *Archives of Neurology* 50:873-880.

Cunnington, R., R. Iansek, J. L. Bradshaw and J. G. Philips
    1995   Movement-related potentials in Parkinson's disease: Presence and predictability of temporal and spatial cues. *Brain* 118:935-950.

Damasio, H., T. J. Grabowski, D. Tranel, R. D. Hichwa and A. R. Damasio
    1996   A neural basis for lexical retrival. *Nature* 380:409-505.

Darwin, C.
    1859   *On the Origin of Species* (facsimile ed. 1964). Harvard University Press, Cambridge, MA.

Duchin, L. E.
    1990   The evolution of articulate speech: comparative anatomy of the oral cavity in *Pan* and *Homo*. *Journal of Human Evolution* 19:687-697.

Elman J., E. Bates, M. Johnson, A. Karmiloff-Smith, D. Parisi and K. Plunkett
   1997 *Rethinking Innateness: A Connectionist Perspective on Development*. MIT Press/Bradford Books, Cambridge MA.

Engen, E. and T. Engen
   1983 *Rhode Island Test of Language Structure*. University Park Press, Baltimore.

Falk, D.
   1975 Comparative anatomy of the larynx in man and the chimpanzee: implications for language in Neanderthal. *American Journal of Physical Anthropology* 43:123-132.

Fant, G.
   1960 *Acoustic Theory of Speech Production*. Mouton, The Hague.

Fodor, J.
   1983 *Modularity of Mind*. MIT Press, Cambridge, MA.

Gall, F. J.
   1809 *Recherches sur le Système Nerveux*. B. Baillière, Paris.

Gardner, R. A. and B. T. Gardner
   1984 A vocabulary test for chimpanzees (*Pan troglodytes*). *Journal of Comparative Psychology* 4:381-404.

Geschwind, N.
   1970 The organization of language and the brain. *Science* 170:940-944.

Goodall, J.
   1986 *The Chimpanzees of Gombe: Patterns of Behavior*. Harvard University Press, Cambridge, MA.

Graziano, M. S. A., G. S. Yap and C. G. Gross
   1994 Coding of visual space by premotor neurons. *Science* 266:1054-1057.

Greenberg, J.
   1963 *Universals of Language*. MIT Press, Cambridge MA.

Grossman, M., S. Carvell, S. Gollomp, M. B. Stern, G. Vernon and H. I. Hurtig
   1991 Sentence comprehension and praxis deficits in Parkinson's disease. *Neurology* 41:160-1628.

Grossman, M., S. Carvell, M. B. Stern, S. Gollomp and H. I. Hurtig
   1992 Sentence comprehension in Parkinson's Disease: The role of attention and memory. *Brain and Language* 42:347-384.

Hirsch, I. and C. E. Sherrick
   1961 Perceived order in different sense modalities. *Journal of Experimental Psychology* 62:423-432.

Holloway, R. L.
   1995 Evidence for POT expansion in early *Homo*: a pretty theory with ugly (or no) paleoneurological facts. *Behavioral and Brain Sciences* 18:191-193.

Houghton, P.
   1993 Neanderthal supralaryngeal vocal tract. *American Journal of Physical Anthropology* 90:139-146.

Howells, W. W.
   1989 *Skull Shapes and the Map: Craniometric Analyses in the Dispersion of Modern Homo*. Papers of the Peabody Museum of Archaeology and Ethnology 79. Harvard University, Cambridge MA.

Jellinger, K.
   1990 New developments in the pathology of Parkinson's disease. In *Advances in Neurology. Vol. 53: Parkinson's Disease: Anatomy, Pathology and Theraphy*, edited by M. B. Streifler, A. D. Korezyn, J. Melamed and M. B. H. Youdim, pp. 1-15. Raven Press, New York.

Just, M. A. and P. A. Carpenter
   1992 A capacity theory of comprehension: Individual differences in working memory. *Psychological Review* 99:122-149.

Kuhl, P. K.
   1978 Speech perception by the chinchilla: identification functions for synthetic VOT stimuli. *Journal of the Acoustical Society of America* 63:905-916.

Laplane D., M. Baulac, D. Widlocher
   1984 Pure psychic akinesia with bilateral lesions of basal ganglia. *Journal of Neurology, Neurosurgery and Psychiatry* 47:377-385.

Leiner, H. C, A. L. Leiner and R. S. Dow
   1993 Cognitive and language functions of the human cerebellum. *Trends in Neuroscience* 16:444-447, 453-454.

Liberman, A. M., F. S. Cooper, D. P. Shankweiler and M. Studdert-Kennedy
   1967 Perception of the speech code. *Psychological Review* 74:431-461

Liberman, A. M. and I. G. Mattingly
   1985 The motor theory of speech perception revised. *Cognition* 21:1-36.

Lichtheim, L.
   1885 On aphasia. *Brain* 7:433-484.

Lieberman, D. E.
   1995 Testing hypotheses about recent human evolution from skulls. *Current Anthropology* 36:159-198.

- 1998 Sphenoid shortening and the evolution of modern human cranial shape. *Nature* 393:158-162.

Lieberman, P.
- 1968 Primate vocalizations and human linguistic ability. *Journal of the Acoustical Society of America* 44:1157-1164.
- 1975 *On the Origins of Language: An Introduction to the Evolution of Speech.* Macmillan, New York.
- 1984 *The Biology and Evolution of Language.* Harvard University Press, Cambridge, MA.
- 1991 *Uniquely Human: The Evolution of Speech, Thought, and Selfless Behavior.* Harvard University Press, Cambridge, MA.
- 1994a Biologically bound behavior, free-will, and human evolution. In *Conflict and Cooperation in Nature*, edited by J. I. Casti, pp. 133-163. John Wiley and Sons, New York.
- 1994b Hyoid bone position and speech: Reply to Arensburg et al. (1990). *American Journal of Physical Anthropology* 94:275-278.
- 1994c Functional tongues and Neanderthal vocal tract reconstruction: A reply to Houghton (1993). *American Journal of Physical Anthropology* 95:443-452.
- 1998 *Eve Spoke: Human Language and Human Evolution.* W. W. Norton, New York.
- 2000 *Human Language and our Reptilian Brain: The Subcortical Bases of Speech, Syntax, and Thought.* Harvard University Press, Cambridge, MA.

Lieberman, P. and E. S. Crelin
- 1971 On the speech of Neanderthal man. *Linguistic Inquiry* 2:203-222.

Lieberman, P., E. S. Crelin and D. H. Klatt
- 1972 Phonetic ability and related anatomy of the newborn, adult human, Neanderthal man, and the chimpanzee. *American Anthropologist* 74:287-307.

Lieberman, P., J. Friedman and L. S. Feldman
- 1990 Syntactic deficits in Parkinson's disease. *Journal of Nervous and Mental Disease* 178:360-365.

Lieberman, P., J. T. Laitman, J. S. Reidenberg and P. Gannon
- 1992a The anatomy, physiology, acoustics and perception of speech: Essential elements in analysis of the evolution of human speech. *Journal of Human Evolution* 23:447-467.

Lieberman, P., E. T. Kako, J. Friedman, G. Tajchman, L. S. Feldman and E. B. Jiminez
- 1992b Speech production, syntax comprehension, and cognitive deficits in Parkinson's disease. *Brain and Language* 43:169-189.

Lieberman, P. and Tseng, C.-Y.
- 1994 Subcortical pathways essential for speech, language and cognition: Implications for hominid evolution. *American Journal of Physical Anthropology*, Suppl. 16:93:130.

Lieberman, P. and C-Y Tseng
- n.d. Some speech production and syntax comprehension deficits of Parkinson's disease in Chinese-speaking subjects

Lieberman, P., B. G. Kanki, A. Protopapas, E. Reed and J. W. Youngs
- 1994 Cognitive defects at altitude. *Nature* 372:325.

Lieberman, P., B. G. Kanki, A. Protopapas
- 1995 Speech production and cognitive decrements on Mount Everest. *Aviation, Space and Environmental Medicine* 66:857-864.

Lindblom, B.
- 1996 Role of articulation in speech perception: Clues from production. *Journal of the Acoustical Society of America* 99:1683-1692.

Lisker, L. and A. S. Abramson
- 1964 A cross language study of voicing in initial stops: acoustical measurements. *Word* 20:384-442.

MacDonald, M. C., N. J. Perlmutter and M. S. Seidenberg
- 1994 Lexical nature of syntactic ambiguity resolution. *Psychological Review* 101:676-703.

Marsden, C. D. and J. A. Obeso
- 1994 The functions of the basal ganglia and the paradox of sterotaxic surgery in Parkinson's disease. *Brain* 117:877-897.

Martin, A., J. V. Haxby, F. M. Lalonde, C. L. Wiggs and L. G. Ungerleider
- 1995 Discrete cortical regions associated with knowledge of color and knowledge of action. *Science* 270:102-105.

Mayr, E.
- 1982 *The Growth of Biological Thought.* Harvard University Press, Cambridge, MA.

McCarthy, R. C. and D. E. Lieberman
- n.d. Reconstructing vocal tracts from cranial base flexion: an ontogenetic comparison

of cranial base angulation in humans and chimpanzees. *Journal of Human Evolution* in press.

McGurk H. and J. MacDonald
    1976 Hearing lips and seeing voices. *Nature* 263:747-748.

Mega, M. S. and M. F. Alexander
    1994 Subcortical aphasia: The core profile of capsulostriatal infarction. *Neurology* 44:1824-1829.

Mesulam, M. M.
    1990 Large-scale neurocognitive networks and distributed processing for attention, language, and memory. *Annals of Neurology* 28:597-613.

Middleton. F. A. and P. L. Strick
    1994 Anatomical evidence for cerebellar and basal ganglia involvement in higher cognition. *Science* 266:458-461.

Naeser, M. A., M. P. Alexander, N. Helms-Estabrooks, H. L. Levine, S. A. Laughlin and N. Geschwind
    1982 Aphasia with predominantly subcortical lesion sites; description of three capsular/putaminal aphasia syndromes. *Archives of Neurology* 39:2-14.

Natsopoulos, D. G. Grouios, S. Bostantzopoulou, G. Mentenopoulos, Z. Katsarou and J. Logothetis
    1994 Algorithmic and heuristic strategies in comprehension of complement clauses by patients with Parkinson's Disease. *Neuropsychologia* 31:951-964.

Nearey, T.
    1978 *Phonetic Features for Vowels*. Indiana University Linguistics Club, Bloomington.

Negus, V. E.
    1949 *The Comparative Anatomy and Physiology of the Larynx*. Hafner, New York.

Palmer, J. B., N. J. Rudin, G. Lara and A. W. Crompton
    1992 Coordination of mastication and swallowing. *Dysphagia* 7:187-200.

Peterson, G. E. and H. L. Barney
    1952 Control methods used in a study of the vowels. *Journal of the Acoustical Society of America* 24:175-184.

Pickett, E. R., Kuniholm, A. Protopapas, J. Friedman and P. Lieberman
    1998 Selective speech motor, syntax and cognitive deficits associated with bilateral damage to the head of the caudate nucleus and the putamen. A single case study. *Neuropsychologia* 36:173-188.

Pinker, S.
    1994 *The Language Instinct: How the Mind Creates Language*. William Morrow, New York.

Savage-Rumbaugh, S., D. Rumbaugh and K. McDonald
    1985 Language learning in two species of apes. *Neuroscience and Biobehavioral Reviews* 9:653-665.

Schepartz, L. A.
    1993 Language and modern human origins. *Yearbook of Physical Anthropology*. 36:91-126.

Senecail, B.
    1979 *L'Os Hyoide; Introduction Anatmique à l'Étude de Certains Méchanismes de la Phonation*. Mémoires du Laboratoire d'Anatomie de la Faculté de Médecine de Paris 36.

Sereno, J., S. R. Baum, G. C. Marean and P. Lieberman
    1987 Acoustic analyses and perceptual data on anticipatory labial coarticulation in adults and children. *Journal of the Acoustical Society of America* 81:512-519.

Stephan, H, H. Frahm and G. Baron
    1981 New and revised data on volumes of brain structures in insectivores and primates. *Folia Primatologia* 35:1-29.

Stevens, K. N.
    1972 Quantal nature of speech. In *Human Communication: a Unified View*, edited E. E. David Jr. and P. B. Denes, pp. 51-66. McGraw Hill, New York.

Stromswold, K., D. Caplan, N. Alpert and S. Rausch
    1995 Localization of syntactic processing by positron emission tomography. *Brain and Language* 51:452-473

Strub, R. L.
    1989 Frontal lobe syndrome on a patient with bilateral globus pallidus lesions. *Archives of Neurology* 46:1024-1027.

Stuss, D. T. and D. F. Benson
    1986 *The Frontal Lobes*. Raven Press, New York.

Warden, C. J. and L. H. Warner
    1928 The sensory capacities and intelligence of dogs, with a report on the ability of the noted dog "Fellow" to respond to verbal stimuli. *Quarterly Review of Biology* 3:1-28.

Wernicke, C.
    1874 The aphasic symptom complex: a psychological study on a neurological basis.

Breslau: Kohn and Weigert. Reprinted in *Boston Studies in the Philosophy of Science*, Vol. 4, edited by R. S. Cohen and M. W. Wartofsky, pp. 34-97. Reidel, Boston.

Xuerob, J. H., B. E. Tomlinson, D. Irving, R. H. Perry, G. Blessed and E. K. Perry
    1990   Cortical and subcortical pathology in Parkinson's Disease: Relationship to Parkinsonian dementia. In *Advances in Neurology. Vol. 53: Parkinson's Disease: Anatomy, Pathology and Theraphy*, edited by M. B. Streifler, A. D. Korezyn, J. Melamed and M. B. H. Youdim, pp. 35-39. Raven Press, New York.

Zubrow, E.
    1989   The demographic modeling of Neanderthal extinction. In *The Human Revolution: Behavioral and Biological Perspectives on the Origin of Modern Humans*, vol. 1, edited by P. Mellars and C. B. Stringer, pp. 212-31. Edinburgh University Press, Edinburgh.

13. Discovering the Symbolic Potential of Communicative Signs: The Origins of Speaking a Language

William Noble and Iain Davidson

Abstract

The problem of the origin of language may be framed as an issue in 'behavioral evolution'. That is to say, speaking a language can be considered as a form of communicative behavior, hence its origin can be understood as a change in such behavior. The change in question is one in which the ancestors of modern humans discovered they could use their communicative signs to refer to things, even in the absence of those things; that is, they could use those signs as symbols. One of the key elements required for such discovery is the *coordination* of behavior among two or more group members. We analyze the nature of inter-individual coordination, as preliminary to identifying features of hominid evolution which might support the coordination that led to discovery of the symbolic potential of communicative signs. Many of those evolutionary features, by themselves, have nothing to do with communication: taken together, they provide conditions which might affect it in the way we prescribe. Throwing missiles (stones or spears) with deliberate aim paves the way for pointing with the forelimb. Decrease in body hair combined with increase in post-natal brain growth promotes active carrying of infants, hence increased opportunity for joint attention—including inter-individual coordination of pointing and other gestures. Making a trace of such a gesture in a plastic medium allows it to be witnessed as an object, hence to be useable as a symbol.

Introduction

Contributors to this volume, engaged in interdisciplinary research, are asked to indicate the relationship between whichever neighboring disciplines they exploit. In our case, the relationship is between psychology and archaeology, and a critical form of that relationship is as follows: To understand the place of artifacts (the province of archaeology) in grasping the nature of human intelligence (psychology's interest), one must take account of the fact that artifacts with 'imposed form' (Mellars 1989:365) 'spring from interactive behavior based on the use of communicative signs symbolically' (Noble and Davidson 1996:227). Artifacts with imposed form differ from artifacts without it by showing signs of planning by their makers to achieve the forms they have. Many other artifacts have forms which are purely contingent upon the interaction between physical forces applied by the producer or user, and the mechanical properties of the raw material (Davidson and Noble 1993). Planning to achieve particular forms indicates the existence of conventions: following a convention and articulating a plan, are behaviors that among other things rely on symbolic communicative interaction—speaking a language. It follows that artifacts without imposed form do not imply speaking a language as a behavior while those with imposed form do.

We rely on the somewhat labored phrase, 'speaking a language' so as to emphasize the point that with the concept 'speaking' we are at all times, and only, referring to the activity of uttering strings of words and phrases identifiable (by speaker and hearer) to a particular language, and that 'language' always entails that sort of activity. As argued in more detail elsewhere (Noble and Davidson 1996:2-3), it is preferable to understand speaking as a behavior, and to actively avoid the assumption that 'language' is some abstract thing, of which speaking is its perceived expression (see also Dewart 1989).

In considering 'speaking a language', we will emphasize that this is a form of human social interaction, a form of coordinated attention. How might this form of activity have arisen in evolution? We begin by discussing interactive behavior generally, with the aim of showing the conditions necessary for the evolutionary emergence of symbol-based communication. In focusing on interactive behavior we argue that 'human intelligence' (mindedness) is another way of describing behavior that involves language use. In the present treatment, we write as if speaking uses the voice, but our points can as validly cover 'speaking' using manual signs (Groce 1985).

Some activity can be seen and appreciated as *coordinated* between two or more organisms. How does this activity get identified in this way? Behavior can be said to be coordinated if the activity of one organism affects that of another. The particular form of coordination we want to distinguish, after considering various different kinds, is one that humans routinely express among themselves when engaged in conversation or other speaking bouts. There are ways to analyze this sort of activity in terms of technical issues to do with articulation and sound production (Ladefoged 1975), or to do with how the coordination is achieved as a form of practical action: turn-taking signals, for example, or adjacency pairing rules, as in questions and answers (Schenkein 1978). These are not the focus of analysis here. It is the social character of coordination in speaking that we are concerned with.

As a starting position: it is straightforward to identify speaking as *referential* communication. To speak is to exercise a capacity to identify objects or events by the use of signs (verbal utterances) which users of those signs understand as referring to those objects or events. Such a capacity differs from the making of signs simply in response to events. When a user understands that a sign refers to an object or event, among other things this enables use of the sign in the absence as well as the presence of the object or event to which the sign refers, what Hockett and Altmann (1968) describe as 'displacement'. The word used for a sign that stands for something other than itself is *symbol*. Symbols occur as a larger class than the referential signs we have been talking about so far. The word 'gazelle' is a symbol that may be used in reference to a number of entities, including a particular sort of living (or dead) animal. The word may be said to represent the entity; so can a picture, an iconic mage, of the same thing. The picture, therefore, is also a member of the class 'symbols'.[1]

The argument we have developed here and elsewhere is that symbol use is unique to humans, or to contexts in which non-human primates acquire the use of symbols through inclusion in human interactive behavior. Part of the discussion in this chapter will address that argument.

Our final aim is to suggest ingredients which were needed in the course of the evolution of human beings to allow referential, symbol-based communication to arise. In this argument, referential communication is understood as a form of coordination between two or more organisms. It arose in evolutionary terms through a particular combination of behaviors and contexts, none of which in themselves (or in another combination) amounted to referential communication. Occurring piecemeal, and in some part for reasons that may have had nothing to do with communication, these behavioral and contextual elements accrued over generations. The last aspect of the argument is that eventual selection of a capacity for referential communication was on the basis of a biobehavioral package that conferred advantage in terms of information control.

We spell out these behavioral and contextual ingredients, and say something about that advantage, when describing a speculative prehistoric scenario in which they may have arisen. We recognize that speculative scenarios are not to be encouraged in this domain, because they are always at risk of being or becoming 'just so' stories (Kipling 1902/1987)—accounts unconstrained by evidence. Our speculation attempts not to violate the limits of empirical evidence about human and other primate behavior, nor that of the archaeological record of hominid evolution. We have tried to keep ourselves on the straight and narrow, as it were. In the end, the strong claim for the story we have developed is this: future speculation will not be able to avoid the conceptual issues that our theorizing identifies, even if every element in our scenario turns out to be wrong.

Coordinated Activity and Communication

As planning entails speaking, and as speaking, we argue, is a uniquely human form of symbol use, it follows that behaviors claimed as showing signs of planning, in the absence of signs of symbol use, have to be questioned. Boesch and Boesch (1989) have advanced a claim that chimpanzees hunt down monkeys in a manner that implies planning of their coordinated actions. By contrast, Goodall (1986) concluded that chimpanzee hunting is opportunistic, occurring only after the sighting of prey. There is no sure evidence of chimpanzees relying on, for example, tracks or other environmental signs, to seek out prey not immediately in view. Thus, the chimpanzees all seem to respond to the sight or sound of the monkeys without any evidence for planning ahead on the basis of cues previously left by the monkeys. The chimpanzees' actions can be called *coordinated* because moves by one affect moves by the others, and it is to the analysis of this sense of coordination that we now turn. The issue is the indisputable place of symbol use in the coordination of many human affairs, even in contexts where it is not immediately obvious that symbol use is involved, as in the following example.

Consider an experiment in which two people are seated in sight of each other, and each is oscillating the lower part of one of their legs, in time to a metronome beat, whilst also trying to keep their oscillations in parallel with or in opposition to the other person's (Schmidt et al. 1990). This is not necessarily as odd an experiment as it might sound. Turvey (1990) argues that certain structural features of individual's and pairs of individual's motor coordinations are identifiable by means of such activity. Our point is that, although the activity in this case may itself be non-communicative, referential and other forms of communication are needed throughout such an experiment in order that it happen at all. First, reference is needed to a purpose to be pursued in carrying out the exercise, and to the body parts to be used in its procedure. This is required so as to establish what is to happen. It need not be an elaborate account explaining, say, the abstract principles that lie behind the stated purpose. Indeed, the explanation will be limited to the practical action of swinging the lower limbs, so as to carry out the task. Something referential is entailed nonetheless to get the thing going. A mix of verbal and gestural (indexical) communication will be included so as to show participants what is involved (Schmidt et al.1990:230). In carrying on with the task, participants in this kind of experiment must maintain mutual *non*-verbal communication. Each must keep an eye on, and be seen by the other to keep an eye on, what the other is doing. This is vital for the action to be continued. Finally, an instruction must be communicated by the experimenter so that the activity is brought to a halt.

Communication is unavoidable as a feature of actions that are coordinated between or among individuals, even actions such as in the above experiment, whose organizers would probably consider it to be purely an example of eye, ear and limb coordination across pairs of participants, and not something involving much in the way of communication at all. It is precisely the communicative element, though, which allows such action to be witnessed *as* coordinated, rather than as coincidental.

The contrast between coordination and coincidence can be seen in the following example. Imagine circumstances in which two animals are in each other's vicinity, yet unaware of that fact. Two cats, for instance, may be engaged in individual actions, and incidentally heading closer to each other. At some point cat A becomes aware of the presence of cat B, and cat A's behavior immediately changes. The change may be radical or uneventful, antagonistic or cooperative. If only one cat has noticed the other, its action may be limited to attending to the other's movement. If cat B also notices cat A, the actions of both may be simply to attend to each other, or may be more than that. Whatever the outcome *following* each or only one of them noticing the other, prior to that moment of awareness it would not be coherent to describe any actions of the two animals as inter-individually coordinated. Their co-presence could not be described as other than coincidental—they would be said simply to have been in each other's ambit.

Is the case of cat A monitoring cat B's movements, when cat B is unaware of this, an instance of coordinated activity? We would say 'yes': the coordination in question comprises the monitoring movement of the eyes and head of cat A, linked with the movement of cat B. Is this also an example of communication between these animals? Ethologists are divided on this matter. Some would argue that no communication is going on in such a case, because for coordinated action to be recognized as communicative it has to be mutual (Krebs and Dawkins 1984). Others would argue that to be communicative it has to be mutually beneficial (Marler 1967). However, you could say cat B is (unwittingly) communicating its presence to the other.

The notion which attends this debate is that of intentionality—do animals make signs to each other on purpose? If no intent can be established, should the behavior be considered communicative? The position we support begs no questions about a role for intentionality in behavior that could nonetheless be understood as communicative. We rely on Goffman's (1963) distinction between information 'given', and information 'given *off*'. In the course of a public performance, information may be given by a human performer to an audience. At the same time, the performer may be unable to avoid 'giving off' information about, for example, his/her gender, or his/her social and ethnic background. The information given can be said to be intentionally produced and hence under the performer's control. The information 'given off' is not an intentional product, and may be considered to be broadcast in ways the performer does not control—what Ekman and Friesen (1969a) call 'nonverbal leakage'. (Part of successful play-acting is to fashion a controlled performance that reveals features of the character being played but which looks or sounds as though it is information 'given off'.) In this way of considering things, cat B in the earlier example is 'giving off' information about its presence and heading to cat A, even when it has not noticed the other animal.

One way to decide on what is to count as communication is provided by Mead (1934). He preferred to consider communication as a form of *behavior*, and argued that we can only recognize behavior *as* communicative because of its effects. A signal does not communicate if it does not affect a partner in the communication. Mead did not seek to judge about intentionality nor to assess whether benefit is mutual. He does leave matters unclear about certain forms of life. Behavior is usually only said to be a feature of *animate* life. What about the communicative powers of plants? Mead would probably say that as plants do not typically join in any game of responding to communicative signs, they are outside his scheme of things. On this issue we can borrow a point from Gibson (1966), who speaks about objects *broadcasting* information about their presence. Plants exude odors or reflect light in ways that attract animals, but, for the most part, plants are not thought to 'respond' to signals in their vicinity.

Bees attracted by the light or odor of plants do communicate, certainly with each other but do they communicate with the rest of the animate world? In Mead's terms, the buzz of a bee within earshot of humans communicates because it brings about a range of effects. The chirps of frogs may have the effect of attracting mates, but at the same time may attract predators (Slater 1983). This second outcome, under Mead's scheme, is communicative, though it may be neither intended by nor beneficial for the frog. It can be thought of as another example of information being 'given off' or 'leaking out'. To another frog, the chirp of the first frog is information 'given', and indicates mating status or rivalry. The indication to the predator of the frog's presence is information 'given off' or 'leaked out'. While the content of the utterance does not change the nature of the communication does, depending on the type of audience.

Some birds make distraction displays. For example, the ground plover communicates a 'broken wing' display at the time when it is nesting (Ristau 1991). It walks away from the nest, dragging one wing, whenever a potential predator comes too close. The effect is to draw the intruder's attention away from the nest and on to itself—to what appears to be easy pickings. Such communication may or may not be intentional; and it does not confer benefit to the potential predator, distracted by the display and led away from the nest. The actions occurring in all these examples are, without doubt, coordinated: the behavior of one organism changes in relation to that of the other.

Is the ground plover acting intentionally, with known purpose? When human observers give accounts of this behavior, they are acting with known purpose—*they* know they are observing something and that they are giving an account of it. There is no sign, though, that the bird 'knows what it is doing' in this sense. To use Ryle's (1949) distinction, the plover knows *how* to produce a broken wing display: does it know *that* this is what it is doing? As before, we prefer to leave the question of intentionality in this sort of case unresolved. Consistent with Mead's approach, the question of intent has no particular significance in the context of behavior like the 'broken-wing' display, which, while obviously communicative, is ultimately enigmatic when considered from the perspective of the plover's state of knowledge.

Some people (e.g., Dennett 1987) may deeply desire to see if the plover's behavior can be described using intentional language. The wellspring of that desire is the everyday knowledge people have about intentions they themselves and others routinely express. We are clear about the intentional nature of human behavior, because of what human speech achieves. By its very nature, speaking is intentional, it displays the characteristic of '*about*ness'. When anyone speaks they speak *about* something to someone (if only themselves). As Wittgenstein (1958) noted, we are in the dark about the intentionality and, indeed, about the nature of the consciousness, of creatures that do not speak.

As an index of this contrast between speakers of a language and others, we note the matter of emotional expression. In Darwin's (1872) extensive treatment of this topic he takes care not to uncritically extend to other life-forms his description of certain seemingly universal features of human emotional expression. Even so, he finds it straightforward that all sentient creatures experience emotions of the sort that can be described in human terms. 'Even insects express anger, terror, jealousy, and love by their stridulation' (1872:349). A problem with such across-the-board application of categories of emotion is that it ignores human awareness of and constructions of meanings for emotion states. Such awareness and socially-derived meanings lead to forms of social and personal 'management' of emotional responses—what Ekman and Friesen (1969b) termed 'display rules'. Confirming Darwin's analyses, there is evidence in support of universality in human emotional expression (e.g. Ekman 1977). But the existence of display rules—a mix of implicit and explicit conventions—shows the distinctiveness of the contrast between speakers of a language versus others: speakers of a language can conceal or otherwise modify their emotions.

Gesture Calls

While we need to honor the contrast just outlined, it does remain universally the case—across human communities and across species—that communication underpins coordinated activity between organisms, whatever their intentions and however we construe them. Most communicative signs serve to attract or ward off other organisms. Some signs have a further property, that of calling attention to something else in the environment independent of sign-maker and perceiver of the sign. Take the case of vocal calls made by vervet monkeys in response to different types of predator (Struhsaker 1967). The calls are made in the presence of other monkeys and differ depending on the type of potential threat (leopard vs. eagle vs. snake, for example). The cries cause the others to reorient in the direction of the threat associated with the cry. This is a significant form of coordinated activity. The coordination is not just with respect to each other's conduct, but with respect to that of the whole group relative to an external event or object.

Research by Cheney and colleagues (e.g., 1990) has included questions about the nature of the monkeys' calls. Are these calls *names* for things? Or are they, perhaps, expressions of emotion, or propositions about what actions to take? Those investigators are not certain what sort of status to give them. Burling (1993) has commented on the issue and is less circumspect. The closest form of human utterance to the monkey calls that he sees is the class of signs he terms 'gesture calls'. Gesture calls are expressions of emotion, such as the cries we make on being startled by an unexpected object or event, or the 'wows' we produce in the face of something spectacular. These have no semantic content or referential force, but they are specific to classes of events, and can function to make others aware of something in the landscape besides the gesture-caller. One may gloss the overall effect of such expressions as: 'Something marvelous or threatening is going on over there'.

Referential Communication

If it is incorrect to identify the vervets' calls as referential, then a good question is: what would be needed to transform such calls into signs having referential function? What does 'referential function' entail? In referring to something, one must use its commonly understood name (Terrace 1985). The analytic and experimental work of Olson and others (e.g., Olson 1989), involving children in throes of mastering their own language, shows that to refer to things by name is to know that signs so used stand in symbolic relation to the objects they refer to. There comes a point in the widening experience of the child where the words and phrases that make up the content of conversations are *themselves* appreciated as the means whereby objects and people get referred to and organized. At that point we witness the flowering of reference to what is not in the immediate vicinity, because the language-user is starting to see that words themselves have a reality. Fantasies and lies abound at that point, as children experiment with this new-found fact of life.

Modern human children are guided by older family members and others in achieving reference. Among ethnographies of this achievement are those by Lock (1980) and Zukow (1990; see also Zukow-Goldring and Ferko 1994; Zukow-Goldring 1997), which represent studies of the communicative interactions that form the everyday agenda of children in a variety of different communities and contexts. The coordinations which make up the guidance in question are in the form of deliberate showings of how words, actions and objects relate to each other. Older family members recruit the young child's attention in quite explicit ways, for example, in play sequences in which they make salient for the child both the play-action and the words to describe it. Emphasizing how to refer to things by their names calls the child's attention to the *fact* of reference, in the act of reference. By doing this, the signs themselves are made to be objects for the child. Making the sign itself an object for perception is at the heart of what is going on, and this sort of achievement may be argued to hold the evolutionary key (Greenfield 1991).

One feature of behavior critical to the successful transmission of knowledge is the capacity for imitation. It is doubtful that other primates express such a capacity routinely, although chimpanzees have been shown to imitate in the context of social interaction with humans (Tomasello et al. 1994). In their natural ecology, the actions of adult chimpanzees, such as in the use of stones to crack nuts (Boesch 1993), can be witnessed by younger members. There are rare instances of actions by adult chimpanzees which are more oriented to those infants than to the food-getting (nut-cracking) task as such. If observed among humans, such actions would be seen at least as attempts to teach. Humans routinely shape the actions of their infants, including acknowledging and encouraging imitation (Meltzoff and Gopnik 1993). When humans interact with infant chimpanzees they, being speakers of a language, appreciate the aims of teaching and the separate parts of what is being taught. In consequence, as they interact with the chimpanzees, the human teachers reinforce any action by the chimpanzee that they construe as imitative, in the same way as they do with human infants (Butterworth 1991). This teaching is a direct consequence of the practice of reference, both to objects and their names, in normal human child-rearing.

The Evolutionary Emergence of Referential Communication

The argument, then, is for the occurrence, in some prehistoric context, of some means for turning communicative signs into objects, so as to enable their use as symbols. If we look at the case of the vervet monkeys we see that the signs they utter give rise to coherent responses by the other monkeys, but we do not see evidence that the monkeys notice their own signs. Rather, through the signs the objects or events in the landscape are drawn to the attention of the other animals. The signs do not feature in the monkeys' awareness; they are not witnessed as objects in the way that human observers and recorders witness those signs. If the monkeys used the signs *outside* of contexts of immediate association with the objects which provoke them, that would be one form of evidence showing the signs are witnessed as objects in their own right. There is presently no persuasive evidence for that.[2]

Humans, by contrast, routinely use signs outside of contexts of immediate association (the 'displacement' mentioned earlier). For this difference to occur, a prehistoric context would need to have arisen in which signs themselves became objects of coordinated attention. We promised to describe a scenario in which this happened. The scenario involves, as prerequisite, coordinated actions among hominids, in relation to predatory or scavenging behavior.

One feature of hominids is the lengthy record, over about 2.5 million years, of behavior involving the carrying and flaking of stone. The traditional story about the use of this material is that its modification was a cultural invention which was either guided by language or exhibited features akin to language (Holloway 1969). Holloway argued that the removal of flakes from stone cobbles had the aim of fashioning those cobbles into a range of implements for a range of purposes. He suggested that the sequence of removal had rule-governed structure, just as sentences in a language do. Following Toth (1985) we have argued that the flakes themselves appear to be the items that were used more regularly (Davidson and Noble 1993), rather than the stone from which they were removed. If the reduced cobbles, from which the flakes were struck, were not the objects to be used, the sequence of removal of flakes becomes incidental rather than structured.

The evidence of use and transport is consistent with a scene in which the sharp-edged flakes struck from stone cobbles were used to hack off pieces of flesh from the bones of animals. It used to be taken for granted that organized hunting, also requiring the conceptual resources language affords, was part of that use. But those animals were not necessarily felled by the hominids. They may have been, or they may have either died naturally or been downed by other predators (Binford 1983). The flaked stone cores (the Oldowan choppers), left at what became archaeological sites, may in fact be material from which no further flakes either were or could easily be removed.

The situation is slightly more complex with Acheulian handaxes. This might be an appropriate name for reasons other than normally appreciated. Whilst 'handaxes', like flakes, can be used to slice meat from bones (Mitchell 1995), if you tried to use one of these stones for the sorts of things a present-day axe is more commonly used for, the only thing you would likely sever would be parts of your hand. Although they are difficult to use for that sort of purpose, it turns out that the form of 'handaxes' gives them interesting aerodynamic properties (O'Brien 1981), such that they can be thrown with relative ease. Based on a proposal originally made by Darlington (1975), Calvin (e.g., 1993) has argued that throwing stones is a behavior offering a range of advantages for hominids competing increasingly for high energy food. (There is some archaeological evidence to support such a use for handaxes, since there are sites where many of them have been found in waterholes, where prey animals would likely congregate [e.g., Olorgesailie, see Isaac 1977]).[3]

A critical aspect about competition for food is the capacity to run, as a form of bipedal locomotion (Carrier 1984). A constraint on running is hairiness, and the lesser ability to lose heat which goes with that (Wheeler 1985). Where running aids in the competition for getting meat, there might have been selection for greater hair<u>less</u>ness combined with sweating. In addition, the brain consumes energy and oxygen and requires regulation of its temperature. Any improvement in the transport of blood around and within the brain, as well as any alteration that enables increased airflow in the nasal cavities (White and Cabanac 1995), improves the radiation of heat from the head. There is fossil evidence for changes in blood flow (Falk 1986). With the improvement in temperature regulation this implies, natural selection against large brains is reduced, and that ties in with documented prehistoric changes in cranial capacity. Being able to run faster may offer selective advantage by itself, and be an exaptive consequence of larger brains, through the recruitment of greater neural resources dedicated to lower limb control.

The brain is very energy-hungry (Foley and Lee 1991), so it can be presumed that there are costs as well as benefits associated with increasing brain size (Aiello and Wheeler 1995; see Weaver et al., this volume, for a detailed discussion of this topic). Furthermore, it is harder to give birth to an organism with a large head (to encase a large brain) (see Hogan and Gallup, this volume). Increase in the size of the birth canal to try to accommodate that may compromise a running gait. One way for these competing pressures to resolve is limitation on the amount of head growth before birth. Human heads and brains grow outside the mother's body to a greater extent than the heads of other primates (Foley and Lee 1991). A consequence is that humans at birth are more immature and hence dependent on mothers' sustenance for a longer period than is the case with other apes, a phenomenon called secondary altriciality. The timing of emergence of secondary altriciality is uncertain (and probably not the result of a single change). It seems to have been absent in *A.afarensis* (Tague and Lovejoy 1986) 2.9 million years ago, but Walker and Ruff (1993) argued from the pelvic bones of the very complete skeletal remains found at Nariokotome (WT-15000) that it was present in *Homo erectus* 1.6 million years ago. Smith and Tompkins (1995) are more cautious, pointing to the fact that the Nariokotome individual was male, and therefore not ideal for examining the relation between neonate head-size and pelvic dimensions. Nevertheless, Smith and Tompkins acknowledge that secondary altriciality had emerged by the time of Neandertals. Smith (1993) showed that the life history parameters of the Nariokotome individual were not completely matched to modern humans, and in particular, lacking an adolescent growth spurt, it had features that were more ape-like than human-like.

None of the foregoing characteristics—increasing use and flaking of stone, increasing hairlessness and improved thermoregulation of the brain—has anything directly to do with communication. The argument, ultimately, is that their combination may nonetheless have certain critical consequences for the coordination of activity including communication. Hairlessness, for example, increases the likelihood of carrying the offspring in ways that allow greater face-to-face interaction. One behavior observed in the bonobo 'Kanzi' is a forelimb gesture that has the characteristics of pointing (Savage-Rumbaugh 1984). Because, in infancy, Kanzi clung in typical chimpanzee fashion to the hairy back of his foster-mother 'Matata', this gesture went unobserved by her, and it was not part of a coordinated communication system. Hairlessness (and, indeed, speedier gait) may promote the carrying of infants in front, thus their behavior would be more visible and they would become subject to greater opportunities for adult-infant interaction—the kind of coordinated attention described earlier in relation to contemporary adult-infant play episodes. Prolonged carrying during a period when the brain was growing at fetal rates would presumably have important consequences for learning in very young infants, particularly as it resulted from coordinated attention.

If we pick up the stone-throwing part of the story, two matters may be noted. First, Calvin has argued that throwing stones farther and faster is just going to be good for those who can do it. This is because, as he suggests, it improves the chances of downing prey, or because, as we additionally note (Noble and Davidson 1991), it helps in chasing off rival scavengers and predators. Chimpanzees throw missiles, but, in general, they do it badly by human standards (Goodall 1964). They are not well adapted to it in virtue of their relative non-use of bipedal gait, and their limited actions with stone or other potential missile material. Nonetheless, we can accept throwing as a behavior likely to find expression in hominids. The Nariokotome *Homo erectus* individual, however, had a small spinal canal, of proportional size similar to other non-human primates, in the thoracic region (MacLarnon 1993). Weaver et al. (this volume) conclude that the Nariokotome evidence indicates "that a significant portion of [subsequent] encephalization...was related to aspects of upper limb innervation". It has been suggested (Walker 1993), that the small spinal canal of Nariokotome might have restricted the muscular control associated with throwing and with the production of vocal utterances. Thus, consistently effective aimed throwing may be of relatively recent origin.

The second matter is that to improve the precision of throwing as a behavior requires many orders of increased neural timing control so that the missile gets to its target with more force and accuracy. Increased timing control is plausibly only delivered by an increase in neural populations dedicated to muscle and joint coordination (Georgopoulos et al. 1986; Kalaska and Crammond 1992). Here is a possible further source of selective pressure for increased brain size and such pressures can be responded to, given the release of constraints consequent upon thermoregulation changes.

We have argued that aimed throwing is a basis for pointing as a novel behavior (Noble and Davidson 1989, 1996). This is proposed on the ground that, at the end-phase of an aimed throw, the forelimb releasing the missile must be aligned with the target of the throw. For an observer of the throw, that limb position is salient in incidentally indicating the bearing of the target. Infant bonobos spontaneously make gestures that humans read as pointing: infant hominids probably made similar gestures that were also likely to have been attended to and shaped up into a repeated action through the coordinated communication between adults and infants. This will have been especially so if, for its caregivers, the gesture was saliently linked with throwing. Here is one way in which adults may take over a behavior through observing it done in non-utilitarian ways by infants. To round out this matter: wild chimpanzees and bonobos do not, as adults, point. They hardly ever engage in one-handed aimed throwing. Thus the speculated bridge that enables passage from throwing to pointing is not part of their behavioral/perceptual experience. Thus, further, even if Kanzi's 'pointing' had been seen by Matata, she might not have responded to it in the way humans respond to such gestures. It is relevant to note, however, that humans can make pointing salient for certain other primates in conditions of 'encultured' rearing (Call and Tomasello 1994); hence its emergence in hominids is plausible.

By whatever means, pointing is feasibly selected for as a behavior because it serves to indicate the whereabouts of predators and prey. What is more, pointing can be done silently, thus not revealing the presence of the pointer. At this stage, matters of communication, hence inter-organism coordination, have crept into the story. Such communications were presumably commonplace as soon as bipedal hominids began to pick up their babies and walk. We are talking about the expression of gestures that get imitated, or that inform a conspecific about something else in the environment. This is by no means remote from the vervet monkeys and their gesture calls, so we cannot be accused of idleness in this speculation.

From here, the speculation is unabashed, but not so reckless as merely to invite dismissal. Infants' opportunities for observation of adults' activities using stone, adults' reinforcement of infants' movements that have salience, in turn allow the prospect of infants elaborating and refining their control of forelimb activity. Such elaborations are observed in deaf infants of non-native sign-language-using human adults (Newport and Supalla 1980; Volterra et al. 1990). Some of these actions may be incorporated in adult repertoires of gesture because they add to success in predation or improve chances of survival by alerting others to potential threats. Thus, controlled elaborations of pointing might evolve. These could take the form of tracking the path and imitating the gait or form of a prey or predatory animal. And this may be maintained as a behavior because it delivers information to others, not just about the bearing, but also the approach or retreat, and the category of the indicated creature. As with the vervet cries, the gestures serve to orient others to critical features of the environment and to elements of the nature of those features.

These are speculated products that result from certain prior adaptations concerning gait, from changes in carrying/holding habits, from the more regular use of stone, from thermoregulatory adaptations, and from changing constraints on brain size, including the emergence of secondary altriciality. The products are identifiably communicative, and thus entail inter-organism coordination. But there is nothing in what has been described that carries us to a specific alteration in the consciousness of the sign users, such that they start to attend to the signs themselves. After all, we are talking about hominids subsisting in semi-open terrain, in competition with a range of other, hungry animals, and with nothing to go on but their ability to run, provide themselves with stone flakes to remove pieces of meat, and to throw the stones that can no longer yield flakes. Gestures that assist in this risky trade are not likely to be themselves the object of any attention, when what they signify could cause death.

Signifiers and signifieds

It is necessary at this juncture to distinguish *signifier* and *signified*. We have talked about signs and their use as symbols to refer to entities or events. Another way to characterize this is to identify the sign as the signif*ier*, and the object or event as the signif*ied*. Saussure (1916/1983) is usually credited with deriving these terms, although he used them in a different sense, ours being closer to Voloshinov's (1929/1973). As we stressed at the start, behavior falling under the heading of 'linguistic' is expressed by creatures who are aware of signifiers as items that refer to signifieds. This is what Terrace (1985) emphasizes as the essence of 'naming': understanding that this X stands for that Y. Grasping this link makes X (the signifier) as prominent as Y (the signified). Savage-Rumbaugh and Lewin (1994) describe how difficult it is to create conditions to achieve this linkage for apes in language experiments, and to be sure they have achieved the linkage. How hominids got to notice their signs *as* signifiers, *as* objects in the landscape, required, in our argument, the recording of these signs in a persistent medium. The signs had to be transferred from the transient context of visible gesture to something more permanent. A gesture made in the course of food-getting had to persist as a *trace* due to that same gesture, which then became visible in and of itself in the environment. By some such transfer the signifier became an object of perceptual attention.

It is striking that vervet monkeys are wide awake to the *direct* visible signs of predatory snakes, and to the significance of the cries of conspecifics in response to that visible appearance. Yet those same monkeys show no signs of noticing what to humans are the highly salient tracks those snakes leave in the dust after they have wound their way out of sight (Cheney and Seyfarth 1990:285-6; Noble 1993).

It requires a particular mosaic of adaptations and practices, in particular contexts of interaction, to bring about the chance that a trace of a hand-arm gesture could be made by a hominid (or any other animal). Furthermore, that hominid must engage in gestural practices in the presence of others before the trace arising from it will be attended to as a salient appearance in the landscape. This is not utterly implausible if the trace in question resulted from the repetition of a highly salient gesture in the course of attending to a prey or predatory animal. Were a trace to be made, and were the result to be noticed, there would occur for the first time a sign that could become, itself, an object of attention. A trace deriving from such a sign, when *re*-traced, when the trace itself is imitated, is the first occurrence of 'representation'.

Something of this order is needed, we suggest, for attention to be turned from signifieds to signifiers, so that, in effect, the reality of signifiers is *discovered*. A discovery of this kind cannot happen to any animal which makes communicative signs; it needs the sort of background conditions we have been outlining. And it cannot be a meaningful event outside of contexts of co-action; that is to say, coordinated attention is required to what it is that signs signify, to what they may be made to refer to. By some means such as this, signs became objects of perception, and became useable as symbols.

Vocal gestures associated with visible hand and arm gestures that functioned as symbols presumably became symbols also. Furthermore, selection for greater vocal control could occur in contexts that offered an advantage for ongoing communication during coordinated use of the forelimbs for other or for related purposes. The intimacy of the link, neurally, between vocal and forelimb control (e.g., Peters 1990) might be explained on the basis of a takeover of functions by the upper laryngeal airways that originated with the other primary bodily articulators.

The discussion of the prehistory of hominid vocalizations has been dominated by attempts to understand the vocal abilities of hominids based upon the shape of their basicranial anatomy. Lieberman, Crelin and Klatt (1972; see also Lieberman this volume) claimed that the vocal range of Neandertals was restricted, but subsequent work has shown that changes in the human direction were already in place by the time of the ER-3733 *Homo erectus* (Laitman 1985) about 1.8 million years ago. Reassessment of Lieberman and colleagues' work suggests that their generalizations about the larynx cannot be sustained (Gibson, personal communication; Davidson 1997a).

Whatever the nature of vocal flexibility and control for any hominid, it remains the case that the issue concerning language is about how vocal utterances are used in coordinated communication, rather than about the range of utterances. Speaking a language might constitute a selective context favoring increased vocal range (Noble and Davidson 1996:211-212), but first, language as communication using symbols is not restricted to vocal utterances, and second, it need not have emerged as a behavior in hominids with modern human vocal ability.

The argument we have presented previously (Noble and Davidson 1996) points to the first appearance of symbolic objects in different regions after the time of first colonization of Australia (Davidson and Noble 1992), and the implausibility of claims for symbolism any earlier (Davidson 1990). The colonization of Australia is the earliest event in the record of hominid evolution that implies the planning derived from communication using symbols—from the behavior of speaking a language.

Selection for symbol use

The survival of the phenomenon of symbol-making cannot be accounted for just because it happened. It is often taken as satisfactory that linguistic behavior is, as it were, inherently self-preserving. Once it emerged it is surely self-evident that it was too good to lose. But more analytic care is needed here.

Behavior involving the deliberate use of signs survives because of the power it confers in the control of information between in-groups and out-groups (Noble and Davidson 1993; Sherif 1966). Once the potential of signs to be used as symbols is realized, their use in erroneous ways is bound to happen, hence their deliberate use to mislead rivals is inevitable. Human groups without the capacity to use symbols would lose out very thoroughly to those which could. Human groups would also quickly develop means to distinguish those whose symbols were trustworthy from those who were batting for a different team. Nettle and Dunbar (1997) have shown that simulations can be constructed which demonstrate the advantage of distinctive codes in making reciprocal exchange more stable in large groups, with implications for the emergence of linguistic boundaries when group sizes increased. We have suggested it is no coincidence that the earliest archaeological sign of symbols is accompanied by signs of objects that can plausibly be interpreted as personal decoration or identifiers (Davidson 1997b). It is thus also not coincidental that genetic divisions correlate with language divisions (Barbujani and Sokal 1990; see also Lieberman this volume).

This summary account of the evolutionary emergence of linguistic behavior depends at critical points on key elements that have nothing whatever to do with coordinated attention and communication among conspecifics. But other critical points in the story are necessarily due to such joint action. Speaking a language survives at all times in virtue of its commonality of use within human groups, whilst that commonality also inevitably leads to division between such groups.

Notes

[1] Some authors (e.g., Chase 1991) argue that iconic images may be distinguishable from symbols. We offer a counter-argument to such a view (Noble and Davidson 1996:72-80).

[2] One monkey was observed to make the call associated with the appearance of a leopard when there was none present (Cheney and Seyfarth 1990:213-215). The usual response of vervet monkeys hearing this call is to move away from any open terrain where they would be vulnerable to a leopard attack, and to head for the trees. But this monkey continued to make the leopard call whilst moving across open terrain suggesting that this (single) example of unusual call use cannot be considered a case of knowingly giving a false indication of danger. No explanation of this anomalous behavior is available. Cheney and Seyfarth (1990:216) suggest that the monkey in question may 'simply have learned that a leopard alarm caused other animals to run away'.

[3] Throwing does not need to be associated with handaxes. We have previously discussed uses of pointed sticks as spears for throwing (Noble and Davidson 1996:203). There is now evidence for pointed sticks about 400,000 years ago at Schöningen (Thieme 1997), hence throwing as a behavior of hominids could be that old.

References Cited

Aiello, L. C. and Wheeler, P.
  1995 The expensive-tissue hypothesis: The brain and the digestive system in human and primate evolution. *Current Anthropology* 36:199-221.

Barbujani, G. and Sokal, R. R.
    1990   Zones of sharp genetic change in Europe are also linguistic boundaries. *Proceedings of the National Academy of Sciences of the United States of America* 87:1816-1819.

Binford, L. R.
    1983   *In Pursuit of the Past*. Thames and Hudson, London.

Boesch, C.
    1993   Aspects of transmission of tool-use in wild chimpanzees. In *Tools, Language and Cognition in Human Evolution*, edited by K. R. Gibson and T. Ingold, pp. 171-183. Cambridge University Press, Cambridge.

Boesch, C. and H. Boesch
    1989   Hunting behavior of wild chimpanzees in the Tai National Park. *American Journal of Physical Anthropology* 78:547-573.

Burling, R.
    1993   Primate calls, human language and nonverbal communication. *Current Anthropology* 34:25-53.

Butterworth, G.
    1991   The ontogeny and phylogeny of joint visual attention. In *Natural Theories of Mind*, edited by A. Whiten, pp. 223-232. Blackwell, Oxford.

Call, J. and M. Tomasello
    1994   The production and comprehension of referential pointing by orangutans (*Pongo pygmaeus*). *Journal of Comparative Psychology* 108:307-317.

Calvin, W. H.
    1993   The unitary hypothesis: a common neural circuitry for novel manipulations, language, plan-ahead, and throwing? in *Tools, Language and Cognition in Human Evolution*, edited by K. R. Gibson and T. Ingold, pp. 230-250. Cambridge University Press, Cambridge.

Carrier, D. R.
    1984   The energetic paradox of human running and hominid evolution. *Current Anthropology* 25:483-495.

Chase, P. G.
    1991   Symbols and Paleolithic artifacts: style, standardization, and the imposition of arbitrary form. *Journal of Anthropological Archaeology* 10:193-214.

Cheney, D. L. and R. M. Seyfarth
    1990   *How Monkeys See the World*. University of Chicago Press, Chicago.

Darlington, P. J.
    1975   Group selection, altruism, reinforcement, and throwing in human evolution. *Proceedings of the National Academy of Sciences of the United States of America* 72:3748-3752.

Darwin, C.
    1872   *The Expression of the Emotions in Man and Animals*. Murray, London.

Davidson, I.
    1990   Bilzingsleben and early marking. *Rock Art Research* 7:52-56.
    1997a   The evolution of language: assessing the evidence from non-human primates. *Evolution of communication* 1:75-95.
    1997b   The power of pictures. In *Beyond Art: Pleistocene Image and Symbol*, edited by M. Conkey, O. Soffer, D. Stratmann and N. G. Jablonski, pp. 125-160. Memoirs of the California Academy of Sciences 23.

Davidson, I. and W. Noble
    1992   Why the first colonization of the Australian region is the earliest evidence of modern human behavior. *Archaeology in Oceania* 27:135-142.
    1993   Tools and language in human evolution. In *Tools, Language and Cognition in Human Evolution*, edited by K. R. Gibson and T. Ingold, pp. 363-388. Cambridge University Press, Cambridge.

Dennett, D.
    1987   *The Intentional Stance*. MIT/Bradford Books, Cambridge, MA.

Dewart, L.
    1989   *Evolution and Consciousness: The Role of Speech in the Origin and Development of Human Nature*. University of Toronto Press, Toronto.

Ekman, P.
    1977   Biological and cultural contributions to body and facial movement. In *The Anthropology of the Body*, edited by J. Blacking, pp. 39-84. Academic Press, London.

Ekman, P. and W. V. Friesen
    1969a   Nonverbal leakage and clues to deception. *Psychiatry* 1:88-105.
    1969b   The repertoire of nonverbal behavior: categories, origins, usage, and coding. *Semiotica* 1:49-98.

Falk, D.
    1986   Evolution of cranial blood drainage in hominids. *American Journal of Physical Anthropology* 70:311-324.

Foley, R. A. and P. C. Lee
  1991 Ecology and energetics of encephalization in hominid evolution. *Philosophical Transactions of the Royal Society of London, B* 334:223-232.
Georgopoulos, A. P., A. B. Schwartz and R. E. Kettner
  1986 Neuronal population coding of movement direction. *Science* 233:1416-1419.
Gibson, J. J.
  1966 *The Senses Considered as Perceptual Systems*. Houghton-Mifflin, Boston.
Goffman, E.
  1963 *Behavior in Public Places*. The Free Press of Glencoe, New York.
Goodall, J.
  1964 Tool-use and aimed throwing in a community of free ranging chimpanzees. *Nature* 201:1264-1266.
  1986 *The Chimpanzees of Gombe. Patterns of Behavior*. Harvard University Press, Cambridge, MA.
Greenfield, P. M.
  1991 Language, tools and brain: The ontogeny and phylogeny of hierarchically organized sequential behavior. *Behavioral and Brain Sciences* 14:531-595.
Groce, N. E.
  1985 *Everyone Here Spoke Sign Language: Hereditary Deafness on Martha's Vineyard*. Harvard University Press, Cambridge, MA.
Hockett, C. F. and S. A. Altmann
  1968 A note on design features. In *Animal Communication*, edited by T. A. Sebeok, pp. 61-72. Indiana University Press, Bloomington.
Holloway, R. L.
  1969 Culture: a human domain. *Current Anthropology* 10:395-412.
Isaac, G. L.
  1977 *Olorgesailie: Archaeological Studies of a Middle Pleistocene Lake Basin in Kenya*. University of Chicago Press, Chicago.
Kalaska, J. F. and D. J. Crammond
  1992 Cerebral cortical mechanisms of reaching movements. *Science* 255:1517-1523.
Kipling, R.
  1902/1987 *Just So Stories for Little Children*. Penguin, Harmondsworth.
Krebs, J. R. and R. Dawkins
  1984 Animal signals: Mind-reading and manipulation. In *Behavioral Ecology: An Evolutionary Approach*, edited by J. R. Krebs and N. B. Davies, pp. 380-402. Blackwell, Oxford.
Ladefoged, P.
  1975 *A Course in Phonetics*. Harcourt Brace Jovanovich, New York.
Laitman, J. T.
  1985 Evolution of the hominid upper respiratory tract: the fossil evidence. In *Hominid Evolution: Past, Present and Future*, edited by P. V. Tobias, pp. 281-286. Alan R. Liss, New York.
Lieberman, P. E., E. S. Crelin and D. H. Klatt
  1972 Phonetic ability and related anatomy of the newborn, adult human, Neanderthal man, and the chimpanzee. *American Anthropologist* 74:287-307.
Lock, A.
  1980 *The Guided Reinvention of Language*. Academic Press, London.
MacLarnon, A
  1993 The vertebral canal. In *The Nariokotome Homo erectus Skeleton*, edited by A. Walker and R. Leakey, pp. 359-390. Harvard University Press, Cambridge, MA.
Marler, P.
  1967 Animal communication signals. *Science* 157:769-774.
Mead, G. H.
  1934 *Mind, Self, and Society*. University of Chicago Press, Chicago.
Mellars, P.
  1989 Major issues in the emergence of modern humans. *Current Anthropology* 30:349-385.
Meltzoff, A. and A. Gopnik
  1993 The role of imitation in understanding persons and developing a theory of mind. In *Understanding Other Minds: Perspectives from Autism*, edited by S. Baron-Cohen, H. Tager-Flusberg and D. Cohen, pp. 335-366. Oxford University Press, Oxford.
Mitchell, J. C.
  1995 Studying biface utilization at Boxgrove: roe deer butchery with replica handaxes. *Lithics* 16:64-69.
Nettle, D. and R. I. M. Dunbar
  1997 Social markers and the evolution of reciprocal exchange. *Current Anthropology* 38:93-99.
Newport, E. L. and T. Supalla
  1980 Clues from the acquisition of signed and spoken language. In *Signed and Spoken Language: Biological Constraints on*

Linguistic Form, edited by U. Bellugi and M. Studdert-Kennedy, pp. 187-211. Verlag Chemie, Weinheim.

Noble, W.
1993 What kind of approach to language fits Gibson's approach to perception? *Theory and Psychology* 3:57-78.

Noble, W. and I. Davidson
1989 On depiction and language. *Current Anthropology* 30:337-342.
1991 The evolutionary emergence of modern human behavior: language and its archaeology. *Man: Journal of the Royal Anthropological Institute* 26:223-253.
1993 Tracing the emergence of modern human behavior: methodological pitfalls and a theoretical path. *Journal of Anthropological Archaeology* 12:121-149.
1996 *Human Evolution, Language, and Mind: A Psychological and Archaeological Inquiry*. Cambridge University Press, Cambridge.

O'Brien, E.
1981 The projectile capabilities of an Acheulian handaxe from Olorgesailie. *Current Anthropology* 22:76-79.

Olson, D. R.
1989 Making up your mind. *Canadian Psychology* 30:617-627.

Peters, M.
1990 Interaction of vocal and manual movements. In *Cerebral Control of Speech and Limb Movements*, edited by G. R. Hammond, pp. 535-574. North-Holland, Amsterdam.

Ristau, C. A.
1991 Before mind reading: attention, purposes and deception in birds? in *Natural Theories of Mind*, edited by A. Whiten, pp. 209-222. Blackwell, Oxford.

Ryle, G.
1949 *The Concept of Mind*. Hutchinson, London.

de Saussure, F.
1916/1983 *Course in General Linguistics* (trans. R. Harris). Duckworth, London.

Savage-Rumbaugh, E. S.
1984 *Pan paniscus* and *Pan troglodytes*: contrasts in preverbal communicative competence. In *The Pygmy Chimpanzee: Evolutionary Biology and Behavior*, edited by R. L. Susman, pp. 395-413. Plenum Press, New York.

Savage-Rumbaugh, S. and R. Lewin
1994 *Kanzi: The Ape at the Brink of the Human Mind*. Wiley, New York.

Schenkein, J. (ed.)
1978 *Studies in the Organization of Conversational Interaction*. Academic Press, New York.

Schmidt, R. C., C. Carello and M. T. Turvey
1990 Phase transitions and critical fluctuations in the visual coordination of rhythmic movements between people. *Journal of Experimental Psychology: Human Perception and Performance* 16:227-247.

Sherif, M.
1966 *Group Conflict and Cooperation: Their Social Psychology*. Routledge and Kegan Paul, London.

Slater, P. J. B.
1983 The study of communication. In *Animal Behavior*, edited by T. R. Halliday and P. J. B. Slater, pp. 9-42. Blackwell, Oxford.

Smith, B. H.
1993 The physiological age of KNM-WT 15000. In *The Nariokotome Homo erectus Skeleton*, edited by A. Walker and R. Leakey, pp. 195-220. Harvard University Press, Cambridge, MA.

Smith, B. H. and Tompkins, R. L
1995 Toward a life history of the Hominidae. *Annual Review of Anthropology* 24:257-279.

Struhsaker, T. T.
1967 Auditory communication among vervet monkeys (*Cercopithecus aethiops*). In *Social Communication among Primates*, edited by S. A. Altmann, pp. 281-324. University of Chicago Press, Chicago.

Tague, R. G. and C. O. Lovejoy
1986 The obstetric pelvis of A.L. 288-1 (Lucy). *Journal of Human Evolution* 15:237-255.

Terrace, H. S.
1985 In the beginning was the name. *American Psychologist* 40:1011-1028.

Thieme, H.
1997 Lower Paleolithic hunting spears from Germany. *Nature* 385:807-810.

Tomasello, M., E. S. Savage-Rumbaugh and A. Kruger
1994 Imitative learning of actions on objects by children, chimpanzees, and enculturated chimpanzees. *Child Development* 64:1688-1705.

Toth, N.
1985 The Oldowan reassessed: a close look at early stone artifacts. *Journal of Archaeological Science* 12:101-120.

Turvey, M. T.
- 1990 Coordination. *American Psychologist* 45:938-953.

Voloshinov, V. N.
- 1929/1973 *Marxism and the Philosophy of Language* (trans. L. Matejka and I. R. Titunik). Seminar Press, New York.

Volterra, V., S. Beronesi and P. Massoni
- 1990 How does gestural communication become language? in *From Gesture to Language in Hearing and Deaf Children*, edited by V. Volterra and C. J. Erting, pp. 205-216. Springer-Verlag, Berlin.

Walker, A.
- 1993 Perspectives on the Nariokotome discovery. In *The Nariokotome Homo erectus Skeleton*, edited by A. Walker and R. Leakey, pp. 411-430. Harvard University Press, Cambridge, MA.

Walker, A. and C. B. Ruff
- 1993 The reconstruction of the pelvis. In *The Nariokotome Homo erectus Skeleton*, edited by A. Walker and R. Leakey, pp. 221-233. Harvard University Press, Cambridge, MA.

Wheeler, P. E.
- 1985 The loss of functional body hair in man. *Journal of Human Evolution* 14:23-28.

White, M. D. and M. Cabanac
- 1995 Physical dilatation of the nostrils lowers the thermal strain of exercising humans. *European Journal of Applied Physiology* 70:200-206.

Wittgenstein, L.
- 1958 *Philosophical Investigations*. Blackwell, Oxford.

Zukow, P. G.
- 1990 Socio-perceptual bases for the emergence of language: an alternative to innatist approaches. *Developmental Psychobiology* 23:705-726.

Zukow-Goldring, P.
- 1997 A social ecological realist approach to the emergence of the lexicon: educating attention to amodal invariants in gesture and speech. In *Evolving Explanations of Development: Ecological Approaches to Organism-Environment Systems*, edited by C. Dent-Read and P. Zukow-Goldring. American Psychological Association, Washington, D.C.

Zukow-Goldring, P. and K. R. Ferko
- 1994 An ecological approach to the emergence of the lexicon: Socializing attention. In *Sociocultural Approaches to Language and Literacy: An Interactionist Perspective*, edited by V. John-Steiner, C. P. Panofsky and L. W. Smith pp. 170-190. Cambridge University Press, New York.